Fractals and Chaos

Springer
New York
Berlin
Heidelberg
Hong Kong
London
Milan
Paris
Tokyo

SELECTED WORKS OF BENOIT B. MANDELBROT

REPRINTED, TRANSLATED, OR NEW
WITH ANNOTATIONS AND GUEST CONTRIBUTIONS

COMPANION TO *THE FRACTAL GEOMETRY OF NATURE,*
FRACTALS AND SCALING IN FINANCE (SELECTA E),
MULTIFRACTALS AND 1/f NOISE (SELECTA N),
AND *GAUSSIAN SELF-AFFINITY AND FRACTALS* (SELECTA H)

Benoit B. Mandelbrot

FRACTALS AND CHAOS

The Mandelbrot Set and Beyond

SELECTA VOLUME C

with a foreword by P.W. Jones
and texts co-authored by
C.J.G. Evertsz and M.C. Gutzwiller

Springer

Benoit B. Mandelbrot
Mathematics Department
Yale University
New Haven, CT 06520-8283, USA
http://www.math.yale.edu/mandelbrot
and
IBM T.J. Watson Research Center
Yorktown Heights, NY 10598-0218, USA

Library of Congress Cataloging-in-Publication Data
Mandelbrot, Benoit, B.
Fractals and chaos : the Mandelbrot set and beyond / Benoit Mandelbrot.
p. cm.
Includes bibliographical references and index.
ISBN 0-387-20158-0 (alk. paper)
1. Fractals. 2. Mandelbrot sets. 2. Differentiable dynamical systems. I. Title.
QA614.86.M23 2004
514′.742—dc22 2003063815

ISBN 0-387-20158-0 Printed on acid-free paper.

Printed in the United States of America.

9 8 7 6 5 4 3 2 1 SPIN 10962329

Springer-Verlag is part of *Springer Science+Business Media*

springeronline.com

My uncle Szolem Mandelbrojt (1899–1983)
initiated me—long ago—to the delights of iteration
To his memory I dedicate this intellectual fruit of mine

BENOIT B. MANDELBROT is Sterling Professor of Mathematical Sciences at Yale University and IBM Fellow Emeritus (Physics) at the IBM Research Center.

Best known as the founder of fractal geometry which is the first broad attempt to investigate quantitatively the ubiquitous notion of roughness.

Author of *Les objets fractals,* 1975, 1984, 1989, and 1995 (translated into Basque, Brazilian, Bulgarian, Chinese, Czech, Italian, Portugese, Rumanian, and Spanish) and *The Fractal Geometry of Nature,* 1982 (translated into Chinese, German, Japanese, Korean, Russian, and Spanish). His multi-volume *Selecta* began with *Fractals and Scaling in Finance: Discontinuity, Concentration, Risk,* 1997, *Fractales, hasard et finance,* 1997, *Multifractals and 1/f Noise: Wild Self-Affinity in Physics,* 1999, and *Gaussian Self-Affinity and Fractals, 2002.* With M.L. Frame, he authored *Fractals, Graphics, and Mathematics Education,* 2002.

Fellow, American Academy of Arts and Sciences; Member U.S. National Academy of Sciences; Foreign Member, Norwegian Academy of Science and Letters.

His awards include the 1985 *F. Barnard Medal for Meritorious Service to Science* ("Magna est Veritas"), granted by the U.S. National Academy of Sciences, the 1986 *Franklin Medal for Signal and Eminent Service in Science* from the Franklin Institute of Philadelphia, the 1988 *Charles Proteus Steinmetz Medal* from IEEE, the 1988 (first) *Science for Art Prize* from Moet-Hennessy-Louis Vuitton, the 1989 *Harvey Prize for Science and Technology* from the Technion in Haifa, the 1991 *Nevada Prize,* the 1993 *Wolf Prize for Physics,* the 1994 *Honda Prize,* the 1996 *Médaille de Vermeil de la Ville de Paris,* the 1999 *John Scott Award,* the 2000 *Lewis Fry Richardson Award* of the European Geophysical Society, the 2002 *William Procter Prize* of Sigma Xi, and the 2003 *Japan Prize for Science and Technology.*

He also received a *Distinguished Service Award for Outstanding Achievement* from the California Institute of Technology, and a *Humboldt Preis* from the Alexander von Humboldt Stiftung.

Graduate of the Paris École Polytechnique; M.S. and Ae.E. in Aeronautics, California Institute of Technology; Docteur ès Sciences Mathématiques, Université de Paris. Doctor *honoris causa:* Syracuse University, Laurentian University (Canada), Boston University, State University of New York, University of Guelph (Canada), University of Dallas, Union College, University of Buenos Aires (Argentina), Open University (UK), Athens University of Business and Commerce (Greece), University of St. Andrews (Scotland), Emory University, Universität Bremen (Germany), Pace University, and the University of Tel Aviv (Israel).

Positions before joining IBM were with the CNRS in Paris, Philips Electronics, M.I.T., Princeton Institute for Advanced Study, University of Geneva, University of Lille, and Ecole Polytechnique. Institute Lecturer at M.I.T. Visiting Professor of Economics, later of Applied Mathematics, Mathematics, and the Practice of Mathematics, at Harvard; of Engineering at Yale; of Physiology at the Albert Einstein College of Medicine. Professeur de l'Académie des Sciences à l'École Polytechnique, Paris. Visited Cambridge, UK as G.C. Steward Visiting Fellow at Gonville and Caius College, Scott Lecturer at Cavendish Laboratory and Member at Isaac Newton Institute of Mathematical Sciences.

Mandelbrot had no formal teacher but his early work was strongly influenced by Paul Lévy, Norbert Wiener, and John von Neumann. An eloquent spokesman for "the unity of knowing and feeling," he seeks a measure of order in physical, mathematical or social phenomena that are characterized by abundant data but extreme sample variability.

List of Chapters

*In this List, * and (2003) mark the chapters not published previously. Other chapter titles are followed (in parentheses) by the letter M followed by the year of publication and by the lower case letter that the* Bibliography *uses to distinguish different texts published in the same year. In the* Bibliography, *the items reproduced in this book and in Volumes E, N and H are marked by a star followed by a chapter number, which in some cases is incomplete.*

II NONQUADRATIC RATIONAL DYNAMICS

III ITERATED NONLINEAR FUNCTION SYSTEMS AND THE FRACTAL LIMIT SETS OF KLEINIAN GROUPS

IV MULTIFRACTAL INVARIANT MEASURES

V BACKGROUND AND HISTORY

CF

Foreword by Peter W. Jones, Yale University

IT IS ONLY TWENTY-THREE YEARS SINCE BENOIT MANDELBROT published his famous picture of what is now called the Mandelbrot set. The graphics available at that time seem primitive today, and Mandelbrot's working drafts were even harder to interpret. But how that picture has changed our views of the mathematical and physical universe! Fractals, a term coined by Mandelbrot, are now so ubiquitous in the scientific consciousness that it is difficult to remember the psychological shock of their arrival. A twenty-first-century researcher does not think twice about using a computer simulation to begin the investigation of a problem; indeed, it is now routine to use a desktop computer to search for new phenomena or seek hints about research problems. In 1980 this was very far from the case.

When a paradigm shift hits, it is rarely the old guard who ushers it in. New methods are required, and accepted orthodoxy is often turned on its head.

Thirty years ago, despite the appearance of an avant garde, there was a general feeling in the mathematics community that one should distrust pictures and any information they might carry. Computer experiments had already appeared in the undergraduate physics curriculum, but were almost nonexistent in mathematics. Perhaps this was due in part to the relatively weak computers then available, but there were other aspects of this attitude. Abstraction and generality were seen by many mathematicians as the guiding principles. There were cracks in this intellectual foundation, and the next twenty years were to see many of these prejudices disappear.

In my own field of analysis there had been overblown expectations in the 1950s and 1960s that abstract methods could be developed to solve a large range of very concrete problems. The correct axioms and clever theorems for abstract Banach spaces or algebras would conquer the day. By the late 1960s, groups in France and Sweden, along with the Chicago school in the U.S., had developed entirely new methods of a very concrete nature to solve old conjectures and open new frontiers. The hope of abstract salvation, at least in its most extreme forms, was revealed as naive. Especially for problems of a statistical nature, hard tools needed to be developed. (One should note that in other areas of mathematics, abstract methods have had spectacular success in solving even very concrete problems. What this means for the future of those fields is now a topic of broad speculation.)

How fascinating it is to look back on this period and observe Benoit Mandelbrot. He was looking at pictures, drawing conclusions in many fields, and being largely ignored by all. He was outside every orthodoxy imaginable.

To understand Mandelbrot's contributions to science, one must first give up the tendency to find a disciplinary pigeonhole for every scientist. What should one call someone who works simultaneously in mathematics, physics, economics, hydrology, geology, linguistics... ? And what should one think of someone whose method of entry into a field was often to find puzzling patterns, pictures, and statistics. The former could not be a scientist, and the latter could not be science! But Benoit Mandelbrot was really doing something very simple, at least at the entry point to a problem: He was looking at the pictures and letting them tell their own story.

In the mid 1500s, Galileo peered through telescopes to find astonishing celestial features imperceptible to the human eye. In very much the same spirit, Mandelbrot used the most modern computers available to investigate phenomena not well studied by closed formulas, and out popped strange and unexpected pictures. Furthermore, he worked with the idea that a feature observed in a mathematics problem might be related to "outliers" in financial data or the observed physics of some system. Perhaps these rare events or outliers were not actually so rare at all; perhaps they were even the main feature of the system!

After getting his foot in the mathematical door, Mandelbrot would start the next phase of research, erecting a mathematical framework and doing the hard estimates. Try today to explain to the scientifically literate high-school student that the beautiful fractal pictures on a computer screen are not interesting, at least not to be trusted, and try asserting that the fractals arising in wholly different problems are similar due just to chance.

While the aversion to looking at pictures has faded, there is still confusion as to why Mandelbrot's early works on fractals, e.g., his book *The Fractal Geometry of Nature,* generated such wild popularity in the general scientific community. One does not see on every page the "theorem–proof" methodology of a mathematics textbook. Furthermore, though one can easily find theorems and rigorous proofs in the book, the phenomena and pictures discussed may seem to a mathematician to be unrelated, because there is not necessarily an exact theorem to link any two of them.

What a poor world we would live in if this were the only permitted method to study the universe! Consider the plight facing a working biologist, where all data sets are dirty and causality difficult to determine. Should one demand a theorem in this situation? Should a geologist looking at rock strata search first for a theorem, when the formalism of multifractal measures might be more important? An old tradition in science is to seek first a description of the system at hand; this apparently simpler problem is usually much more difficult than is generally believed. Few doubt that Kepler's laws would have been formulated without his first seeking patterns by poring over reams of data.

Perhaps, however, the pictures studied by Mandelbrot arose randomly, and any connection to interesting science is just a coincidence. The Mandelbrot set *M* offers an instructive example. Despite twenty years of intensive research by the world's best analysts, we still do not know whether *M* is locally connected (the MLC conjecture), and progress on this problem has rather ground to a halt. This is now seen as one of the most central problems of complex dynamics, and the solution would have many deep consequences. The geometry of *M* is known to be devilishly complicated; M. Shishikura proved that the boundary has dimension equal to two.

We know today that the "Sullivan dictionary" provides many analogues between iteration of rational functions and the theory of Kleinian groups, but there is very much that remains open. For example, we do not know whether it is possible for either a Julia set or a limit set (of a Kleinian group) to have positive area unless it is the full sphere. If all Julia sets from quadratic polynomials have zero area, then the Fatou conjecture on density of hyperbolic systems would be proven for quadratics. It is also known that MLC implies both the Fatou conjecture for quadratics and the nonexistence of certain (but not all) Julia sets of positive area.

Another example is furnished by the Brownian boundary that is the subject of Plate 243 of *The Fractal Geometry of Nature.* Arguing by analogy and examination of simulations, Mandelbrot proposed that the Brownian boundary has dimension 4/3 and serves as a model for (continuous) self-avoiding random walks (SARW). The 4/3 conjecture was only recently solved by the spectacular work of G. Lawler, O. Schramm, and W. Werner. Their proof relied heavily on the new processes called SLE that Schramm invented. We now know that SLE (8/3) represents the Brownian boundary. This also proves another prediction of Mandelbrot that the two sides of the Brownian boundary are "statistically similar and independent." One of the major challenges in probability theory is to prove that SARW exists, and the new conjecture is that it can be identified with SLE (8/3).

The study of multifractals is another area where Mandelbrot played a leading role. Through multiplicative measures with singular support were known in certain areas of Fourier analysis and conformal mappings, their fine structure had not been examined, and they were virtually absent in discussions of physical problems until the work of Mandelbrot. He was also the first to write down $f(\alpha)$ in the form of normalized logarithms of large deviation probabilities.

The status of these problems may be open, but the beautiful pictures, now easily reproduced by the aforementioned high-school student, continue to fascinate and amaze. What we see in this book is a glimpse of how Mandelbrot helped change our way of looking at the world. It is not just a book about a particular class of problems; it also contains a view on how to approach the mathematical and physical universe. This view is certain not to fade, but to be part of the working philosophy of the next mathematical revolution, wherever it may take us.

Peter W. Jones, Professor of Mathematics, Yale University
New Haven, Connecticut, October 1, 2003

CP

Preface

THE INTERCONNECTIONS BETWEEN FRACTALS AND CHAOTIC dynamical systems are numerous and varied. But this is neither a monograph on those interconnections, nor a textbook.

The core consists of reprints of the direct technical contributions I made in the 1980s to four great and enduring topics of mathematics: (A) Fatou–Julia iteration of the quadratic map $z^2 + c$, (B) Fatou-Julia iteration of other rational maps, (C) Poincaré's "Kleinian" limit sets, and (D) related singular measures. My contributions are not available at present in any single library. They were few in number, but several became influential, while others are perhaps more rarely quoted than they deserve.

To weave those topics together, new chapters were specially written, and many reprints are clarified by new forewords and annotations. There is a strange but widely held belief that science is a passionless and dull enterprise. This belief is certainly contradicted by the historical and biographical sketches in this book.

An eventful history and newly published pictures might well attract to this book some readers not concerned with mathematics *per se*. To help the pictures catch the interest of those readers, existing expository material that is comparatively "light" has been scattered throughout, especially in Chapters C23 and the first half of Chapter C17.

Sketches of the four main topics

Part I. Quadratic iteration and its Mandelbrot set. In the case of Fatou–Julia iteration, an object now denoted by M and called the "Mandelbrot set" has opened wide new vistas. For the quadratic map, I defined M in the plane of the complex variable c by the condition that the sequence c, $c^2 + c$,

$(c^2+c)^2+c, \ldots$ does *not* diverge. This definition, which may seem haphazard, will be seen to be deeply motivated. In contrast to its extreme simplicity, the complexity and beauty of M provoke wide fascination.

In 1980, paying close attention to computer-generated pictures led me to a number of striking observations that — either immediately or after a short delay — became mathematical conjectures concerning the quadratic Mandelbrot set. Though very simple to state, those conjectures were hard to prove. In fact, the most important of them—the Mandelbrot set is locally connected— remains open and has become notorious under the letters MLC.

My discovery of M consisted of those observations, and the deep contrast between merely seeing and discovering is discussed in Chapter C1.

Fractals and the Mandelbrot set in the classroom. A striking and important broad feature, not only of the Mandelbrot set but of all of fractal geometry, is that unknown territory lurks close to elementary considerations now taught in many high schools. The fact that the boundary of the unknown comes close to every known area has been of great help to many teachers. The bibliography lists two "waves" of material on fractals for the classroom. One was coauthored by Heinz-Otto Peitgen. Another is coauthored by M.L. Frame and me and includes Frame's course notes on the web and a DVD.

Part II: Nonquadratic iterations. Preparing this book brought a delightful surprise. Old archives preserved by my programming assistant in 1977–1979, Mark R. Laff, included my never-before-published illustrations (each imprinted by a date) concerning nonquadratic rational maps. Those pictures reveal that the discoveries I made in 1980 were preceded by a rich and subtle early period of fumbling and bumbling. Until now it could not be documented and therefore I mentioned it rarely. Today, with hindsight, everyone will recognize in Chapter C14 the overall shape and other features of the quadratic Julia and Mandelbrot sets studied in Part I. However, the nonquadratic environment of those early pictures was so complex that there was very little I could do with them in 1979.

The story of what happened in 1980 remains unaffected, but the events of 1979 and 1980 combined into an interesting case of scientific search and discovery that several of the chapters written especially for this book will discuss.

Part III: Kleinian groups' limit sets. I contributed a rapidly converging algorithm that filled a longstanding gap in an old theory. More specifically, my algorithm constructs by successive approximations a set that is self-inverse with respect to a given collection of circles.

Part IV: Exponentially vanishing multifractal measures. Chapter C20 arose when my IBM colleague Martin Gutzwiller and I, coming from thoroughly distinct areas of physics, realized that we were both investigating the same strange singular measure. Our results were easily combined in one paper. That measure then turned out to have been defined long ago by H. Minkowski, but further study was well-deserved.

Motivation and tools of investigation

A strong long-term motivation. To a large extent — in fact, surprisingly so, even to me — my thinking was triggered by being young and adventurous enough to become a master of the use of the computer and old enough to have been immersed in some ancient mathematical traditions. They had arisen in early twentieth-century but by the 1970s were unfashionable and slumbering. Those traditions caused me to begin the study of iteration with the complicated rational maps taken up in Part II.

My involvement with this book's topics was largely independent of "chaos theory," understood as the revival of nonlinearity in the 1970s. While chaos theory favored the real map $x^2 + c$, it was already said that my move to its complex counterpart came late and reluctantly.

The relative roles of primitive or refined pictures, and of the eye. It was near-universally believed among pure mathematicians around 1980 that a picture can lead only to another, and never to fresh mathematical thinking. A striking innovation that helped thoroughly destroy this belief resided in my work's heavy reliance on detailed pictures, in contrast to schematic diagrams. Incidentally, a picture is like a reading of a scientific instrument. One reading is never enough. Neither is one picture.

More precisely, my discoveries of new mathematical conjectures relied greatly on the quality of visual analysis and little on the quality of the pictures. Indeed, Chapter C1 will establish that for discovering the Mandelbrot set, high quality graphics was not necessary, while Chapter C12 will establish that it was not sufficient, either.

Altogether, my lifetime scientific work rescued the verb "to see" from the figurative meaning to which both common usage and hard quantitative science had reduced it, and restored its concrete meaning, whose instrument is the eye.

Some fractal pictures are realistic and proudly called "forgeries" of mountains, clouds, trees, or galaxy clusters. Other pictures are totally abstract, like those of the Mandelbrot set. Moreover, some fractal pictures are perceived as having high aesthetic quality. Enormous numbers of persons have posted pictures of fractals on the Web. But the black-and-white computer pictures in my old files continue to be very valuable. If a suitable environment can be found, I would love to extend the small art portfolio implicit in this book into a "permanent exhibit" on the Web.

Fractal geometry opens up a quantitative theory of roughness

Given the variety of its manifestations, fractal geometry continues to surprise both the technical and the nontechnical audiences. It remains hard to pigeonhole, to classify, and to compare with existing disciplines.

Mostly after the fact, I view fractal geometry as opening up a study of roughness that is parallel to—but distinct from—the studies of brightness/color, loudness/pitch, heaviness, and heat, each of which has long since developed into a science. Compared to the studies of those other basic sensations, the study of roughness came late because it is more complex. Its quantitative measurement demands Hölder exponents and Hausdorff dimension—concepts that arose far later than, for example, periodic oscillations; fractal geometry was first in recognizing that they concern anything "real."

While it is tightly bound by the tools it uses and the flavor of the problems it faces, fractal geometry retains an intrinsic diversity that is rare, amusing and—I think—important. It has survived the childhood diseases and crises that strike intellectual initiatives involving an ambitious synthesis has been described as having changed the view of nature held by many mathematicians, scientists, engineers, artists, other professionals, and even every man and woman.

Open and fortress mathematics. Starting at the latest in ancient Greece with Archimedes and Plato, the views of the nature of mathematics has ranged between two extremes. My self-explanatory words for them are *open* and *fortress mathematics.* The former involves a lively sprawling collection of buildings permanently under construction or reconstruction, with many doors and windows revealing beautiful and varied landscapes. The highest ambition of fortress mathematics, to the contrary, is to wall off all openings but one. Its dwellers believe that their endeavors can evolve on their own steady path and need not interact with society at large.

While mathematics and science are among the highest achievements of humanity, all evidence shows that their history and the history of human civilization have been indissolubly intertwined. The claim that fortress mathematics *has* become independent is wishful thinking, and the notion that it *can* become independent is gratuitous.

Other contributions of fractals to pure mathematics

Following *The Fractal Geometry of Nature* (M 1982F), a series of my *"Selecta"*—selected papers— began with M1997E, M1999N, and M2002H, and continues with this Volume C, M2004C. The style of the preceding references is explained on the first page of the bibliography. Denoting those *Selecta* by nonconsecutive mnemonic letters suggests that they can be examined in any sequence.

The previous three volumes all concern a "state" of randomness and variability that I call "wild." This volume C is unrelated to the previous three, with the following important exception. Not only do Chapters C20 and C21 involve the topic of M1999N, which is multifractals, but those chapters were motivated by statistical physics through diffusion-limited aggregation. This is why M and Evertsz 1991 is reprinted as Chapter C22.

Early plans called for additional *Selecta* volumes. But, the Internet having transformed our world, the further *Selecta* will be "Web books" on my Web site. Each will reduce to a title page, a foreword, a table of contents, and links to papers on my home site. Given the diversity of my work, the web's flexibility is a great asset.

The "Overview" Chapter HO of M2002H presents a partial but nearly-up-to-date status report on fractal geometry. More specifically, what has been its overall impact on mathematics? While fractal geometry was young, it was invidiously observed that it "has not solved any mathematical problems." This is no longer true and in any event was always irrelevant. Indeed, my role in mathematics has been to provide a mass of *new problems and conjectures*. Each opened a new field that continues to prosper as I move to other concerns. Most widely known are the examples discussed in this book, but a few other examples deserve brief mention now.

M1982F (p. 243) introduces and M2002H (Chapter H3) investigates the concepts of Brownian cluster and Brownian boundary, culminating with the conjecture that this boundary's Hausdorff dimension is 4/3. Combined with related conjectured dimensions for percolation and Ising clusters, the diverse occurrences of 4/3 grew into sharp challenges to the analysts and has led since 1998 to widely acclaimed proofs by Duplantier, Lawler, Schramm, Werner, and Smirnov. Earlier, a dozen or so scattered technical conjectures in analysis had been shown to be equivalent to that

4/3. All have now been proven as corollaries and together provide mathematics with a new element of unity that continues to be explored.

M1999N collects many early papers in which I introduced and investigated the random multiplicative singular measures, now called "multifractal," an example of which stars in Part IV. They were not intended as new esoterica but as a model of turbulence and finance. My conjectures created an active and prosperous subbranch of mathematics, they served to organize some features of DLA (as already mentioned), and they underlie the main current branch of statistical modeling of the variation of financial prices. Increasingly rich structures arose as I repeatedly weakened the constraints on the multifractal multiplicands. The papers collected in M1999N took the step from microcanonical to canonical multiplicands. Recent papers coauthored by J. Barral moved on to products of pulses and other functions.

The telling pictures I drew of old standbys like the Koch and Peano curves and the Cantor dust achieved a broader and deep change of perspective. Those sets used to be viewed as "pathological" or "monsters." Quite to the contrary, I turned them around into unavoidable rough models ("cartoons") of a reality that science had previously been powerless to tackle, namely, the overwhelming fact that most of raw nature is not smooth but very rough. For example, I reinterpreted Peano "curves" as nothing but motions following a plane-filling network of rivers.

Norbert Wiener once described his key contributions to science as bringing together — starting from widely opposite horizons — the fine mathematical points of Lebesgue integration and the vigorous physics of Gibbs and Perrin. Also (like Poincaré), Wiener was very committed (and successful) in making frontier science known to a wide public. On both counts, the theory of fractals is arguably a multiple second flowering of Wiener's Brownian motion.

Overall acknowledgments

Let me now proceed beyond the acknowledgments printed after each paper, which were preserved, and the further acknowledgments found in several introductory chapters. Firstly, warm thanks go to the coauthors of joint papers for permission to reprint them.

For over 35 years, the Thomas J. Watson Research Center of the International Business Machines Corporation, in Yorktown Heights, N.Y., provided a unique haven for mavericks and for various investigations that science and society forcefully demanded, but academia and its funding agencies neither welcomed nor rewarded.

The originals of the old texts reprinted in this book were written at that haven. Invaluable programming and graphics support was provided there by Mark R. Laff, V. Alan Norton, and J.A. Given, and at Harvard in 1980 by Peter Moldave. The preparation of the original texts and the long–drawn–out preparation of this book were performed by several long–term secretaries, H. Catherine Dietrich (1933–2003), Janis Riznychok, Leslie Vasta, Premla Kumar, Kimberly Tetrault, Catherine McCarthy, and Barbara White. After retiring from IBM, I continued at Yorktown part time as IBM Fellow Emeritus, largely in order to prepare the *Selecta* books. Short-term assistants far too numerous to list were of great help. The clumsy English of some old papers was copy edited by Helen Muller-Landau, Noah Eisenkraft, and others. Of course, extreme care was taken never to modify the meaning. The originals are available in libraries and are gradually being posted on my Web site.

Never was IBM's pioneering *Script* word-processing language under VM expected to survive (unattended!) for ten years. But the clock is ticking, and this may be the last major project served by *Script*.

I am deeply indebted to the Yorktown of its heyday as a scientific powerhouse. Among long-term friends and colleagues, it will remain most closely associated with Richard F. Voss, Martin Gutzwiller, Rolf Landauer (1927–1999), and Philip E. Seiden (1934–2001). As to management, at a time when the old papers in this book were being written and were widely perceived as a wild gamble, my work received wholehearted support from Ralph E. Gomory, to whom I reported in his successive capacities as Group Manager, Department Director, and finally IBM Director of Research and Senior Vice-President. Gomory reminisced on the old times in a *Foreword* written for M 1997E.

As an adjunct in the Yale Mathematics Department before retiring from IBM, then as a tenured professor, I had the renewed great fortune of being invited, especially by R. R. Coifman and Peter W. Jones, to move on to another haven that also provided my life with welcome balance between industry and academia. The Yale postdocs I supervised include Carl J.G. Evertsz, coauthor of a paper that became Chapter C22.

Last but not least, this book is not solely dedicated to my uncle. As all my work, it is also dedicated to my wife, Aliette. The original papers would not have been written, assembled, and added to without her constant and extremely active participation and unfailingly enthusiastic support.

Decorative design.

PART I: QUADRATIC JULIA AND MANDELBROT SETS

&&&&&&&&&&&&&&&&&&&&&&&&&&&&&&&&&&&&&&&&&&&

First publication

C1

Introduction to papers on quadratic dynamics: a progression from seeing to discovering

THIS CHAPTER DESCRIBES THE CIRCUMSTANCES under which I had the privilege of discovering in 1980 the set that is the main topic of this book. As will be meticulously documented in Chapters C12 and C14, I actually saw this set in 1979 but bumbled and fumbled for about a year.

M1980n{C3} ushered in the modern theory of complex dynamics, specifically of the quadratic dynamics of the maps z^2+c and $\lambda z(1-z)$. *The Fractal Geometry of Nature*, M 1982F, followed and was widely read. Very rapidly, interest in the Mandelbrot set became broad and extraordinarily intense. Many eminent mathematicians immediately took up its study and achieved spectacular results that provoked a historically significant and highly beneficial change in the mood of mathematics. Yesterday, "generality at all cost" was in the saddle. Today, "special" problems are more readily recognized as compelling.

However, my own research moved on to different topics. Therefore later developments of quadratic dynamics are little known to me, and few will be quoted; Urbanski 2003 is a recent survey.

The broad "popular infatuation" with the Mandelbrot set must be mentioned. This social phenomenon continues, astonishes everyone, and of course enchants me. It was largely spontaneous, no committee or organization being involved. It suggests that a strong interest for mathematics is widespread among humans, but only if its links with nature and the eye are not actively suppressed but, instead, brought out and celebrated.

This broad interest may bring to this book some readers from a "general public." For their sake, the introductory chapters interpret the term "historical circumstances" rather broadly and include facts already well known to many professional mathematicians. Earlier accounts such as M 1986p were very incomplete.

Section 1 is a sketch, and Sections 2 and 3 provide fuller discussions. Broader acknowledgments are postponed to the next chapter.

"Nice" illustrations are scattered throughout this book, but this chapter's main point is strengthened by returning to the sources. Figures 1, 2, 3, 4, and 5 are a small sample of the crude illustrations, many of them published for the first time, that led to the actual discovery of the Mandelbrot set during the Harvard spring term of 1980. Their number and variety, which I had forgotten, are significant. The computer did not automatically imprint a date, but they might be roughly ordered in time.

To print those old pictures, it was necessary to enhance them by repeated xeroxing. Since they are well-known today, and contain no fine detail that risks being lost, many were made small.

1. THE PROGRESSION FROM SEEING TO DISCOVERING

1.1 Definition and a key quotation from Adrien Douady

Everyone knows, or so it seems, that the set M is defined in the complex plane of the variable c by the condition that the sequence $c, c^2 + c$, $(c^2 + c)^2 + c, \ldots$ does not diverge. It is Adrien Douady who proposed the term "*Mandelbrot set M* because Benoit Mandelbrot was the first one to produce pictures of it, using a computer, and to start giving a description of it."

Verba volant, scripta manent. The words quoted above are found on the third and second lines from the bottom of page 161 of Douady 1986, which this chapter will quote again.

Those words and their date are important. Despite its brevity, Douady's statement subdivides into two clearly separate issues, one inconsequential, and the other broad and historically important. The fact that I was the first to produce pictures of the Mandelbrot set, in 1979, is nice. But in the context of mathematics, this is not much to be praised for. Section 2 will argue that the issue of the "first picture" is, by itself, unimportant. Section 3 will argue that the actual discovery occurred later in 1980, and consisted in my early description of many fundamental features of M. This discovery mattered a great deal, because it soon triggered important developments.

Between the unbeatable simplicity of the definition of M and its visual and mathematical complexity there is a profound contrast that marks an important discovery of the late twentieth century.

This book's core consists of reprints of papers in which my main observations were first presented in the form of mathematical challenges/conjectures.

1.2 Motivation for investigating the Mandelbrot set and a sketch of key observations

1.2.1 Orbits, their limit points or cycles, and the "filled-in" Julia sets. A rational function of a complex variable z is the ratio of two polynomials in z. Let $f(z, c)$ denote a rational function of z depending on a complex parameter c that can be onedimensional or multidimensional. For fixed c, the orbit of a starting point z_0 is defined as the infinite sequence z_0, $z_1(z_0, c) = f(z_0, c)$, $z_2(z_0, c) = f(z_1, c)$, and generally $z_k(z_0, c) = f\,[z_{k-1}(z_0, c), c]$.

In the late nineteenth century, the notation arose that such sequences provide idealized versions of dynamical systems of a discrete time k. Within that perspective, it is important to classify the points z_0 and c according to the limit behavior of the corresponding orbit. Where does it fail to converge and can be called "chaotic"? Where does it converge to one of several fixed points or finite cycles and can be called "orderly"?

Quadratic dynamics corresponds to the case where f is a second-order polynomial. Changing the variable z reduces f to either $z^2 + c$ or $\lambda z(1 - z)$. In either case, there is one complex parameter, c or λ. For every c, there is a fixed point at infinity to which an orbit converges if its starting point z_0 is far enough from the origin. But there also exist points z_0 such that the orbit starting at z_0 fails to converge to infinity. Those points taken together define the "filled-in Julia set" corresponding to c.

1.2.2 M^0 versus M. For some values of c, the orbits of some points z_0 converge not to infinity but to a finite stable cycle of size $N \geq 1$. In this dynamical-systems perspective, I became interested in 1979 in identifying the set M^0 of those values of c and classifying them according to N.

In all interesting cases, an analytic study of M^0 is impossible. Therefore, I attempted to study it numerically. But the task proved to be extremely hard computationally and the approximate M^0 it yielded was very blotchy. Making the task even harder was the fact that in 1979, I opted to start not with quadratic dynamics but with the far more complicated $f(z, c)$ to be discussed in Chapter C14. This part picks up the story at the point where I turned back to $f(z) = z^2 + c$.

It occurred to me that the existence of limit sets implied domains of convergence separated by curves. Hence the set M^0 I was seeking had to be a subset of the set of c's for which the Julia set is not a dust but connected. Fatou and Julia had given a criterion that is straightforward and particularly easy to program for the quadratic map: c belongs to the set M if and only if the orbit with the starting point $z_0 = 0$ (called "critical point") *fails* to converge to infinity. This set of values of c is identical to the set M as defined above.

I conjectured that M was the closure of M^0 but that in any event, M was relatively easy to investigate, hence well worth exploring. My conjecture is most often restated today as asserting that the Mandelbrot set is locally connected (MLC). Despite heroic efforts it has not yet been proven true or false. How fortunate that I did not try to settle this issue!

1.2.3 A structure made of "atoms" combined in a big "molecule." Douady 1986 continues (p. 162) his description of my key early observations on M as follows: "When you look at the Mandelbrot set, the first thing you see is a region limited by a cardioid, with a cusp at the point .25, and its round top at the point – .75. Then there is a disk centered at the point – 1 with a radius .25, tangent to the cardioid. Then you see an infinity of smaller disk-like components, tangent to the cardioid, most of which are very small. Attached to each of those components, there is again an infinity of smaller disk-like components, and on each of these there is attached an infinity of smaller disk-like components, and so on."

Let me interrupt Douady to describe the circumstances of my discovery of those disc-like components for the quadratic map in the alternative form $z \to \lambda z(1-z)$. After a few iteration stages on a rough grid, we saw that the set M includes very crude outline of the disks $|\lambda| < 1$ and $|\lambda - 2| < 1$. Two lines of algebra confirmed that these disks were to be

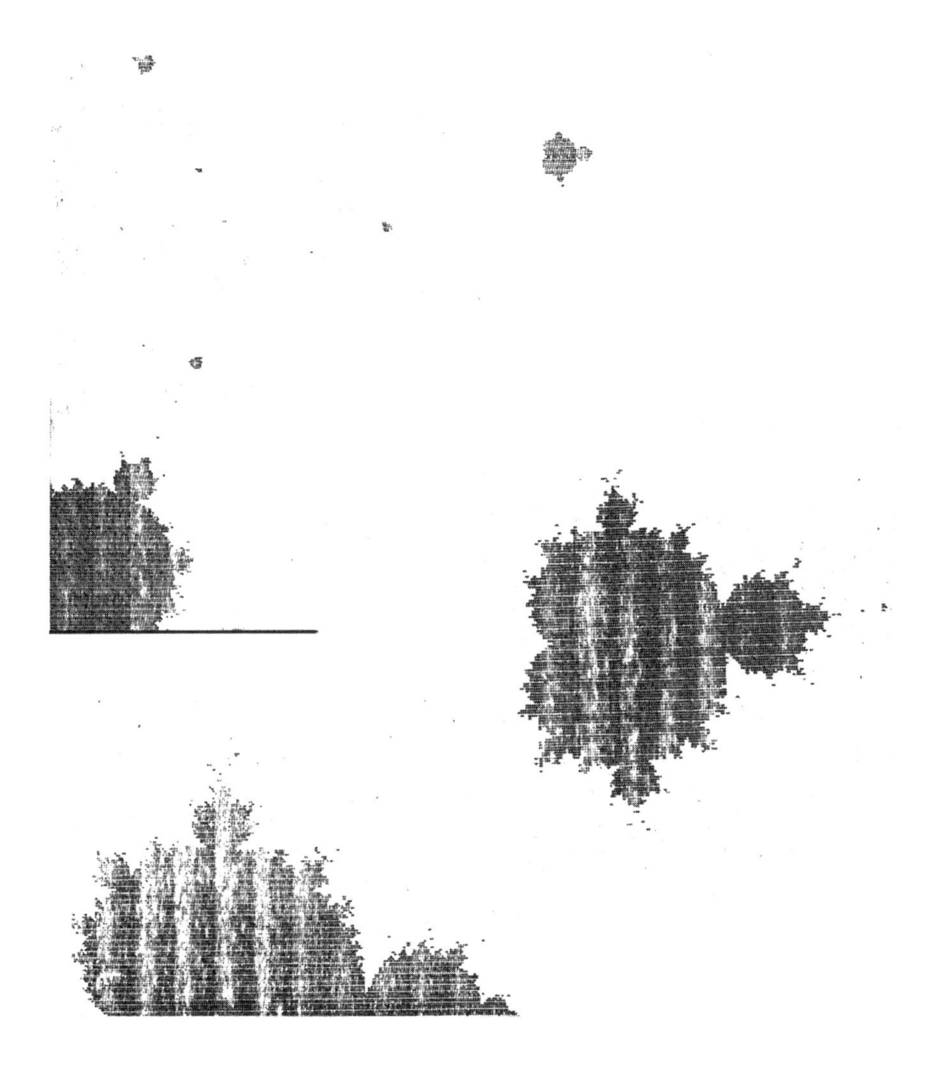

FIGURE C1-1. [Harvard, early 1980] My first picture of the whole *M* set is—unfortunately— either misfiled or lost. The first picture made at IBM in 1980 is reproduced in later chapters. The top panel here is a blow-up of the most conspicuously "messy" corner of *M*, near the bifurcation of order 3. The middle and bottom panels show the oldest preserved blow-ups of the two largest islands, one already seen in the top panel and the other intersected by the real axis (for reasons of economy, only half was computed).

expected. Also, we saw, on the real line to the right and left of the above disks, the outlines of disc-like "atoms." They appeared to be bisected by intervals known in the theory of bifurcations for the real map $x^2 + c$ that was provided in Myrberg 1962. An increasing investment in computation yielded increasingly sharply focused pictures. Helped by imagination, I saw the atoms fall into a hierarchy, each carrying smaller atoms attached to it. We verified that the points where the big disk-shaped atoms carry smaller atoms are as expected. Thus, I saw geometric implementations, not only for the familiar Myrberg sequence of successive binary bifurcations, but for every sequence of bifurcation of arbitrary order.

1.2.4 The "satellites" or "offshore islands." Pearls in the pigsty. We now return to Douady 1986. "But that is not all! If you start from the big cardioid ... and keep going ... to ... the Myrberg–Feigenbaum point, situated at -1.401.... Now, the segment from this point to the point -2 is contained in M and on this segment, there is a small cardioid-like component, with its cusp at -1.75 ... accompanied by its family of disk-like satellites just like the big cardioid. Actually there are infinitely many such cardioid-like components. There are also cardioid-like components off of the real axis. B. Mandelbrot ... showed that there are an infinite number of them."

How were those "island" components discovered? As witnessed by Figure 1, early originals looked awful: filled with apparent specks of dust that the Versatic printer produced in abundance. This complicated matters. But for the skilled, meticulous, and tireless observer that I was, mess was not a reason to complain but a reason to be particularly attentive. Fearing that the mess was the fault of faltering Harvard equipment (as will be explained momentarily), I drove to Yorktown for a day I shall never forget. On the IBM mainframe we obtained an iconic illustration that need not be repeated here. It became Figure 1 of M1980n{C3} then the top of Plate 189 in Chapter 19 of M 1982F, that is, Chapter C4 below.

Not only the mess failed to vanish, but some specks looked suspiciously symmetric in relation to one another. Since bona fide dirt is not expected to be symmetric, both kinds of specks demanded to be blown up for close inspection. Many small specks vanished after we zoomed in. Other specks, however, not only remained, but were revealed to be complex "molecular" structures, miniature versions of the Mandelbrot set-to-be for the map $z \to z^2 + c$.

A thought comes to mind. Had that memorable day at Yorktown been less rushed and better organized, I might have drawn only the upper half of that picture. The symmetry would have been missed—with unknowable consequences.

Next, we focused on the sprouts corresponding to different orders of bifurcation, and we compared the corresponding offshore islands. They proved to lie at the intersections of stellate patterns and of logarithmic spirals!

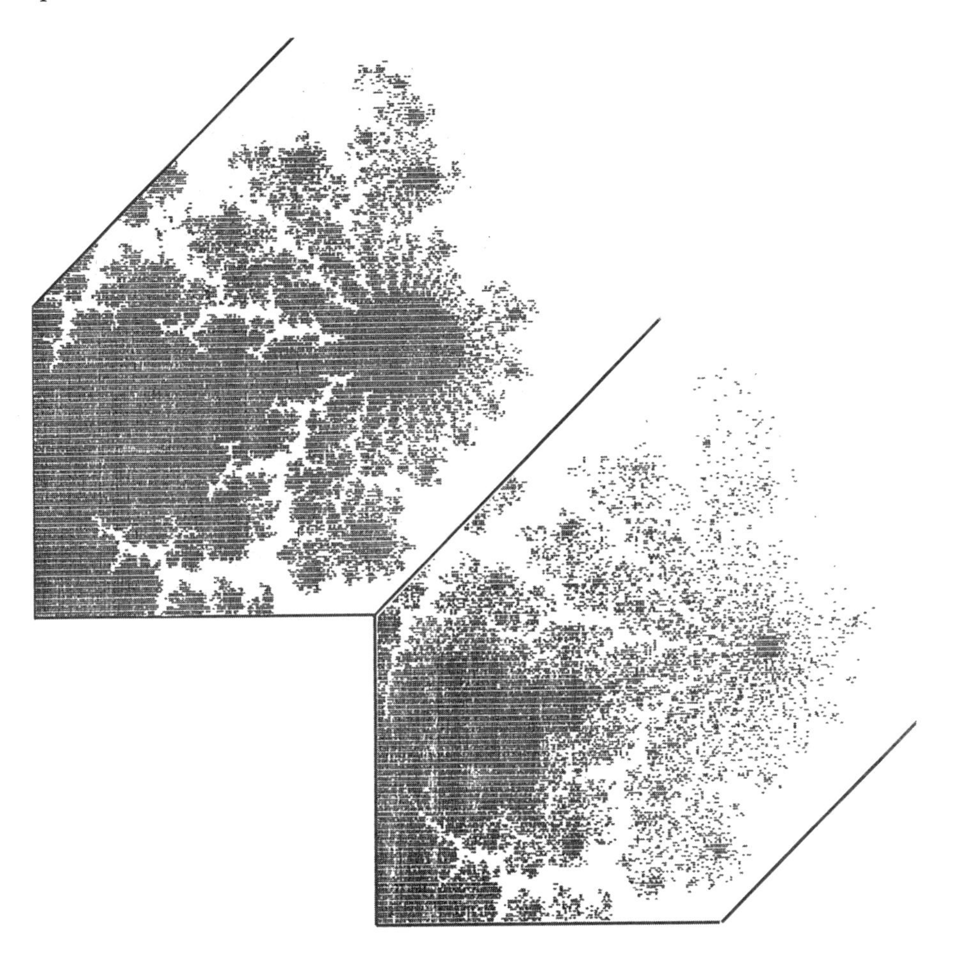

FIGURE C1-2. [Harvard, early 1980] Increasingly sharper pictures of a detail of the Mandelbrot set near a bifurcation of order 11. My concern at that time was whether or not the *M* set is connected. Therefore, pictures were also drawn by quadratic iteration with a starting point other than the critical value $z_0 = .5$. The pictures' crudeness allowed no conclusion to be drawn. A more convincing circumstantial evidence lies in the observation reported in the caption of Figure 4.

Discovering such "pearls" in a pigsty motivated a passionate investigation that led to the first paper on the Mandelbrot set, M1980n{C3}; some of its pictures were prepared after I returned from Harvard to IBM. This text appeared very quickly in the *Annals of the New York Academy of Sciences*, in the widely read proceedings of a major meeting on nonlinearity. The title, "Fractal aspects of the iteration of $z \rightarrow \lambda z(1-z)$ for complex λ

FIGURE C1-3. [Harvard, early 1980] Miscellaneous Julia sets for complex parameter values in the main continental molecule of the Mandelbrot set.

and z, " suffices to show that my goal was to revive experimental mathematics by reporting observations triggering new mathematics.

1.2.5 The web-like structures linking the islands to create a connected set M. The quotation from Douady 1986 continues as follows: "This is not all

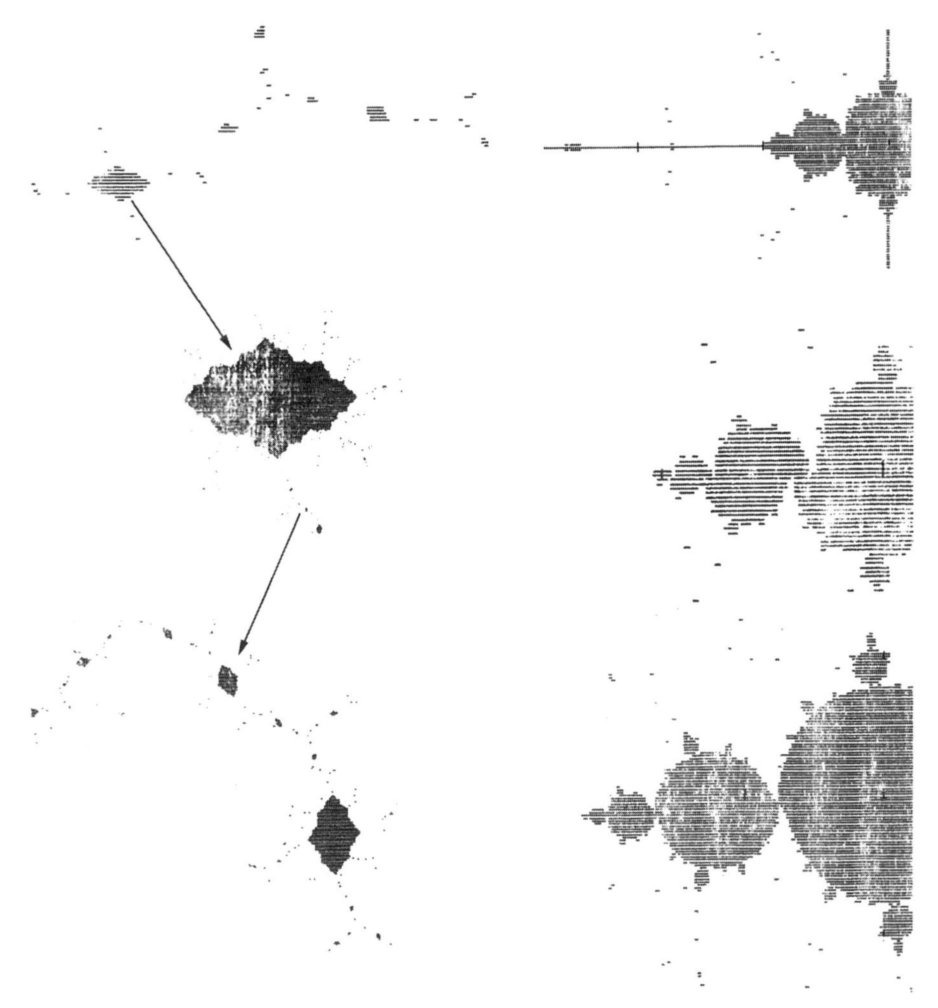

FIGURE C1-4. [Harvard, early 1980]. Miscellaneous Julia sets for parameter values in the "island molecules" of the Mandelbrot set. By a theorem Julia and Fatou, those Julia sets are connected. Therefore the broken-up appearances is necessarily due to the discrete variables used in computation. These graphs were important to my thinking because they sufficed to show that the broken-up early M set pictures were compatible with connectedness.

... All of these cardioid-like components are linked to the main cardioid by filaments, charged with small cardioid-like components, each of which is accompanied by its family of satellites. These filaments are branched according to a very sophisticated pattern."

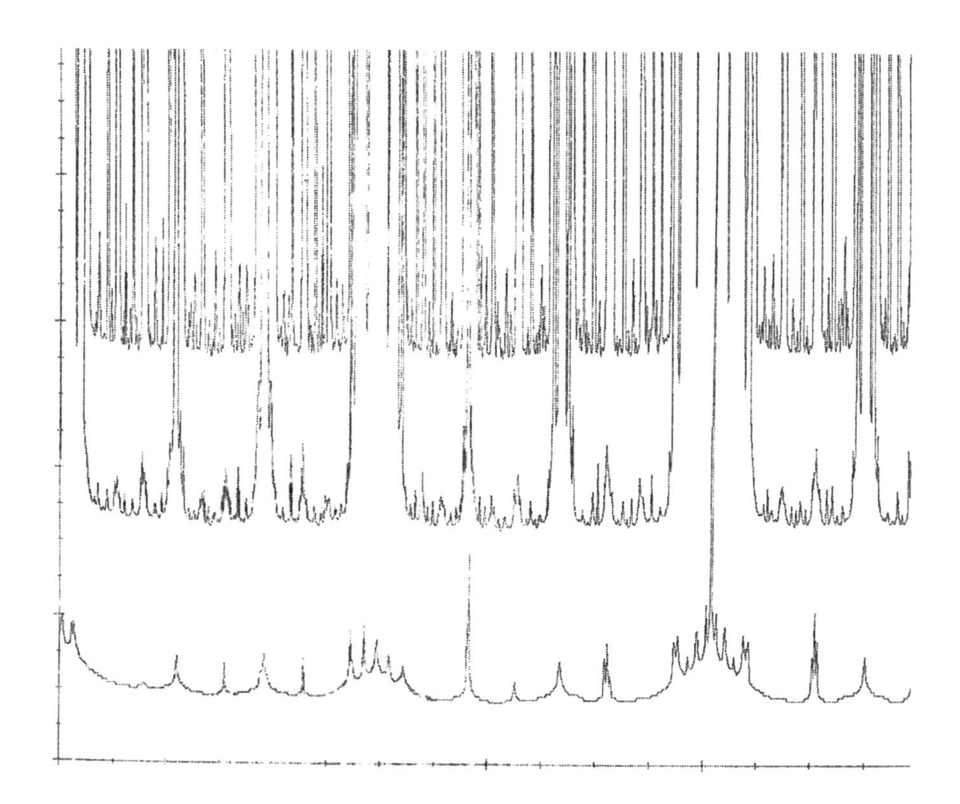

FIGURE C1-5. [Harvard, early 1980] Early on, I imagined the surface that is defined, for each λ, by the function " $H(\lambda)=$ the number of iterations needed to first achieve the inequality $|f_n(1/2)|>2$. " The M set is defined by $H=\infty$. Plate 189 of M1982F (reproduced in Chapter C4) includes level surfaces of $H(\lambda)$ represented in shades of grey. Peitgen & Richter 1986 taught everybody how to represent isolines or perspective views of $H(\lambda)$ in bright colors. But in early 1980, all I could do is to examine vertical cuts of that surface along lines in the λ plane. The abscissa being denoted by t, this figure combines the (non-overlapping original) records of cuts truncated to $H\leq 300$. corresponding to the lines $\lambda=t+0.00001i$, $\lambda=t+0.0001i$, and $\lambda=t+0.001i$. The range of t (not recorded on the originals) clearly corresponds to small islands along the real axis of λ. To make the overlap legible, the first and second cuts are moved up, respectively, by 100 and 50 iteration stages. The result confirmed what one could expect on the basis of real, as opposed to complex, iteration.

Again, let me elaborate. In parallel with blowups of the set *M*, we were also running pictures of the Julia sets for values of *c* that lie within the island molecules. They are exemplified by Figure 4. This was the "Eureka moment" for which decades of work seemed to have prepared me. The shapes we saw also appeared to split into many islands, each of these a reduced-scale version of the Julia set corresponding to a matching value of *c* in the continental molecule of the set *M*. However, we knew from old mathematics that this appearance had to be misleading. While island interiors cannot overlap, Julia's criterion implied—and closeups confirmed—that the gaps between islands had to be partly spanned by a peculiar geometry best illustrated by the following analogy. Imagine a stream so wide that only large beasts can jump from one side to the other. To accommodate smaller beasts, smaller stones are added half way through each gap. And to accommodate devilishly small beasts, the process must continue ad infinitum. Ultimately, the islands connect by their coastlines, adding up to "devil's" polymer, whose "strands" were invisible because actual computation is necessarily limited to a lattice.

Returning to the set *M*, it became very important that Myrberg had developed a theory of the iteration of the real map $x^2 + c$. That theory implies that the real axis pierces a string of islands of the set *M*, and connects them in devil's fashion by their coastlines. This suggested the conjecture that the whole set *M* is a connected devil polymer.

However, I faced this issue with a degree of caution that was out of character and perhaps brought about by living during that year among pure mathematicians. I presented the connectedness of the set *M* as a question to be answered, not a conjecture to be verified. This was a distinction without a difference but it led me in M 1980n{C3} to define an awkward surrogate to *M*, whose properties could be described in mathematically more "firm" fashion. M 1982F snapped back into proper assertiveness. The issue soon became moot: also in 1982, A. Douady and J.H. Hubbard gave a proof of the connectedness of *M* and went on to study it in admirable detail.

1.2.6 "Stellate structures" and uncanny resemblances between Julia sets and the "corresponding" corners of the Mandelbrot set. Continual flipping between the set *M* and selected Julia sets led to another exciting discovery. I saw that the set *M* goes beyond being a numerical record of numbers of points in limit cycles. It also has an uncanny "hieroglyphic" character: It includes within itself a whole deformed collection of miniature versions of all the Julia sets.

My other early observations included "stellate structures" that are common to the Julia set of parameter c and the neighborhood of c in the Mandelbrot set.

1.2.7 The conjecture that the Hausdorff–Besicovitch dimension of the boundary of M is 2. This conjecture was gradually refined and is discussed in Chapter C8.

2. AN INCONSEQUENTIAL ISSUE EXAMINED CLOSELY: WHICH WERE THE "FIRST PICTURES" OF THE MANDELBROT SET?

Once again, Douady's statement in Section 1.1, despite its brevity, touched upon two clearly separate issues that this and the next section will examine separately and closely.

2.1 In mathematics, a picture cannot be interesting in itself, but only for the descriptions and investigations that it triggers

Douady calls me the "first to produce pictures" of M. This is indeed the case. Moreover, his use of the plural "pictures" is accurate and important, but the statement demands careful elaboration. Suppose a future historian finds that Julia and/or Fatou observed the following somewhere, in passing: for $|c| > 2$, the quadratic Julia set is totally disconnected. Such a hypothetical but conceivable finding would imply an "approximate picture" of M as a circle. No conclusion can be drawn from this approximation, hence no one would claim it as a first.

Instead of a hypothetical event, consider two well-documented ones concerning a different object, namely, the diffusion-limited aggregates, which will be the topic of Chapter C22. After "DLA" was discovered in Witten & Sander 1981, a search through the literature revealed many old pictures of DLA-like natural phenomena. But those pictures had led to no insight, because suitable tools became available only after the development of fractal geometry. Fractals were a prerequisite for the brilliant and very influential description provided by Witten and Sander. Those authors are the discoverers of DLA, and the early pictures deserve neglect.

Similarly, Figure 2 of Chapter C19 served in M and Evertsz 1991 to report some very peculiar properties of the Laplacian potential around DLA. A scientist then pointed out a related figure that was older. But that figure had appeared without comment, for the sole purpose of illustrating a computing technique. Maybe it was "first" but it was forgotten, as it deserved.

2.2 Computing at Harvard in 1980

In 1980, the programming skills and minimal computer power needed to produce some kind of picture were widely available to anyone who asked. In terms of exploratory research, everyone was in the same boat. Only a few years earlier, when preparing M 1975O, I had tested my fractal model of coastlines by superposing the letters O, X, M, and W and filling the islands using a pen marker.

Moreover, I discovered the Mandelbrot set in 1980 while at Harvard, at a time when the computer facilities there were among the most miserable in academia. The basement of the Science Center housed its first D.E.C. Vax 50 computer (not yet "broken in"), pictures were viewed on a Tektronix cathode-ray tube (worn out and very faint), and hard copies were printed on an ill-adjusted Versatec device. We could only work at night when we had only one competitor for the machine, chemist Martin Karplus. We alternated in using almost all the machine time.

All this suffices to prove that my permanent position at the Thomas J. Watson Research Center of IBM and the quality of IBM computers were *not* a factor in those early discoveries. The contrary assumption, though widely expressed, is contradicted by the facts.

In 1979, as will be shown in Chapter C14, the quality of the IBM computers allowed me to see the Mandelbrot set, but failed to help me discover it. In any event, IBM was absolutely not graphics-oriented. I had no roomful of up-to-date custom equipment, only good friends who built a custom contraption that established the state of the art in image rendering. The now classic pictures in M 1977F and M 1982F required unheralded heroics and an investment in very rare skills and equipment that could not be justified until *after* discovery.

2.3 How the main picture of M 1980n{C3} became the tack of the town

The messiness of my early pictures was widely noticed. Mindful that offshore "islands" were hard to tell from specks of dirt, I kept warning the printer's assistants that the spots on my picture imprinted with the date 3/27/80 must *not* be "cleaned." However, as can be checked in Chapter CXX, the printer's assistants *did* "clean" the picture, to the widespread merriment of those who read the paper or heard me present its contents. Later, in this book's preparation, my files yielded a better picture imprinted with the date 81/12/10. It repeats the same warning: "Do *not* clean off the dust specks. They are *real*." For a later use of the same picture, this message was covered by a label with the following words in French: "IMPRIMEUR: N'EFFACEZ PAS CE QUI SEMBLE N'ÊTRE QUE DE PETITES SALETÉS SUR CE CLICHÉ ELLES SONT TOUTES RÉELLES ET IMPORTANTES."

Do not clean off
the dust specks
They are real, See Note →

IMPRIMEUR
N'EFFACEZ PAS CE QUI
SEMBLE N'ÊTRE QUE DES
PETITES SALETÉS SUR CE
CLICHÉ. ELLES SONT
TOUTES RÉELLES ET IMPORTANTES

To no avail! Also later in that book's preparation, a horrible thought came to mind. What about the two pictures of the *M* set that are featured in the widely read M1982F? Between the first and the current printing, the *M* set of $z^2 + c$ has been updated to show the notorious "hairs" or "polymer structure." In my early papers, this set was called the μ-*map*, because the physicists used to denote the constant c by $-\mu$, hence used to write the map in question as $z \rightarrow z^2 - \mu$.

But the illustration of the *M* set of $\lambda z(1-z)$ (originally called λ-map) was not willfully changed from the original one. Horrors! It is now free of specks! I rushed to check my archive copies of the first printing. Specks were present. Clearly, gremlins in the printing business had taken advantage of the update and repeated that evil deed.

To conclude, a message was heard far and wide *despite*, not *because of*, the pictures' quality. The discovery of *M* began only after pictures had been used for an inspiring description.

3. THE DISCOVERY OF THE MANDELBROT SET CONTRIBUTED TO THE REBIRTH OF EXPERIMENTAL MATHEMATICS

The demise of the presumed "beautiful IBM graphics" leads one to wonder why the Mandelbrot set was not discovered by someone with access to graphics better than what I had in 1980. The answer is part of the mystery of scientific discovery. A discovery is made when the tools—both intellectual and physical—are available to an individual with the motivation, acuity, and inspiration to use them, and I was motivated to sniff out the ramifications of those specks of dirt.

Later conversions to a belief in the power of observations were triggered by the inspiration that my conjectures about those dust spots awakened among many mathematicians—not by a discontinuity in the availability of computers. To quote from John F. Kennedy's 1962 Yale commencement address, "The great enemy of truth is very often not the lie— deliberate, contrived, and dishonest, but the myth—persistent, persuasive, and realistic." It was a myth—one all too persistent, persuasive, and long realistic—that experiment in mathematics had become useless.

3.1 A key fact that is perennially misunderstood and misinterpreted: many sciences, especially early on, encounter periods during which a well-trained and skillful eye is essential

It is worth recalling at this point some notable examples of the role played by properly interpreted messages from the eye. The telescope was invented, built, and marketed in Holland, as a toy. At least one person, Thomas Harriott (1560–1621), well regarded in his day as an astronomer and mathematician, had the idea of pointing it towards the Moon and making a drawing of what he saw. His drawing was preserved but shows nothing but blobs with no structure.

Galileo Galilei (1564–1642) was not the first to handle the telescope, but the first to change it from a toy to a vital tool. He was a trained painter, a negligibly minor one, to be sure, against stiff competition in Renaissance Tuscany, but equal to the new task of taming the telescope. Once directed to structures he had not discovered, Harriott instantly confirmed their existence. He might have uttered T.H. Huxley's exclamation upon reading Darwin's *The Origin of Species:* "How extremely stupid not to have thought of that."

There is a strong reason why Galileo performed (immensely) better than his contemporaries, invented physics, and became the first physicist.

Modern science arose in his hands from the notion that truth did not reside in ancient books written by men but in the "great book of Nature" opened in front of Man's eyes. In effect, Galileo argued against the New Testament assertion that "in the beginning was the word." Had he dared counter the Scriptures (unlikely, since he did not seek a conflict with the Church), Galileo might have proclaimed (as I do not fear to do) that "in the beginning were the picture and the eye." Even better, "In the beginning, the word joined the picture and the eye."

Let us move on. The microscope and the photographic camera were ancient tools in the time of Santiago Ramon y Cajal (1852–1934) but their availability did not suffice to resolve the complexity—nay, the utter messiness— of the human nervous system. Once again, the reason why "fate" chose Ramon y Cajal for this task is because the task was not "normal," and he—perhaps he alone in his day—was well prepared. Being a trained and infinitely patient artist, he overwhelmed the inadequacies of his miserably outdated microscope. He saw—and revealed through classic pictures—marvels that long remained unsurpassed. Around 1950, my neurologist friends were still relying on pictures first published in near-medieval Spain in the 1890s. That Cajal did not achieve and hold true fame is a disgrace.

A third notable "seer" worth mentioning was the meteorologist and geophysicist Alfred Wegener (1880–1930). The near match between the southwest coast of Africa and northeast coast of South America must have been noticed early. Wegener dared take the next step, from seeing to discovering. He compared fossils separated by the Atlantic and imagined a primordial continent that broke into parts that drifted away.

Galileo and Cajal loom high in my personal Pantheon. In kind, and without any claim concerning relative importance, my experience was like theirs. Quite explicitly, I thought of Cajal while discovering the M set and of Galileo while discovering the Brownian boundary dimension 4/3.

Major differences are obvious: Galileo's story marked a nearly absolute beginning, Ramon y Cajal's marked a deepening, and mine marked a renewal after a long lapse. The "fate" that drove me to revive the theory of iteration, first chose me to reinvent the role of the eye in a field, mathematics, where it and explicit computation had become anathema, about as unwelcome as they could possibly be.

3.2 The culture of mathematics during the 1960s and 1970s

Within that culture the Mandelbrot set could not have been discovered. Hence its discovery marked a historical departure. Today—but not yesterday— only a minority among mathematicians would agree with the opinion due to someone *who did not* discover that set, that the study of *M* reflects "a rather infantile and somewhat dull mathematical sensibility" (Brooks 1989).

The attitudes in which all mathematicians were trained not so long ago is witnessed by Stanislas Ulam (1909–1984). He might have been expected to speak to the power and utility of the pictures produced by his associates. Instead, Ulam 1974 (pages 378 and 490) informs us that "Mathematics is not really an observational science and not even an experimental one. Nevertheless ... computations... were useful in establishing some curious facts ... Fermi expressed ... a belief that it would be useful to attempt practice in the mathematics... of nonlinear systems. The results [described in a famous report by Fermi, Pasta, and Ulam reproduced in Ulam 1974] ... were interesting and quite surprising to Fermi."

Among pure mathematicians, Ulam's lukewarm advocacy of the computer strengthened an antagonism that was obvious well before 1979. For example, a sustained effort brought prominent youngish mathematicians to visit the IBM Research Center to lecture on diverse topics. The unexpressed hope was that the computer's promise would impress some of them. The failure was complete, and (much later) certain communities accepted my "anomalous" manners grudgingly.

How to respond to serious thinkers who do not wish to distinguish between seeing and discovering? Some still seek "pictures of *M*" earlier than either the trove of mine, or the single one found in Brooks & Metelski 1981 and often mentioned after 1988. That single picture is so indistinct that it could not—and did not—lead to any discovery.

To show how counterproductive such a search for "first sighting" can become, let me broaden it from iteration to all fractals. Who provided the first massive collection of many pictures clearly recognizable as fractal? Could it be the marvelous Hokusai Katsushika (1760–1849), to whom I pay homage in Plate C16 of M1982F and again in Chapter C13. His unforgettable pen drawings of *One hundred views of Mt. Fuji* depict clouds and trees admirably, and all those who can see and are familiar with fractal geometry recognize that Hokusai had a perfect "eye for fractals."

But neither he nor earlier or later landscape painters have any claim for fractals as a topic for either mathematics or science. Those credits belong, respectively, to contemporaries of Cantor and Peano, and to me. Yet, Hokusai holds a central role in my current view of fractals as a notion familiar to Man, in one form or another, since time immemorial.

The fractals' prehistory is a long story that must be reserved for a different forum, but it raises a question I commend to the specialists. The time when Cantor, Peano, et al. flourished was the heyday of Japanism, when Hokusai's direct influence on Western European art was widely acknowledged. Did this influence extend—with no acknowledgment — from the artists to the mathematicians?

3.3 A changed mood in mathematics?

An entirely different world is called for in Bourguignon 1999. Here is a free translation. "I think one must distinguish the future of mathematics from the future of those who claim to be mathematicians. The future of mathematics overflows with challenges and promise. But I fear that mathematicians may spoil it by failing to open up and dare, and by exhibiting a high propensity to exclude (I should say "excommunicate" because of religious overtones) from their community whole domains of knowledge. Without fear of fresh air, mathematicians must open up without shivering." I read this text with equal pleasure and surprise.

4 SUMMARY

One must heed the wise words of Whitehead 1974 (p. 127) that "To come very near a true theory and to grasp its precise application are two very different things, as the history of science teaches us. Everything of importance has been said before by somebody who did not discover it." The thought also applies when the word "said" is replaced by "seen."

A contribution to mathematics and/or science does not consist of a picture, but rather of a picture combined with a description. Without words and formulas, a picture can, at best, be praised for artistic quality. Without an interest in pictures and other aspects of "reality," pictures can play no role whatsoever.

This comment will be amplified in the next chapter by a discussion of several individuals and institutions. All affected my life when I was seeking my way.

First publication

C2

Acknowledgments related to quadratic dynamics

THIS CHAPTER BEGAN AS A SINGLE PAGE TO ACKNOWLEDGE my indebtedness to three individuals: Mandelbrojt, Douady, and Hubbard. But it grew and — unavoidably — became increasingly autobiographical. It even extended the scope of the word "acknowledgment" by commenting about Bourbaki, my Nemesis.

When first mentioned, key names are set in bold, **roman** or ***italics***. Some background is provided for those unfamiliar with the history of mathematics, especially in France in the middle of the twentieth century. A more extensive background is found in Chapter 25.

1. Szolem Mandelbrojt (1899–1983)

Above all, to amplify this book's dedication, I am endlessly indebted to my uncle Szolem, a noted mathematician who reached the Collège de France, the top of French academia, when he was thirty-eight and I was thirteen. So I always knew that science was not just recorded in dusty tomes but was a flourishing enterprise, and the option of becoming a scientist was familiar to me as long as I can remember.

Brilliant, bold, and ambitious, he left his native Poland for France at age twenty, as an "ideological" refugee repelled by the excessively abstract "Polish mathematics" then being invented by Waclaw Sierpiński (1882–1969). He was, to the contrary, attracted to the mathematical school that ruled Paris in the 1920s, one linked with ***Henri Poincaré*** (1854–1912). He became close to ***Jacques Hadamard*** (1865–1963) and Vito Volterra (1860–1940), the period's most influential mathematicians in Paris and

Rome, respectively. My father later joined Szolem in France, as a refugee from Polish economic depression and political hopelessness. Clashes between Sierpiński and my uncle concerning mathematics and other matters! saved our lives.

My uncle and I loved each other, had very different scientific tastes, and argued endlessly, especially about the best ways of devoting one's life to matters of the mind. Each of us followed a single star with unwavering fidelity. I criticized his path while understanding its value, but he never understood mine. Be that as it may, both overall and in many specific ways, he was my earliest and foremost mentor; on balance, his influence is increasingly obvious: broad and very positive. I interviewed him at length, and Mandelbrojt 1985 can now be found on my Web site.

2. First encounter with Nicolas Bourbaki

In February 1945, I passed the entrance examination of the prestigious École Normale Supérieure. I would have been ranked number one in a mathematics–physics class that (for the whole of France) counted only fifteen students! This was an "acrobatic feat," since war conditions had reduced my preparation for the exams from two years at least to a few weeks. My success was largely due to a natural ability in geometry.

At that point, my uncle told me that Normale was about to be taken over by a utopian movement toward mathematical abstraction, "purity," and isolation, whose corporate name was ***Nicolas Bourbaki.*** My uncle had been the oldest member of the original group. After World War II he chose to "retire early," while remaining close to his old friends. His reservations are alluded to in Mandelbrojt 1982, 1985 and Chapter C25.

Bourbaki showed extraordinarily wide-reaching concern with political influence across age groups and across disciplines. They acquired the authority to educate children (even though they disavowed the French math that was "new" yesterday), to spread correct thinking among the young, to cow the old, to shun the strong-willed heretics, and to try to export their standards of rigor and taste everywhere, even in areas where they did not belong. They have done untold harm.

Bourbaki gave itself an aura of a "secret society" that continued when they seized power. Many former members say little about the record. This would-be token of "discretion" also controls history; hence it is good to provide examples of their attitudes toward science and geometry.

André Weil (1906–1998) used to be described as only the first among equal original members of Bourbaki, but since his death he is publicly acknowledged as having been the instigator and leader. As a "postdoc," he spent a year in the Göttingen of David Hilbert (1862–1943) where the will of Hilbert and the design of Felix Klein (1849–1925) followed by Richard Courant (1888–1972) placed theoretical physics, numerical methods, and the more esoteric mathematics as neighbors in the same building. The physicists Max Born (1882–1970) and Werner Heisenberg (1901–1976) were inventing quantum mechanics in bright limelight. However, Weil saw only a paradise of abstraction, which he extracted and enforced in Paris and later the USA. In his paradise, mathematicians were trained unspoiled by any impure contact. Altogether, his view of the world was sorely biased, and his own life as described in Weil 1992 related to recorded history in a dishearteningly distant fashion. So did the credit he chose to dole to others.

Another founder, ***Jean Dieudonné*** (1906–1992), often described Bourbaki as the response of a few brilliant individuals to various aspects of France after World War I. Moreover, France of the 1940s and the 1950s abounded in all-consuming ideologies that led my most brilliant contemporaries to serve one of a small list of all-devouring causes. Be that as it may, Dieudonné was happy to view "some mathematical objects, like the Peano curve, as totally nonintuitive ... extravagant." Arguing for the precise contrary of this belief, Chapter 7 of M 1982F pointed out that the lining of the lung's pipes is nearly a space–filling surface, providing an approximate image that instantly changed our "intuition."

A later generation Bourbakiste, ***Laurent Schwartz*** (1915–2002), did express an interest in physics. But Schwartz 1997 reports (p. 58) that "The contrast between my love for geometry and my near–complete lack of geometric vision is truly mystifying. All that I recall is non figurative.... Roughly, I love geometry that is close to algebra or analysis, not visual geometry. This lack of spatial ability explains why I never learned to drive. I can prove the Pappus theorem but do not see the figure."

The preceding paragraphs dealt with Frenchmen prominent in the environment to which I was reacting in 1945. But parallel developments occurred in the USA. In Stone 1962, ***Marshall Stone*** (1903–1989) hailed "a revolution in mathematics" consisting in "the discovery that mathematics is entirely independent of the physical world... . When we stop to compare the mathematics of today with mathematics as it was at the close of the nineteenth century we may well be amazed to see how rapidly our mathematical knowledge has grown in quantity and in complexity, but we should also not fail to observe how closely this development has been involved with an emphasis upon abstraction, and an increasing concern with the perception and analysis of broad mathematical patterns. Indeed, upon close examination we see that this new orientation, made possible only by the divorce of mathematics from its applications, has been the true source of its tremendous vitality and growth during the present century."

Once again, mentioning those persons and opinions gives the word "acknowledgment" an unusually broad meaning. Their influence on me has been enormous, since keeping away from them, and later assisting in their ideological demise, became an important aspect of my life and in particular of the work described in this book.

3. Two days at Normale and the continuing fallout

The quotation from M. Stone in Section 2 reminds me of an attractive and revealing autobiographical essay. In Hewitt 1990 we read that "From Stone and his fellow mathematicians at Harvard, I learned vital lessons about our wonderful subject: • Rule #1. Respect the profession. • Rule #2. In case of doubt, see Rule #1."

My uncle also had a deep respect toward his profession. His private reservations about Bourbaki did not in the least extend to École Normale. He passionately wanted me to go there. I did, but the next day walked out. He was bitterly disappointed, constantly worried about my future, and only when close to death did he stop asking me, "But why?"

Because, giddy as I was with surviving the war and passing those exams, prestige and authority did not affect me sufficiently. That strong-willed institution was absolutely the wrong place for a strong-willed person who dreamt of helping unscramble the messiness of nature. To me, the eye and geometry were not mere keys to acrobatic exam scores. I worshipped them, therefore turned down a golden opportunity to "outgrow" them so as to become a "true mathematician." Instead — or so it seems — I went on to prepare myself to sniff specks of dust on bad quality computer pictures.

In Paris, my leaving École Normale for École Polytechnique created a durable scandal. Had I never registered, my life might have been easier. On the other hand, the Directeur and another alumnus of École Normale attended my seventieth birthday celebration in Curaçao. As a joke, they "appointed" me as an "honorary freshman for life."

What I wanted to do in life was incomprehensible to others but surprisingly clear to me in 1945. Thus, as I near eighty, it is a deep privilege to observe the following. Having failed to prove any difficult theorems is not for me a source of any regret, and whatever I accomplished is roughly what I had hoped. Also, in recent years, I devote quite some effort to the teaching of mathematics. We painstakingly revert André Weil's blindness to science and Laurent Schwartz's blindness to pictures, and find — marvelous to say — that the average person actually likes mathematics.

An afterthought: Bourbaki took over Normale two years later than my uncle had been led to expect. In addition to being inappropriate for me, Normale would have been boring. *Felix culpa.*

4. Early encounter with Fatou–Julia

A few years after 1945, I found myself in a state of great certainty. At this point, my uncle made a fresh attempt to save me for pure mathematics of the kind he practiced and loved, whose flavor is sketched in Chapter C25. One of his all-time favorites, due to ***Pierre Fatou*** (1878–1929) and ***Gaston Julia*** (1893–1978), dated to about 1917 and concerned the iteration of rational functions. His personal relations with Julia were dismal, but he agreed with the converse of the opinion of Plutarch (*Pericles* 2.1) that "It does not of necessity follow that, if the work delights you with its grace, the one who wrought it is worthy of your esteem."

After an early period of glory, with no clue of what to do next, the mathematical theory of iteration became dormant. To be sure, there was a trickle of articles, and ***Paul Montel*** (1876–1975), kept the flame alive. Both Fatou and Julia has relied heavily on Montel's theory of "normal families of functions" — arguably providing it with its most beautiful application. Montel (who became my uncle's Ph.D. advisor a few years later) devoted several courses to iteration, but his continued proselytizing was low-key and had no effect. Neither did Chapter VII of Montel 1927 (titled *Itération des fonctions rationnelles*) and Chapters XXIII, XI, and XII of Montel 1957 (respectively titled *Récurrences du premier ordre, Itération,* and *Itération des fonctions à cercle fondamental*). Arguably, no one else cared.

Julia (my professor of differential geometry at Polytechnique) was as "unfashionable" as can be. Abstraction was in the saddle and, as shown in Chapter C25, the subfield of mathematical analysis had my uncle as its "captain" in France but was, throughout the world, very much on the defensive.

Recklessly overconfident in my abilities, my uncle introduced me to Montel and suggested "Fatou–Julia" as the topic of a Ph.D. dissertation. He thought a marvelous career would open if I could find a new "angle" to awaken iteration from what was already at that time a thirty-year-long slumber. He even lent me very valuable original reprints given to him by the authors themselves.

I tried hard but met utter frustration. My uncle implicitly expected the new angle to guide me to originate within mathematics itself. A new angle did arise thirty years later, but it originated in respectful examination of mounds of computer–generated graphics.

The altogether different thesis I defended in 1952 had nothing to do with Fatou–Julia. But in an odd and very indirect fashion it, too, was indebted to my uncle: It was triggered by a paper that he retrieved from his wastebasket and gave me to read. That thesis grew into fractal and multifractal geometry as they stood in the mid-seventies. Building that geometry gave me experience that no one else had in using graphics in the search for mathematical facts to discover, ponder, and perchance prove. And in due time I returned to Fatou–Julia.

5. How Hadamard (posthumously) triggered my return to iteration

My return to diverse forms of iteration was unplanned, but after the fact, it looks inevitable. The trigger was the unique eulogy of Poincaré in Hadamard 1912, an extraordinary historical document that came to my attention after M 1975O but before M 1977F. It made me recall Bourbaki's hostility to Poincaré and the awkward centennial in 1954. As a result, Hadamard 1912 made me move on from linear to nonlinear fractals. My uncle's mentor and hero also influenced me in many ways.

I was bowled over — as noted in M 1982F, p. 415a and retold in Chapter C15 — by Hadamard's beautiful story of how proto-fractal dusts appeared nearly simultaneously in two contexts. By far the best-advertised context was the search for counterexamples that had led Cantor to objects I later used as "cartoons" of nature and culture. But Poincaré had independently uncovered a little-known second source that Hadamard found more admirable, and so did I. It consisted of the limit sets of certain Kleinian groups, the topic to which Part III of this book will be devoted. This made me recall that Julia specifically motivated his study of rational iteration by dynamical arguments that referred to Poincaré.

Reading Hadamard made me realize that all the fractals I had studied were invariant under *linear* transforms. Each could, in a deep way, be described as being a "linear fractal." The linear models of the Earth's relief or turbulence are extraordinarily effective but the underlying phenomena are nonlinear. Hadamard made me view the effectiveness of my models (depending on the mood of the day) as either miraculous or suspect.

Having been (briefly) an aeronautics graduate student at the California Institute of Technology, I had become familiar with the raw beastliness of the Navier–Stokes equations and had read von Karman 1940, titled "The engineer grapples with nonlinear problems." Hence nonlinearity per se had been familiar to me well before chaos theory.

6. The Julia–Fatou theory, revived and brought to fuller glory

After a side excursion into Fuchsian and Kleinian groups described in Chapter C15, I rejoined the scientific bride selected for me by my uncle. This time, I was "seeking by looking."

The search stumbled along for a while in a very interesting manner that, as already mentioned, is best withheld to Chapter C12. It converged in 1979–1980 when I was a visiting professor in the Mathematics Department at Harvard. I am grateful to ***Barry C. Mazur*** for proposing and arranging this visit, which made possible one of the most fruitful years of my life, and to the then chairman ***Heisuke Hironaka*** for welcoming and befriending me. Mine was the first course centered on fractals at any institution, and that year also made possible Gefen, M, and Aharony 1980.

More important from the present viewpoint, my work on iteration moved on, assisted by an excellent programmer, ***Peter Moldave***, Harvard class of 1980. On my arrival in 1979, my new colleagues had dismissed my pictures, but their mood changed, gradually but palpably. Hot from the computer printer, the properties of the emerging Mandelbrot set-to-be were presented daily to colleagues in front of the Mathematics Department mailroom. They were summarized in late spring 1980 in well-attended lectures and in seminars around town. I also described it privately to Jean Leray (1906–1998) who was visiting MIT. I had once viewed him as a possible Ph.D. advisor. He remembered the work of Julia, was amazed that it could spring back to life, and observed that "tout arrive."

In some early papers, the set I investigated was called *separator*. Elsewhere, it was called *μ-map*, because physicists used to write the quadratic map as $z \rightarrow z^2 - \mu$. In the same papers, the locus for the equivalent map $z \rightarrow \lambda z(1-z)$ was called the *λ-map*.

The next step was a visit to the Institut des Hautes Études Scientifiques (IHES) located in Bures–sur–Yvette, near Paris. I am grateful to ***David Ruelle*** for inviting me. Since *verba volant, scripta manent,* as already mentioned, my IHES lecture on November 13, 1980 followed closely the text of M 1980n {C3} and I handed out photocopies of the galley proofs. But I also showed additional illustrations, both pitiful originals from Harvard and better later renderings from IBM.

The lecture hall was packed and filled with excitement. There was a palpable feeling of an "event" that I (for one) shall never forget. On this occasion, the Fatou–Julia theory "officially" came back to life.

7. First encounters with Adrien Douady and John H. Hubbard

Douady and Hubbard earned my gratitude for three reasons:

• For the enthusiasm for my work that they showed immediately after the 1980 Bures lecture. On that account, Douady 1986 reported the following: "I must say that in 1980, whenever I told my friends that I was just starting with J.H. Hubbard a study of the form $z \rightarrow z^2 + c$, they would all stare at me and ask, 'Do you expect to find anything new?,'"

• For denoting a set that I had discovered and described by the term "Mandelbrot set," which they coined for this purpose. A quotation from Douady 1986 is found near the beginning of the preceding chapter.

• For encouraging acceptance of this term by the mathematical community.

Douady's name had long been familiar, not only to me but to all mathematicians. But we met only at the Université de Paris-Sud (Orsay) after I lectured on my "Kleinian" algorithm described in Part III. The day was either November 10 or 12, 1982. I could not check, but November 11 is excluded, being a holiday in France. The quotation from Douady 1986 are best appreciated with the knowledge that their author was a member of Bourbaki (the official leader at that time.) Hence, the sudden (and permanent) change in his interests blurred the image and self-image of Bourbaki and was widely noted and commented upon.

November 15, 1980, is a date in my old appointment book that I remember fondly. Douady and his family invited my wife and me to their house and surrounded me after dinner. I patiently described to them some of the riches that I had uncovered, and they showered me with questions that I was prepared to answer instantly. More generally, I discussed a skill that the development of fractal geometry, specifically, the preparation of M1975O and M1977F, had honed over the preceding fifteen years. Once again, that skill consists in respectful and fruitful examination of those and other pictures. When seeking new insights, I look, look, look, and play with many pictures. (One picture is *never enough*!)

Not surprisingly, Douady, like other eminent "pure" mathematicians, had never examined a picture in that way. In the form of conjectures, most of them already fully specified, I offered him a crop of observations. He was "converted" very quickly and permanently.

Having heard of Hubbard a bit earlier, I invited him to lecture at IBM Research in 1979. He showed and commented upon an image of a Julia set relative to the Newton method for the roots of the equation $z^3 - 1 = 0$ —an ancient favorite of E. Schröder, A. Cayley, and Fatou and Julia. Programmers at the Université de Paris-Sud (Orsay) had prepared it to illustrate his course there as visiting lecturer. He had brought along four letter-size printouts with burned edges. I taped them together and gave him a clean reduced copy that he could show in lectures.

I saw Hubbard next in 1980 when he briefly stopped at Harvard and heard echoes of the above-mentioned seminar. I sketched my findings and discussed my forthcoming visit to IHES.

After my November 1980 lecture at IHES, Douady and Hubbard took up my conjectures and pioneered in translating them rapidly into organized mathematics. They (and also Nessim Sibony, who left this result unpublished) found a way to prove that the Mandelbrot set M is connected. They ran a seminar on this topic and on one occasion received me very warmly as a speaker.

Their deeper conjecture that M is also *locally* connected, "MLC," restates the conjectured relation to M^0 that had originally motivated me to study M. It remains open, and partial results receive high praise, and they helped both J.-C. Yoccoz and C. McMullen receive Fields medals.

Follow-up

Very rapidly, my work and the term for it coined by Douady and Hubbard became widely known in Paris. Back at IBM and unaware of those developments, I completed *The Fractal Geometry of Nature,* M 1982F. It came out in mid-1982, with Chapter 18 being devoted to iteration. Marvelous pictures had become feasible because I knew what was needed, and software that greatly helped M 1982F soar was developed by ***Richard F. Voss*** and ***V. Alan Norton***. They deserve to be thanked once again.

The broad "popular" acceptance of M was largely spontaneous, as already mentioned, but it needed a catalyst. Society is deeply in debt to ***Heinz-Otto Peitgen*** and ***Peter H. Richter***, whose beautiful color graphics were collected in Peitgen and Richter 1986 and previewed by Dewdney 1985 in *Scientific American.* They put my own efforts to shame. I decided immediately not to compete, but to cooperate, and those brilliant disciples became close friends.

In September 1987, the National Science Foundation sponsored a week-long regional conference on fractals at the University of Cincinnati. I am grateful to ***John W. Milnor*** for his support and his lectures.

8. Change of mood and "fracas"

By 1988, the Mandelbrot set had grown from an exotic novelty into the topic of many talks at the centennial celebration of the American Mathematical Society. By then, unfortunately, some persons became unhelpful. Early events stopped being mentioned and writers who had neither witnessed the events nor checked the record provided imaginative but dull anecdotes. Not to be "indicted" without reason would have been preferable, but the fracas subsided when all claims collapsed, proved self-contradictory, or failed to be documented and were silently withdrawn. This cleared up early an issue often left murky. For credit for the discovery of the first and most striking properties of M — including the existence of atoms and islands, the question of connectedness, and the dimension 2 of the boundary — no documented "competitor" is left. The historical background is expanded in later introductory chapters.

9. Postscript

In the last twenty years, mathematics has changed so deeply that to younger persons this chapter's story might be simply incredible. They wonder how it could be that from the 1920s until 1980, the Fatou–Julia theory saw only one significant advance. The renowned Carl Ludwig Siegel (1896–1981) corrected a technical error in Julia's work and discovered "Siegel disks." His work, which M1985g{C10} attempts to motivate, had a deep impact on analysis but did not reawaken Fatou–Julia. In addition, a few Scandinavians worked on the theory of iteration.

The key reason for the neglect of the Fatou–Julia theory of iteration of functions was mentioned repeatedly and elaborated in Chapter C25: it belonged to mathematical analysis, a deeply "unfashionable" field for many decades. Hence, the interest that many brilliant mathematicians took in my observations/conjectures concerning the Mandelbrot set represented a historical change and a return to a far less ideological view of their craft. I regret that the ride was rough, but it is a delight that my work helped.

Annals of the New York Academy of Sciences, 1980.

Fractal aspects of the iteration of $z \rightarrow \lambda z(1-z)$ for complex λ and z

• ***Chapter foreword concerning the illustrations, especially the "missing specks" of Figure 1 (2003).*** As described in Chapter C1, this paper boasts many "firsts" and was instrumental in reviving the theory of iteration. The many new observations it contains concern the set in the μ-plane for which A. Douady and J.H. Hubbard soon proposed the term "Mandelbrot set." Each observation was stated as a mathematical conjecture or became the source of one. Thus, the figures in this paper played a fundamental historical role.

Nevertheless, due to events also described in Chapter C1, Figure 1 of the original was crude beyond necessity, flatly contradicting the accompanying text. The latter spends significant time presenting my identification of apparent "specks of dirt" as "islands." But a last-minute visual "checker" somehow ordered those specks to be erased.

Dire necessity forced me, whenever I handed out a reprint, to add a few specks. Everybody had a good laugh, but the problem continues throughout the literature—while some continue to believe that my graphics were always flawless.

The mishandled original having been preserved, it can at long last be made generally available by changing Figure 1 into a composite. The bulk reproduces my original illustration. Next to it is a piece of the speck-free figure, as "improved" by the printers.

Needless to say, the original does not show the "hairs" that, to use the terminology of Chapter C1, weave the islands into a "polymer." An algorithm that exhibits them was discovered only much later by W.P. Thurston.

Significant historical importance of a different kind attaches to Figure 4, which later became Plate 198 of M 1982F{*FGN*}. To the best of my knowledge, this is the earliest published example of a pattern generated by an "iterated function system" (IFS), a notion discussed in Part X. Asymptotically, the IFS of inverse quadratic dynamics traces an underlying Julia curve made of many loops. Preasymptotically, however, this curve is outlined very imperfectly: It is overexposed at some points and underexposed elsewhere. This irregular exposure—also observed for the IFS of Kleinian groups (Chapter C15)—is a token of multifractality.

In this reprint, the sections have been numbered. •

GIVEN A MAP $z \to f(z, \lambda)$, WHERE f IS A RATIONAL FUNCTION OF the complex numbers z and λ, consider the iterated maps $z_n = f(f(\ldots f(z_0)))$ of the starting point z_0. To achieve a global understanding of these iterates' behavior, Cayley, Fatou and Julia saw was necessary to allow λ and z_0, hence z_n, to be complex variables. To the contrary, the extensive recent studies of the map $z \to \lambda z(1-z)$, for example those found in Gürel & Rössler 1979 and Helleman 1980 are largely restricted to λ real $\in [1, 4]$ and z real $\in [0, 1]$. Hence, they are powerful but local and incomplete. The global study for unrestricted complex λ and z throws fresh light upon the results of these restricted studies, and reveals important new facts.

In this light, a change of emphasis to investigate even more general maps $\{x \to f(x, y, \lambda); y \to g(x, y, \lambda)\}$, appears premature.

1. Introduction

The present paper stresses the role played in the unrestricted study of rational maps by diverse fractal sets, including λ- fractals in the λ plane and z- fractals in the z plane. Some are fractal curves (that is, of topological dimension 1) and others are "dusts" (that is, of topological dimension 0). The z- fractals are of special interest, because they can be interpreted as the fractal attractors of appropriately defined (generalized) discrete dynamical systems, based upon *inverse* maps. This role is foreshadowed in Fatou 1906, 1919-1920 and Julia 1918 (and even earlier in H. Poincaré's work on the related context of Kleinian groups {P.S. 2003. See M 1983m{C18}). However, an explicit and systematic concern with fractals only came with M 1975O and M 1977F, where the notion itself is first defined and given a name: A fractal set is one for which the fractal (Hausdorff-Besicovitch) dimension strictly exceeds the topological dimension. This paper's illustrations are fresh (and better than ever) examples of what this definition implies intuitively. The text is a summarized excerpt from M 1982F{*FGN*}.

This paper's Section 5 comments on "strange" attractors.

2. THE λ-FRACTAL Q

We denote by Q the set of values of λ with the property that the initial points z_0 for which $|\lim_{n\to\infty} z_n| < \infty$ include a closed domain (that is, a set having interior points). It is well known that it suffices that $|\lim_{n\to\infty} z_n| < \infty$ hold when the initial point is the "critical" point $z_0 = 0.5$ The portion of Q for which $Re(\lambda) > 1$ is illustrated on Figure 1, the remainder of Q being symmetric of this figure with respect to the line $Re(\lambda) = 1$.

A striking fact, which I think is new, becomes apparent here: Figure 1 is made of several disconnected portions, as follows.

2.1. The domain of confluence $\mathcal{L}$, and its fractal boundary

The most visible feature of Figure 1 is a large connected domain $\mathcal{L}$ surrounding $\lambda = 2$. This $\mathcal{L}$ splits into a sequence of subdomains one can introduce by successive stages.

The first stage subdomain, $\mathcal{L}_0$, is constituted by the point $\lambda = 1$, plus the open disc $|\lambda - 2| < 1$. This disk is left blank on Figure 1 to clarify the remainder's structure. If and only if $\lambda \in \mathcal{L}_0$, there is a finite stable fixed point. [Proof: when $Re\lambda > 1$, a stable limit point, if it exists, is $1 - 1/\lambda$; the condition $f|$ (1-1/ lambda) lor lt 1 boils down to $|\lambda - 2| < 1$].

The remaining, and truly interesting portion of $\mathcal{L}$ is shown in black on Figure 1. The fact that this black area is "small" means that the map $z \to \lambda z(1-z)$ is mostly not bizarre. However, many interesting and bizarre behaviors (some of them unknown so far, and others thought to be associated with much more complex transformations) are obtained here in small but nonvanishing domains of λ.

Each of the second stage subdomains of $\mathcal{L}$ is indexed by one or several rational numbers α/β. The subdomain $\mathcal{L}(\alpha/\beta)$ is open, except that we include in it the limit point where it attaches to $\mathcal{L}_0$ like a "sprout;" this is the point $\lambda - 2 = -\exp[-2\pi i\alpha/\beta)]$. When $\lambda \in \mathcal{L}(\alpha/\beta)$, the sequence z_n has a stable limit cycle of period β. This cycle can be obtained through a single β-fold bifurcation, by a continuous change of λ that starts with any stable fixed point, for example with the stable fixed point $z_0 = 0.5$ corresponding to $\lambda_0 = 2$.

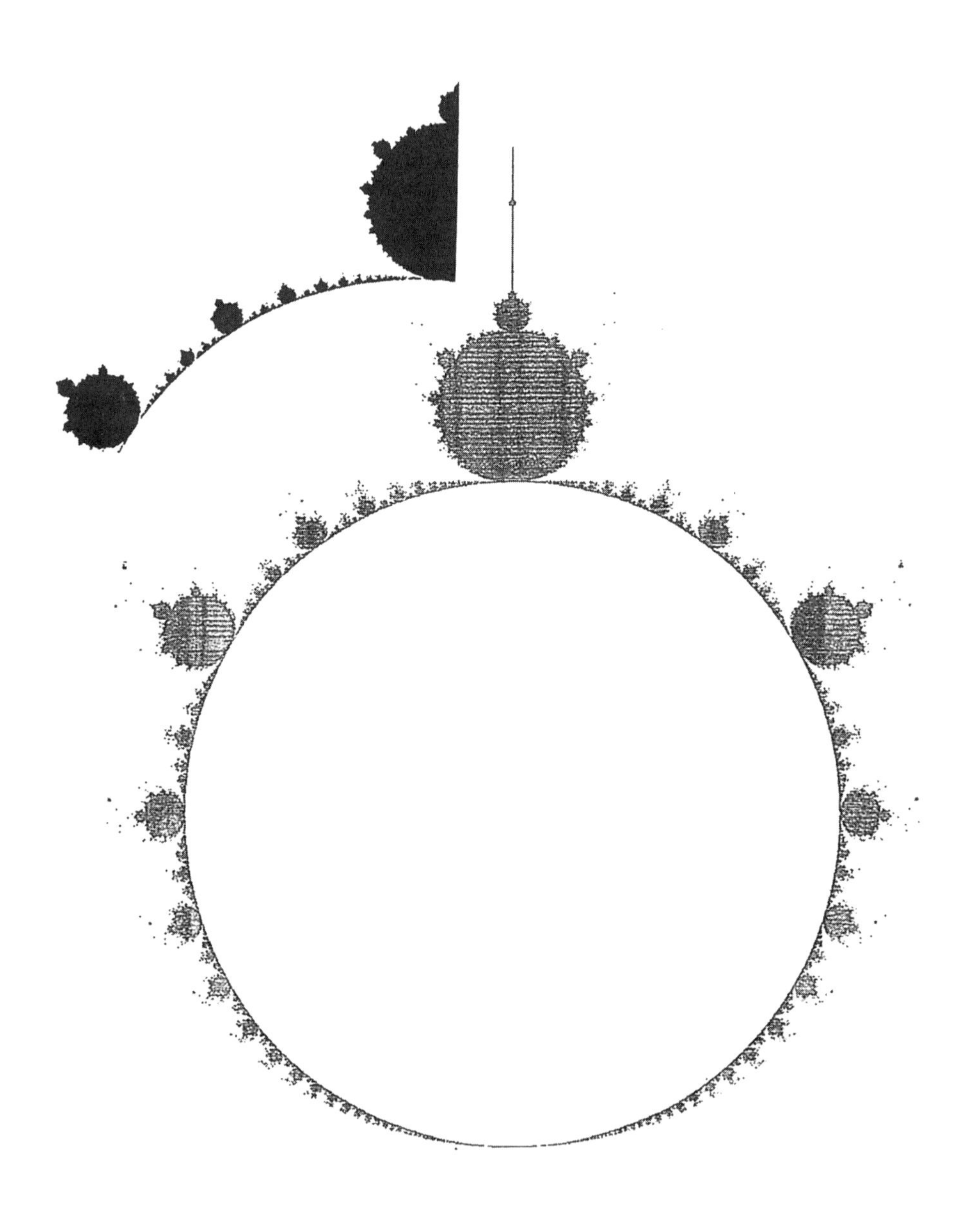

FIGURE C3-1. Complex-plane map of the λ-domain Q. The real axis points up starting at $\lambda = 1$. The circles center is $\lambda = 2$ and the tip is $\lambda = 4$. {PS 2003: please check this chapter's Foreword.}

Each of the third stage subdomains of $\mathcal{L}$ is indexed by two rational numbers: $\mathcal{L}(\alpha_1/\beta_1, \alpha_2/\beta_2)$; it is open, save for the point where it attaches, sprout-like again, to $\mathcal{L}(\alpha_1/\beta_1)$. When $\lambda \in \mathcal{L}(\alpha_1/\beta_1, \alpha_2/\beta_2)$, the sequence z_n has a stable limit cycle of period $\beta_1\beta_2$, resulting from two successive bifurcations, respectively β_1 – fold and β_2 – fold, starting with a stable fixed point in $\mathcal{L}_0$.

Further series of subdomains are similarly indexed by increasingly many rational numbers $\alpha_1\beta_1 \ldots \alpha_g/\beta_g$. $\mathcal{L}$ combines all the values of λ that lead, either to stable limit points of z_n or to stable limit cycles that can be reduced to stable limit points by a process that is the inverse of bifurcation. I propose for this process the term, *confluence*, which is why I call $\mathcal{L}$ the *domain of confluence*.

The domains $\mathcal{L}(\alpha/\beta)$ etc. are nearly disc shaped, but not precisely so. More generally, the boundary of each sprout is nearly a reduced scale version of the whole boundary of $\mathcal{L}$. Recalling the classic construction of the snowflake curve (M 1977F), there is little doubt that the boundary of $\mathcal{L}$ is a fractal curve.

2.2. The transformed domain $\mathcal{M}$ using the often-invoked transformed variable $w = (2z - 1)\lambda/2$

This transformation re-expresses the map $z \to \lambda z(1-z)$ into $w \to \mu - w^2$, where $\mu = \lambda^2/4 - \lambda/2$. This leads to the replacement of the λ-set $\mathcal{L}$ by a μ-set $\mathcal{M}$. The counterpart to the disks $|\lambda - 2| \leq 1 (\equiv \mathcal{L}_0)$ and $|\lambda| \leq 1$ (= the symmetric of $\mathcal{L}_0$ with respect to $Re(\lambda) = 1$) is a shape $\mathcal{M}_0$ bounded by the fourth order curve of equation $\mu = e^{2i\gamma}/4.e^{i\gamma}/2$. The sets $\mathcal{M}$ and $\mathcal{M}_0$ will be needed momentarily. Hence, the scholars' known hesitation between the notations involving λ or μ is not resolved here: the shape $\mathcal{L}_0$ is by far simpler than $\mathcal{M}_0$, but $\mathcal{M}$ is more useful than $\mathcal{L}$.

2.3. The domains of non confluent, or K-confluent, stable cycles

In addition to $\mathcal{L}$ the domain Q is made of many smaller subdomains. Indeed, I discovered that at least some of the apparent specks of dirt or ink on Figure 1 are indeed real: more detailed maps reveal a well defined "island" whose shape is like that of $\mathcal{M}$, except for a nonlinear one-to-one deformation. Each island is, in turn, accompanied by sub islands, doubtless ad infinitum.

When λ lies in an island's deformed counterpart to $\mathcal{L}_0$, then z_n has a stable limit cycle of period $w > 1$.. When λ lies in an island's deformed counterpart to $\mathcal{L}(\alpha_1/\beta_1, \ldots \alpha_g/\beta_g)$, then z_n has a limit stable cycle of period $\omega\beta_1. \ldots \beta_g$. Again, one would like to be able to reduce these cycles through successive confluences provoked by continuous changes in λ, to the fixed point $\lambda_0 = 2$. But this is impossible. *None* of these fixed cycles is confluent to a fixed point.

Some islands of $\mathcal{L}$ that intersect the real axis create intervals that have been previously recognized and extensively studied. It was clear that a cycle with λ in such an interval is not confluent to $\lambda = 2$ through real values of λ. We see it is not confluent through complex λs, either.

2.4. The radial patterns in the distribution of the domains of non confluence

The islands that intersect the real axis can be called "subordinate" to the value of $\lambda = 3.569$, which is known to mark the right-most point of $\mathcal{L}_0$, and corresponds to an infinite sequence of successive 2-bifurcation. More generally, I observe that every island is subordinate to a λ corresponding to an infinite sequence of successive bifurcation. The subordination is spectacular (on a detailed λ-map) when the first of these bifurcations is of high order, that is, when $\theta_1/2\pi = \alpha_1/\beta_1$ with a high value of β_1. But the subordination is already apparent for on Figure 1 for the outermost point of the sprout linked to $\mathcal{L}_0$ at $\theta_1/2\pi = 1/3$. Moreover, the islands are arrayed along β_g directions radiating from an "offshore point" In particular, if λ corresponds to several successive bifurcations, the other β_1 *do not affect the number of radii.* For details, see M 1982F{*FGN*}.

3. THE z-FRACTAL $\mathcal{F}(\lambda)$

We proceed now to a family of z-plane fractals associated with $z \to \lambda z(1-z)$.

3.1. Definition

First, recall that $z = \infty$ is a stable fixed point for all λ. {*Proof:* in terms of $u = 1/z$, the map is $u \to g(u) = u^2/\lambda(u-1)$ and we see that g (0) = 0 lt 1 .} For each λ, the z-fractal $\mathcal{F}(\lambda)$ is defined as the (closed) set of points z_{01}, such that $|\lim_{n\to\infty} z_n| \neq \infty$, that is, of points whose iterates fail to converge to ∞. This set $\mathcal{F}(\lambda)$ is never empty: it includes $z_0 = 0$. which is an unstable fixed point, all of whose iterates also satisfy $z_n = 0$, plus all the finite pre-images of z_0, and their limit points. Furthermore, $\mathcal{F}(\lambda)$ is always bounded: it is easy to see that it is contained—with room to spare!—in the circle $|z - 0.5| = 2.5$. The boundary of $\mathcal{F}(\lambda)$ is to be denoted by $\mathcal{F}*(\lambda)$.

Figure 2 shows an example involving a 7-fold bifurcation.

3.2. Exceptional values of λ for which $\mathcal{F}(\lambda)$ is a standard shape

Those values of λ reduce to $\lambda = 4$, $\lambda = 2$, and $\lambda = \infty$.

For $\lambda = \infty$, hence $\mathcal{F}*(\infty)$ reduce to the points 0 and 1. Obviously, $z_n = \infty$ except conceivably for $z_0 \neq 0$ and $z_0 = 1$: these values yield an indeterminate expression $z_1 = 0\infty$. The indeterminacy is lifted by noting that the inverse transform leaves these points invariant. The relevance of the inverse transform will transpire momentarily.

For $\lambda = 4$, $\mathcal{F}*(4)$, and hence $\mathcal{F}(4)$, both reduce to the segment $[0, 1]$. Indeed, introducing the new variable $w = -(2st - 1)$ changes $z \rightarrow z(1-z)$ into $w \rightarrow 2w^2 - 1$, and the further new variable $u = \cos^{-1} w$ yields $u \rightarrow 2u$, hence $u_n = 2^n u$. When $Im(u_0) \neq 0$, $|Im(u_n)| \rightarrow \infty$ and $|z_n| \rightarrow \infty$. Hence, the u coordinate representation of $\mathcal{F}$ (4) is the real axis, implying that $w \in [-1, 1]$ and $z \in [0, 1]$.

For $\lambda = = 2$, the same new variable w changes the map $z \rightarrow 2z(1-z)$ into the map $w \rightarrow w^2$, meaning that $\mathcal{F}*(2)$ is the circle $|w| = 1$, i.e., $|z - 0.5| = 0.5$ and the closed disc bounded by this circle. Clearly, $w_n \rightarrow 0$, hence $z_n \rightarrow 0.5$ iff $z_0 \in \mathcal{F}(2) - \mathcal{F}*(2)$, and $w_n \rightarrow \infty$, hence $z_n \rightarrow \infty$ iff **C** denoting the complex plane) $z_0 \in$ **C** $- \mathcal{F}(2)$. When $z_0 \in \mathcal{F}*(2)$, so that $z_0 = \exp(2\pi i y)$, then z_n is ergodic on $\mathcal{F}*(2)$ iff $y/2\pi$ is irrational; if $y/2\pi = \alpha/\beta$, then z_n follows an unstable cycle of period β.

The preceding examples show that $\mathcal{F}(\lambda)$ can be of topological dimension 0 (isolated points or dusts), 1 (curves) or 2 (domains). For all other values of λ, $\mathcal{F}(\lambda)$ is a nonstandard set, namely a fractal, but examples of every topological dimension continue to be encountered. Due to lack of space, only a few can be described here (see M 1982F{*FGN*}).

3.3. The shape of $\mathcal{F}(\lambda)$ when $\lambda \in \mathcal{L}$

As λ moves a little away from $\lambda = 2$, the circle $\mathcal{F}*(2)$ "crumples" locally, then bigger folds appear gradually. As long as $\lambda \in \mathcal{L}_0$, $\mathcal{F}*(\lambda)$ remains topologically a circle. As λ reaches a point of β-fold bifurcation, the topology of $\mathcal{F}*(\lambda)$ changes: it becomes "pinched" at an infinity of points, to each of which converge β points of $\mathcal{F}*(\lambda)$.

FIGURE C3-2. Map of $\mathcal{F}(\lambda)$ for λ near a 7-fold bifurcation. {P.S. 2003: M1980n featured the filled-in Julia set to the top left, drawn at IBM. It replaced the faint original to the bottom right, drawn early on at Harvard at the same time as the filled-in Julia sets featured in Figure C1.3.}

FIGURE C3-3. Superposed $\mathcal{F}(\lambda)$ for several real-valued λs: $\lambda = 1$ (light grey), $\lambda = 1.5$ (middling grey), $\lambda = 2.5$ (dark grey) $\lambda = 2.9$ (black). The shape for $\lambda = 3$ is called *San Marco* shape in the text. Two point common to all four diagrams are $z = 0$ and $z = 1$. {P.S. 2003: See also Figure C4.4.}

For example, as λ follows the real axis to the right and $\lambda \rightarrow 3$, $\mathcal{F}(\lambda)$ converges to the characteristic shape shown on Figure 3. (I call it the *San Marco shape* in honor of the Basilica in Venice plus its reflection in a flooded Piazza, ... and an infinite extrapolation.)

When a β_1– fold bifurcation is followed by a β_2–fold bifurcation, the β_1–fold and the pinches in the β_2–fold generally occur at different points. In any event, $\mathcal{F}(\lambda)$ remains connected as long as λ lies in the domain of confluence $\mathcal{L}$.

3.4. The shape of $\mathcal{F}(\lambda)$ when $\lambda \in Q - \mathcal{L}$

I discovered that a totally different shape of $\mathcal{F}(\lambda)$ prevails when λ lies in a domain of non confluence.

(A) The interior of $\mathcal{F}(\lambda)$ ceases to be connected.

(B) components of the interior of $\mathcal{F}(\lambda)$ have a common shape, except for deformations induced by the map $z \rightarrow \lambda z(1-z)$ itself.

(C) This common shape is close to that of the (connected) interior of $\mathcal{F}(\lambda*)$, where $\lambda*$ is the point that is mapped upon λ when the domain of confluence $\mathcal{L}$ is mapped non-linearly on the island under consideration.

(D) There is strong evidence that $\mathcal{F}(\lambda)$ itself is connected because, in addition to the components of its interior and their boundaries, it includes a web of fractal filaments.

3.5. The shape of $\mathcal{F}(\lambda)$ when (λ) is an irrational boundary point of $\mathcal{L}$

An irrational boundary point of $\mathcal{L}$ is defined as a boundary point of $\mathcal{L}$, other that the rational points where a "sprout" is attached to either $\mathcal{L}_0$ or another sprout. Letting λ tend from an interior point of $\mathcal{L}$ to an irrational boundary point provokes *either* an infinite sequence of finite-fold bifurcations, *or* (as when $\lambda \rightarrow \exp(2\pi iy$, with irrational y), an immediate ∞-fold bifurcation *or* a finite sequence of finite-fold bifurcations ending on ∞-fold bifurcation.

I conjecture that, in either case, $\mathcal{F}(\lambda)$ tends to a branched curve (a set without interior points). So does $\mathcal{F}*(\lambda)$. In the case of the real irrational points, which lie on the λ segment $[3.569, 4]$ this conjecture is known to be true, since $\mathcal{F}(\lambda)$ is a subset of the real line. It is known that $\mathcal{F}(\lambda)$ is, or is not, connected, according to the value of λ. An interesting open question arises: to relate the connectedness of $\mathcal{F}(\lambda)$ to the properties of non real values of λ).

3.6. The shape of $\mathcal{F}(\lambda)$ when $\lambda \notin Q$

The very first publication on the global properties of iteration, Fatou 191 remarks that, when λ is real and > 4, $\mathcal{F}(\lambda)$ is nearly a Cantor set. In the new terminology of M 1977F, it is a linear fractal dust.

To make this result perspicuous, let us digress to note that the classic Cantor ternary set is the $\mathcal{F}$-set corresponding to the special "tent" map defined by $z \rightarrow 1.5 - 3 \mid z - 0.5 \mid$. Under this map, indeed, all the points in the open mid-third, $z \in [1/3.2/3]$ yield $z_1 > 0$ hence $\lim z_n \rightarrow -\infty$. All the points in the side thirds' open mid-thirds, $z \in [1/9.2/9]$ or $z \in [7/9.8/9]$, yield $z_1 > 0$ but $z_2 < 0$, hence again $z_n \rightarrow -\infty$. Etcetera. The complement of these mid-thirds is $\mathcal{F}(\lambda)$. Its k-th ternary points, defined as the endpoints of a k-th stage mid-third, yield $z_n = 0$ for $n \geq k$, hence converge to an unstable limit point. Among the non-ternary points of $\mathcal{F}(\lambda)$, some converge to an unstable cycle of period >1, while others are ergodic.

Using the definition in M 1977F, this $\mathcal{F}(\lambda)$ is a *fractal dust*. It is a dust because it is totally discontinuous; so that its topological dimension is $D_T = 0$. On the other hand, its fractal dimension is $D = -\log 2/\log 3 > 0$. Because $D > D_T$, this set is a fractal.

4. THE z-FRACTAL $\mathcal{F}(\lambda)$ AS FRACTAL ATTRACTOR

Today, a map such as $z \rightarrow \lambda z(1 - 1)$ is routinely viewed as a dynamical system. Its attractor is dull (it may be a single point or a finite number of points). However, since $\mathcal{F}(\lambda)$ is the repeller set for $z \rightarrow \lambda z(1 - z)$, it is by the same token the attractor set for the inverse map $z \rightarrow 1/2 + \varepsilon(1/4 - z/\lambda)^{1/2}$ with $\varepsilon = \pm 1$.

More precisely, the last statement holds only after the notion of dynamic system is extended appropriately. An extension is required because, the above inverse map is not unique, but depends upon a parameter ε, to be called "label;" hence it is a 1 to 2 map, and the k-th iterate is a 1 to 2^k map. Considering all these iterates together, Julia 1918 showed that they are everywhere dense on $\mathcal{F}*(\lambda)$. But this is not a satisfactory result, because the intuitive notion of dynamical systems demands a single valued map. To achieve this goal, I propose that one set a discrete dynamical system in the product space of the complex plane by the label-set made of two points, called + and −. The ε_n sequence proceeds according to its own rules, independently of the z_n sequence, while the z_n sequence is ruled by the ε_n sequence. For example, the ε_n sequence may a

Bernoulli process of independent random throws of a fair coin, or a more general ergodic random sequence. The conclusion seems inescapable (though I have not tested the detail) that any ergodic sequence ε_n generates a trajectory whose projection on the complex plane is dense on $\mathcal{F}*(\lambda)$.

For $\lambda = 2$, when $\mathcal{F}*(\lambda)$ is a circle, the invariant measure is known to be uniform. For $\lambda = 4$ when $\mathcal{F}*(\lambda)$ is $[0, 1]$, the invariant measure is readily seen to be the real axis projection of a uniform measure on a circle: hence, it has the "arc-cosine" density $\pi^{-1}[x(1-x)]^{-1/2}$. Both are found empirically to be very rapidly approximated by sample dynamical paths. So is the San Marco shape seen on Figure 4. On the other hand, in the most interesting cases, $\mathcal{F}*(\lambda)$ is extremely convoluted, like on Figure 2; they involve a complication. The limit (invariant) measure on $\mathcal{F}*(\lambda)$ is extremely uneven. The tips of the deep "fjords" require very special sequences of the ε_n to be visited, hence are visited extremely rarely compared to the regions near the figure's outline.

5. Digression concerning "strange attractors"

The term "attractor" has often become associated with the adjective "strange," and the reader may legitimately wonder whether strange and fractal attractors have anything in common. Indeed, they do.

5.1. First point

The fractal (Hausdorff–Besicovitch) dimension D has been evaluated for many strange attractors, and found to strictly exceed their topological dimension. Hence, these attractors (and presumably other ones, perhaps even all strange attractors) are fractal sets. The D of the Smale attractor is evaluated in M 1977b. And the Saltzman–Lorenz attractor with v=40, $\sigma = 16$ and b=4 yields $D \sim 2.06$; this result was obtained independently by M. G. Velarde and Ya. G. Sinai, who report it in private conversations but neither of whom has to my knowledge published it. [Last minute addition: A preprint by H. Mori and H. Fujisaka confirms my value of D for the Smale attractor, and the Velarde–Sinai value of D for the Saltzman–Lorenz attractor. For the Hénon map with a = 1.4 and b = 0.3, it finds that D = 1.26.]. The fact that $D \sim 2.06$ is very close to 2, but definitely above 2, means that the Saltzman–Lorenz attractor is definitely not a standard surface, but that it is not extremely far from being one.

Since the relevance of D in this context may puzzle those who only know of fractal dimension as a measure of the irregularity of continuous curves, let me point out, that, in this instance, D is *not* a measure of irregularity but of the way smooth surfaces pile upon each other – a variant of the notion of fragmentation, which is also studied in M 1975O and M1977F.

Let us also recall from M 1975O, M 1977F that the Hausdorff Besicovitch dimension was not the sole candidate for fractal dimension, but was selected because: (a) it is the most thoroughly studied, (b) it has theoretical virtues, and (c) in most instances, the choice does not matter, because diverse reasonable alternative dimensions yield identical values.

In an interesting further development in the same direction, a relation has recently been conjectured to exist, and verified empirically on examples, between a strange attractor's fractal dimension and its Lyapunov numbers (preprints by H. Nori and H. Fujisaka, and by D. A. Russell, J. D. Hanson and E. Ott).

5.2. Second point

Conversely, are the fractal attractors I study "strange?" The answer depends on the meaning given to this last word. Under its old-fashioned meaning, as a mild synonym to "monstrous", "pathological" and other epithets once addressed to fractals, the answer is "yes, but why bother to

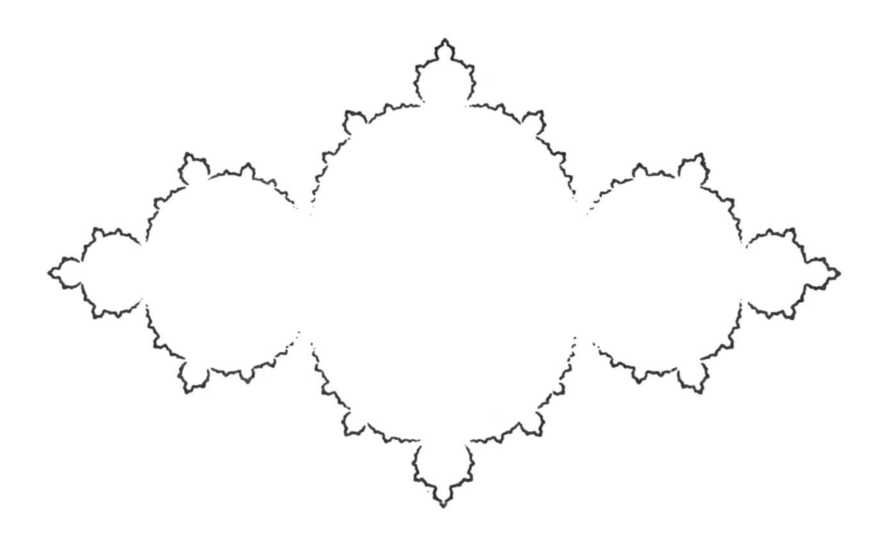

FIGURE C3-4. The 64 000 first positions of a dynamical system attracted to the San Marco fractal shape.

revive a term whose motivation has vanished when M 1975O, M 1977F showed that fractals are no more strange than coastlines or mountains".

Unfortunately, the term "strange" has since then acquired a technical sense, one so exclusive that the Saltzman-Lorenz attractor must be called "strange-strange". In this light, many fractal attractors of my dynamic systems are not strange at all. Indeed "strangeness" reflects non-standard *topological* properties, with the non-standard fractal properties mentioned in Section 5.1 coming along as a seemingly inevitable "overhead". In this sense, (a) a topological circle (intuitively, a closed curve without double points) is strange, however crumpled it may be; hence, (b) the fractal attractors $\mathcal{F}*(\lambda)$ for $|\lambda - 2| < 1$ are surely not strange.

However, the fractal attractors associated with other rational maps I have studied, M 1982F{*FGN*}, *are* topologically peculiar. Thus, the answer to our question is confused. But the question itself is not important. As applied to attractors, the term "strange" has, in my opinion, exhausted its usefulness and ought to be abandoned.

Discussion (omitted in the printed text)

Dr. H. D. Greyber. Could you comment upon the relationship between fractals and the study of physical systems with many interacting particles?

Dr. Mandelbrot. This is an issue my book *Fractals* did not face squarely, but {PS 2003, M1982F} will. The main difficulty is that the existing applications constantly increase in numbers, hence are hard to organize.

(A) In many branches of physics and astronomy (new ones every month, or so it seems), the analytic relationships between observables are expressed by scaling laws. I claim that the geometric shapes involved in all such cases must also be scaling. In the usual geometry of Euclid the only scaling shapes are straight lines, while the shapes found in scaling physical phenomena are anything but straight. Fractal geometry makes scaling compatible with extreme irregularity. I found it effective wherever I tried it, and every one is encouraged to play with it. Anyhow, there is parallelism between everyone else's scaling analytic phenomenologies and my scaling geometric phenomenology; the latter exhibits greater diversity but has no counterpart to the renormalization group.

(B) A question arises for the broken dimensions in physics (e.g., in $4-\varepsilon$ approximations), which are very useful but whose mathematical meaning is completely unclear. Can they be understood fractally? The verdict is not yet in. {P.S. 1980. Gefen, M, & Aharony 1980 study the Ising model on diverse fractal lattices of identical fractal dimension. We establish that the critical temperature satisfies $T_c > 0$ for some lattices while other yield $T_c > 0$, depending on fine topological properties of ramification. P.S. 2003. See also Gefen, Meir, M, & Aharony 1983.}

(C) At a deeper level, the scaling fractals' success in modelling nature is paradoxical. While they are invariant under *linear* transformations, the corresponding natural patterns are typically strong *non-linear*. This discrepancy motivated me to move on to fractals generated through non-linear transformations. Many scaling analytic formulas are justified as linear approximations near fixed points in renormalization groups, and the scaling fractals may eventually find a similar justification.

(D) Until successive improvements of the fractal models may come to be motivated by physics, they are mainly motivated as being the simplest that fit the facts. Consider my successive models of the distribution of galaxies. The first reduces an impossible N-body problem to an infinite collection of 2-body problems; the approximation is unjustified but yields theoretical 2- and 3-point correlations that are identical to those which Jim Peebles and his friends were independently obtaining through mere curve fitting. Later Peebles found that his early empirical 3-point correlation fits the data poorly and my 4-point theoretical correlation does not fit the data at all. As a result, M 1975u went on to the next simplest model, which has identical fractal properties but differs from the first on several non-fractal viewpoints. The new model confirms Peebles' fitted 4-point correlations, and predicts a theoretical 3-point correlation that fits the data with no corrective terms! But it does not fit the data's observed "lacunarity." Fortunately, the third simplest model (M 1979u) does. I am baffled by the success of mathematical models selected primarily on the basis of formal simplicity and invariance properties.

(E) Ultimately, many parameters are needed in most cases. By good chance, which again I do not understand but for which we must be grateful, the new parameters impose themselves one by one on our attention, as the models continue to be refined.

(F) The main parameter always remains the fractal dimension. It is very precisely known in many cases, for example, $D = 1.23$ for galaxies, $D = 1.89$ for critical percolation clusters and $D = 2.06$ for the Saltzman–Lorenz attractor. These are not orders of magnitude, but accurately known constants. All precedents suggest that they cannot fail to become increasingly important.

C4

Cantor and Fatou dusts; self-squared dragons

• *Chapter foreword (2003).* This chapter reproduces Chapter 19 of M 1982F, with a few additions clearly marked as such: Plate C5 of M 1982F (identical to its jacket) and a paragraph and a few illustrations from M 1985g that fit here better than they would elsewhere in this book.

Parts of the 1982 original—and no other text— used "separator" to denote the object that—unknown to me—Douady had already called "Mandelbrot set" (see Chapter 1). To avoid both confusion and anachronistic use of my name, this reprint uses either the specific 1982 terms μ-map and λ-map, or—if unavoidable—the generic term "M set."

In most chapters of M 1982F, the figures had long captions and were collected after the text. This format was a compromise. New and striking illustrations demand to be as large as possible, but large illustrations interrupt the text. Today, those illustrations are familiar, and hence there is no loss in making them smaller, while planning for full versions to be posted on my web site.

The design of Chapter 19 of M 1982F was especially complicated. It may or may not have been as effective as it was artful, but it could not be preserved in this reprint. More extreme rearrangements had to be made than for Chapter C16, the former Chapter 18 of M 1982F. A few overly long captions were integrated into the text. But the captions on pages 188 to 191 were very tightly bound with one another and with the text on pages 183–184. Reshuffling this "multidimensional" material into a "linear" story felt like separating several Siamese twins in a single operation. Some parts were duplicated, a few found no sensible resting place and were discarded, and a few brief minimal parts, as already said, were "borrowed" from publications of the same vintage, clearly acknowledged as the source.

Had I known now how tedious this linearizing was to be, I might not have undertaken it. For absolute historical accuracy, the 1982 original is widely available.

At one point, I felt the urge to acknowledge that Chapter 19 of M 1982F was an afterthought, hence could not be rewritten a hundred times. Also, I could not add several pages to allow the last rewrite to "breathe," because the book's index and table of contents were ready; etcetera, etcetera...

However, this chapter's original exposition was so far below the threshold of easy readability that the benefit of improvement would have been surpassed by the cost of delay. Not only did the figures play their intended role—in fact, became iconic—but the topic took off with extraordinary rapidity. For the overwhelming majority of "readers," the text hardly mattered.

The surprise lies elsewhere, in the fact that Chapter 18 of M 1982F, reproduced below as Chapter C16, met a very different fate. The text and figures were free of obvious blemishes; yet it took forever for Kleinian limit sets to attract the attention they deserve.

The many reprints of M 1982F have brought a few changes. In the original printing, Plate 185 was carried over from M 1980n{C3}, and using it here would have been repetitive. Instead, Figure 2 reproduces a revised Plate 185 introduced between 1983 and 1985. •

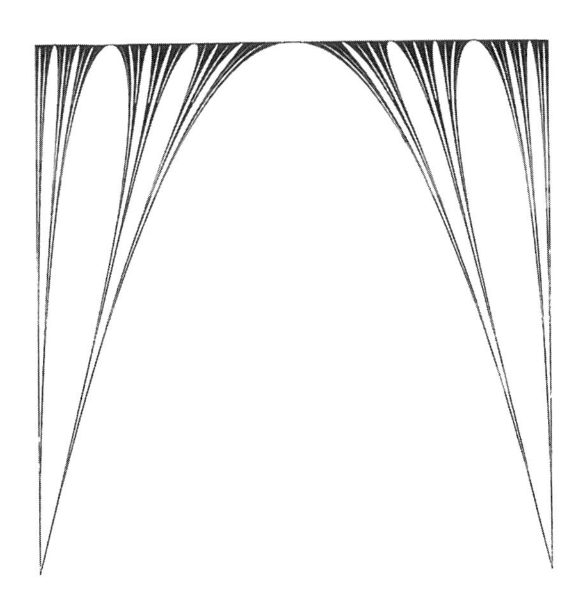

FIGURE C4-1. [Plate 192 of M 1982F]. A composite, explained in the text, of self-squared Fatou dusts on [0,1].

FOR TWO VERY SIMPLE FAMILIES OF NONLINEAR TRANSFORMATIONS, this chapter investigates the fractal sets which these transformations leave invariant and generate.

First, a broken-line transformation of the real line deepens our understanding of an old acquaintance, the Cantor dust. Furthermore, this deepened acquaintance helps appreciate the effect of the real and complex quadratic transforms of the form $x \rightarrow \tilde{f}(x) = x^2 - \mu$, where x and μ are real numbers, or $z \rightarrow \tilde{f}(z) = z^2 - \mu$, where $z = x + iy$ and μ are complex numbers.

The elementary case $\mu = 0$ is geometrically dull, but other values of μ involve extraordinary fractals, many of them first revealed in M 1980n{C3}.

The invariant shapes in question are best obtained as a by-product of the study of iteration, that is, of the repeated application of one of the above transformations. The initial values will be denoted by x_0 or z_0, and the k times iterated transforms by f or $\tilde{f}$ will be denoted by x_k or z_k.

Iteration was studied in three rough stages. The first, concerned with complex z, was dominated by Pierre Fatou (1878–1929) and by Gaston Julia (1893–1978). Their publications are masterpieces of classic complex analysis, greatly admired by the mathematicians, but exceedingly difficult to build upon. In my work, of which this chapter is a very concise sketch,

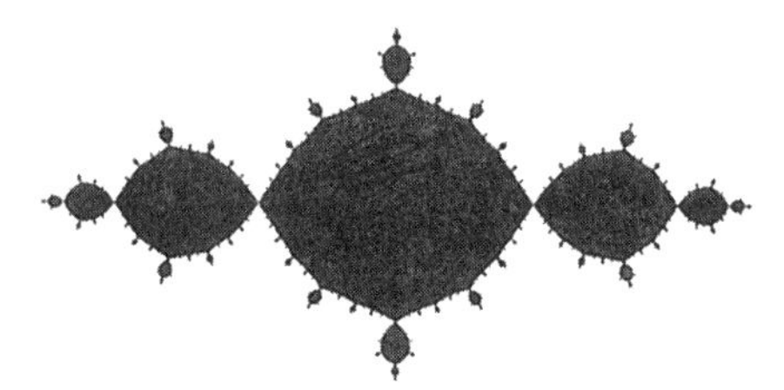

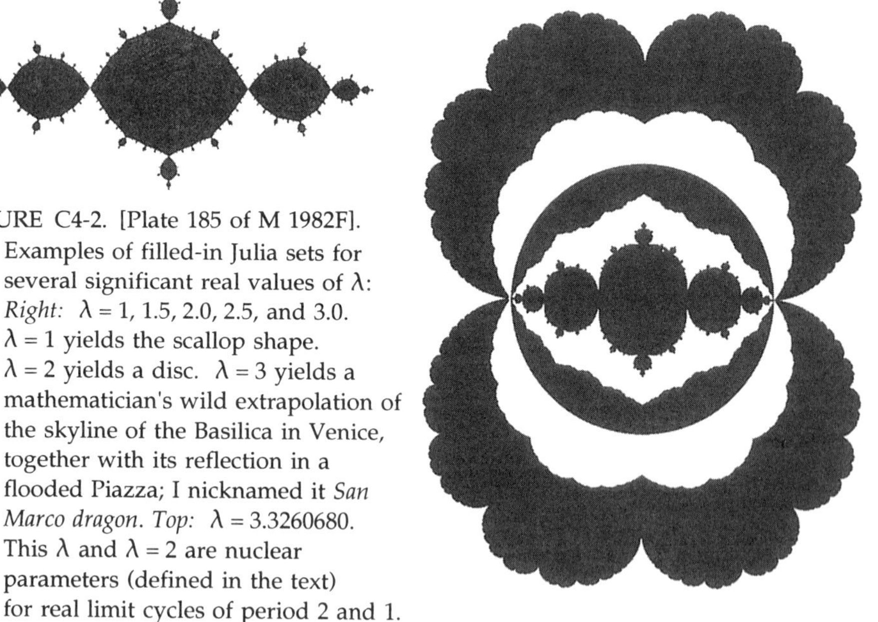

FIGURE C4-2. [Plate 185 of M 1982F]. Examples of filled-in Julia sets for several significant real values of λ: *Right:* λ = 1, 1.5, 2.0, 2.5, and 3.0. λ = 1 yields the scallop shape. λ = 2 yields a disc. λ = 3 yields a mathematician's wild extrapolation of the skyline of the Basilica in Venice, together with its reflection in a flooded Piazza; I nicknamed it *San Marco dragon. Top:* λ = 3.3260680. This λ and λ = 2 are nuclear parameters (defined in the text) for real limit cycles of period 2 and 1.

some of their basic findings are made intuitive by combining analysis with physics and detailed drawing. And innumerable new facts emerge.

The resulting revival makes the properties of iteration essential to the theory of fractals. The fact that the Fatou–Julia findings did *not become the source* of this theory suggests that even classical analysis needs intuition to develop and can be helped by the computer.

FIGURE C4-3. [Plate 187 of M 1982F]. Composite of self-squared Julia fractal curves for real λ ranging from 1 (bottom) to 4 (top). An alternative rendering is found on page 136.

The intermediate stage includes P.J. Myrberg's studies of iterates of real quadratic mappings of $\mathbb{R}$ (e.g., Myrberg 1962), Stein & Ulam 1964, and Brolin 1965.

The current stage largely ignores the past, and concentrates on self-mappings of [0,1], as surveyed in Gurel & Rössler 1979, Helleman 1980, Collet & Eckman 1980, Feigenbaum 1981, and Hofstadter 1981. This chapter's last section concerns the exponent δ due to Grossmann & Thomae 1977 and Feigenbaum 1978: the existence of δ is proven to follow from a more perspicuous (fractal) property of iteration in the complex plane.

The Cantor dust can be generated by a nonlinear transformation

We know from Chapter 8 [of M 1982F] that the triadic Cantor dust $\mathcal{C}$ is invariant by similitudes whose ratio is of the form 3^{-k}. This self-similarity is a vital property, but it *does not* suffice to specify $\mathcal{C}$. In sharp contrast, $\mathcal{C}$ is *entirely* determined as the largest bounded set that is invariant under the following nonlinear "inverted V" transformation:

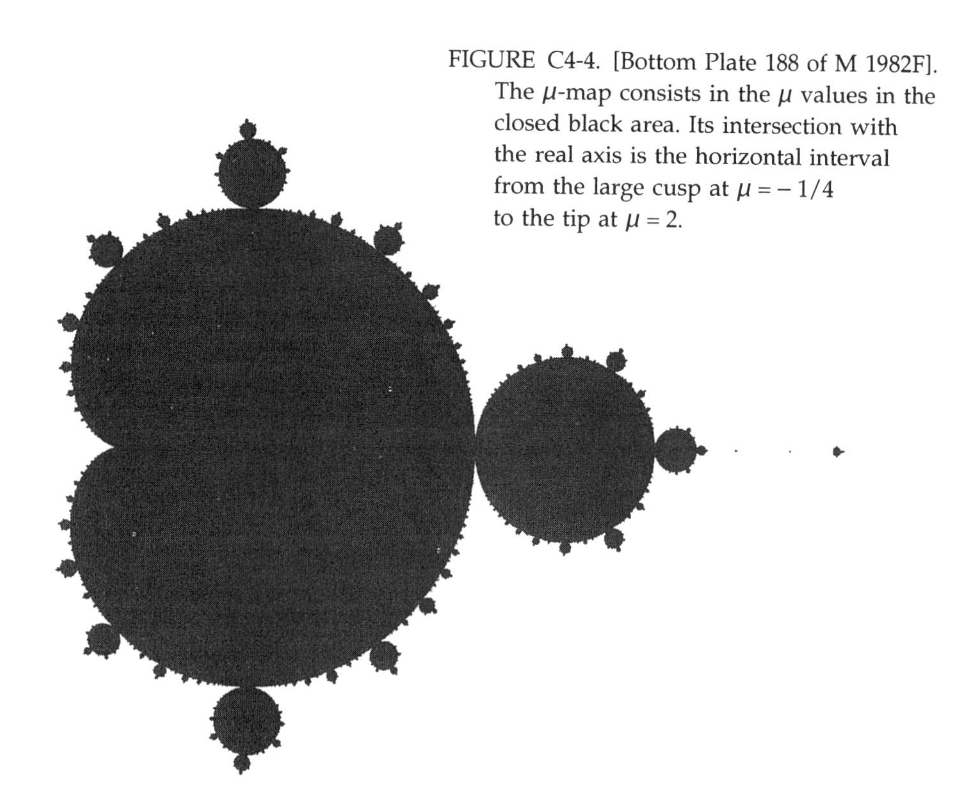

FIGURE C4-4. [Bottom Plate 188 of M 1982F]. The μ-map consists in the μ values in the closed black area. Its intersection with the real axis is the horizontal interval from the large cusp at $\mu = -1/4$ to the tip at $\mu = 2$.

$$x \rightarrow f(x) = \{1/2 - |x - 1/2|\}/r, \text{ with } r = 1/3.$$

More precisely, we apply this self-mapping of the real axis repeatedly, with x_0 spread out over the x-axis, and the final values reduce to the point $x = -\infty$, plus the Cantor dust $\mathcal{C}$. The fixed points $x = 0$ and $x = 3/4$ belong to $\mathcal{C}$.

Sketch of a proof that $\mathcal{C}$ ***is invariant.*** Since $f(x) = 3x$ when $x < 0$, the iterates of all the points $x_0 < 0$ converge to $-\infty$ directly, that is, without ceasing to satisfy $x_n < 0$. For the points $x_0 > 1$, direct convergence is preceded by one preliminary step, since $x_k < 0$ for all $k \geq 1$. For the points in the gap $1/3 < x_0 < 2/3$, where are 2 preliminary steps, since $x_1 > 0$ but

FIGURE C4-5. [Figures 1 and 2 of M 1985g]. Construction of the μ-map $z \rightarrow f*(z, \mu) = z^2 - \mu$ as the limit of monotone decreasing approximations, each a lemniscate of equation $|f_k*(0, \mu)| = 2$. Those curves are shown as boundaries between different shades of gray. Top half: $k = 1, 2, 3, 4,$ and 5. Bottom half: $k = 7, 9,$ and 14.

$x_k < 0$ for all $k \geq 2$. For the points in the gaps $1/9 < x_0 < 2/9$ or $7/9 < x_0 < 8/9$, there are 3 preliminary steps. More generally, if an interval is bounded by a gap that is sent to $-\infty$ after k preliminary steps, this interval's (open) mid third will proceed directly to $-\infty$ after the $k+1$ st step. But *all* the points of $\mathcal{C}$ are found to fail to converge to $-\infty$.

Finiteness of the outer cutoff

To extend these results to the general Cantor dust with $N=2$ and r between 0 and 1/2, it suffices to plug in the desired r in $f(x) = \{1/2 | x - 1/2 | \}/r$. To obtain any other Cantor dust, the graph of $f(x)$ must be an appropriate zigzag curve.

However, no comparable method is available for the Cantor dust extrapolated to the whole real axis. This is a special case of a very general feature: Typically, a nonlinear $f(x)$ carries *within itself* a finite outer cutoff Ω. To the contrary, as we know well, all linear transformations (similari-

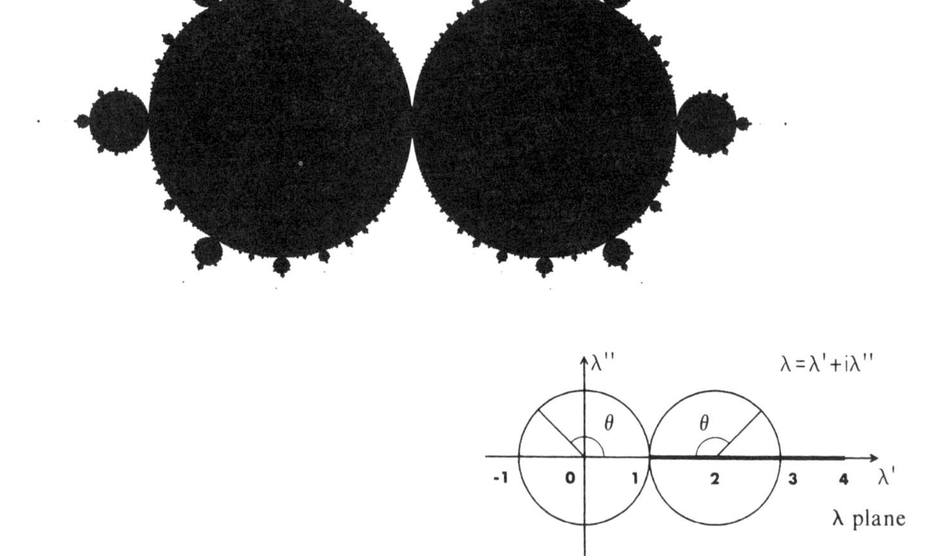

FIGURE C4-6. [Top right Plate 189 of M 1982F and Figure 4 of M 1985g]. The black shape is a tight approximation of the λ-map for $z \rightarrow \lambda z(1-z)$ by the lemniscate $|f_k 1/2, \mu)|$ = constant for a very large value of k. The small diagram identifies the coordinate axes and the abscissas of key values of λ.

ties and affinities) are characterized by $\Omega = \infty$, and a finite Ω (if one is required) must be imposed artificially.

Anatomy of the Cantor dust

We know from Chapter 7 of M 1982F that $\mathcal{C}$ is a very thin" set, yet the behavior of the iterates of $f(x)$ leads to a better understanding of fine distinctions between its points.

Everyone must be tempted, at first acquaintance, to believe that $\mathcal{C}$ reduces to the end points of the open gaps. But this is very far from being the case, because $\mathcal{C}$ includes by definition all the limits of sequences of gap end points.

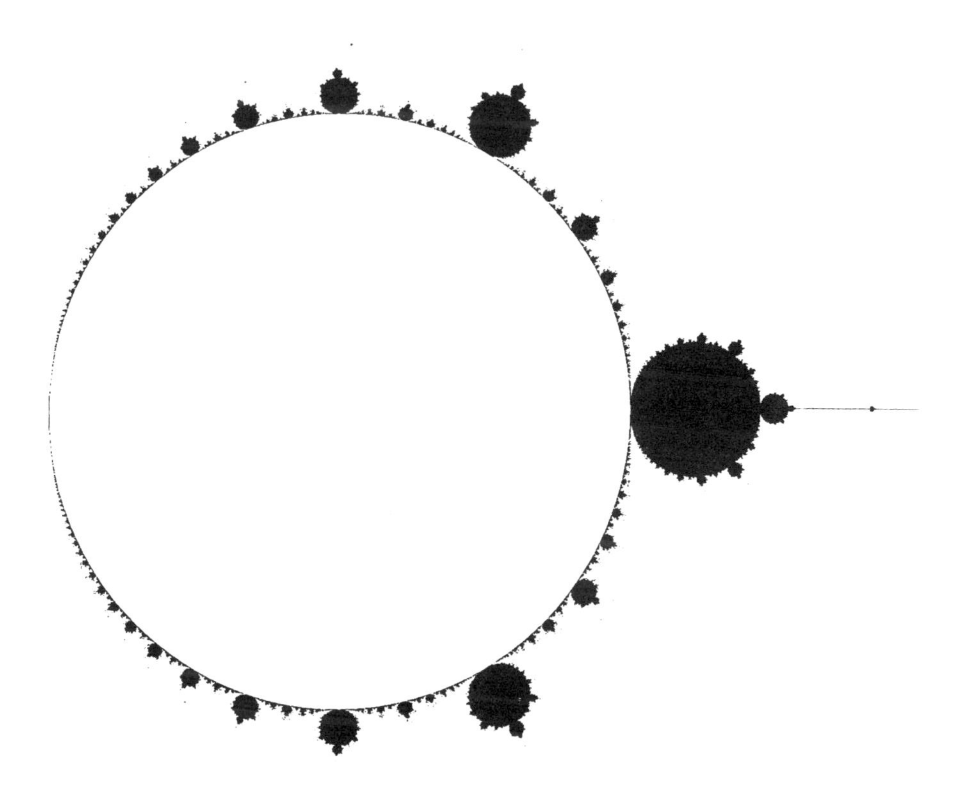

FIGURE C4-7. [Top Plate 189 of M 1982F]. The λ-map's "corona." The full λ-map in Figure 6 is symmetric with respect to the line $\lambda = 1 + i0$. The half λ-map for $\lambda \geq 1$ consists in the values of λ within the closed black area shown here, plus the disc $|\lambda - 2| \leq 1$. When λ is in the corona, the repeated transformation $z \rightarrow \lambda(1 - z)$ *fails* to move the point $z_0 = 1/2$ to *any* limit point, either bounded or at infinity. The coordinates are shown in the small diagram of Figure 6.

This fact is *not* reputed intuitive. Hans Hahn had provided M 1982F with many opinions demonstrating that what he called "intuition" had no geometric content. Long ago, with many fellow students, I would have agreed if he had listed these limit points among the concepts whose existence must be imposed by cold logic. But the present discussion yields *intuitive* proof that these limit points have strong and diverse personalities.

For example, the point $x=3/4$, which $f(x)$ leaves unchanged, lies neither within any mid third interval, nor on its boundary. Points of the form $x=(1/4)/3^k$ have iterates that converge to $x=3/4$. In addition, there is an infinity of limit cycles, each made up of a finite number of points. And $\mathcal{C}$ also contains points whose transforms run endlessly around $\mathcal{C}$.

The squaring generator

The inverted V generating function $f(x)$ used in the preceding sections was chosen to yield a familiar result. But it makes the Cantor dust seem contrived. Now we replace it by

$$x \to f(x) = \lambda x(1-x),$$

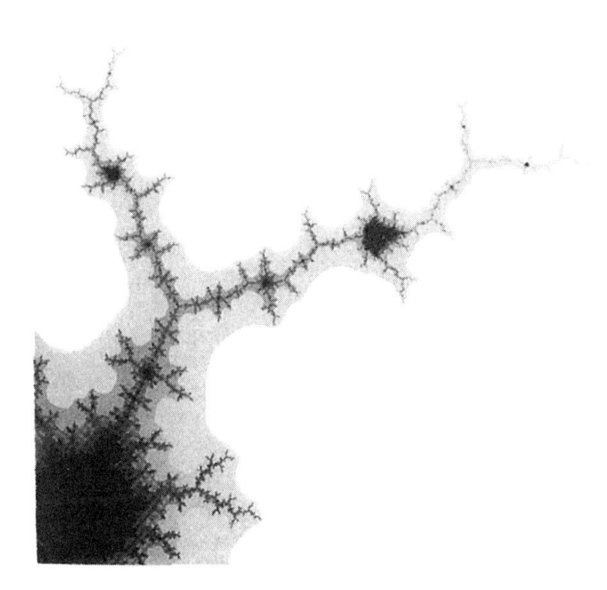

FIGURE C4-8. [Bottom left Plate 189 of M 1982F]. Detailed λ-map near $\lambda = 2 - \exp(-2\pi i/3)$. The M-set is the limit of domains of the form $|f_n(1/2)| < R$, of which a few are shown in superposition. For large n, these domains seem disconnected, and so does the λ map. In fact, they connect outside the computation grid.

whose unexpected wealth of properties was first noted in Fatou 1906. Changing the origin and the scale of the x, and writing $\mu = \lambda^2/4 - \lambda/2$ changes the function f into

$$x \rightarrow \tilde{f}(x) = x^2 - \mu.$$

Depending upon the context, it will be more convenient to use λ and $f(x)$ or μ and $\tilde{f}(x)$.

It is nice to call $f(x)$ or $\tilde{f}(x)$ the *squaring generator*. Squaring is, of course, an algebraic operation, but it is given a geometric interpretation here, so that the sets it leaves invariant can be called *self-squared*. Strict squaring replaces the point of abscissa x by the point of abscissa x^2. Thus, the self-squared points on the line reduce to $x = \infty$, $x = 0$, and $x = 1$. The addition of $-\mu$ may seem totally innocuous, but in fact it introduces totally unexpected possibilities we now consider.

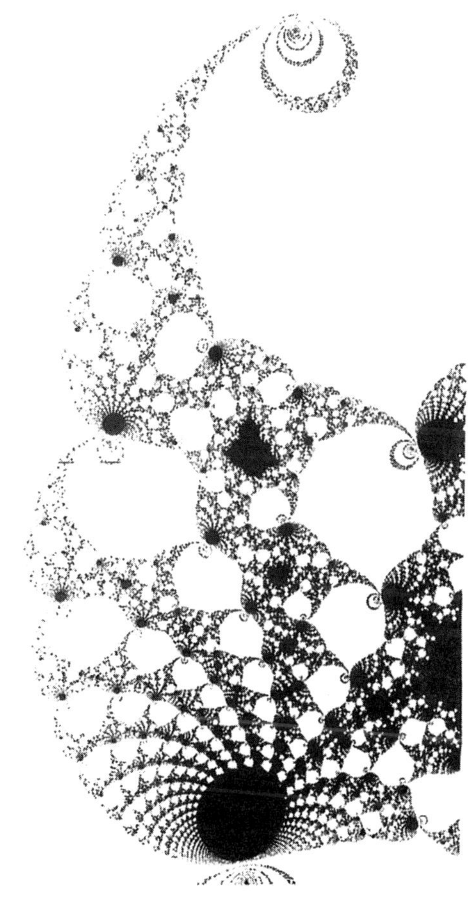

FIGURE C4-9. [Bottom right Plate 189 of M 1982F]. Detailed λ-map near $\lambda = 2 - \exp(-2\pi/100)$. This hundred-fold branching tree shares striking features with Figure 13.

Self-squared Fatou dusts on the real line

The background of the familiar Cantor dust makes an extraordinary discovery of Pierre Fatou easier to state. Fatou 1906 is a masterpiece of an odd literary genre: the *Notes aux Comptes rendus* of the Paris Academy of Sciences. In many cases, the purpose is to reveal little, but to squirrel evidence that the author had thought of everything.

Among other marvelous remarks best understood after long self-study, Fatou 1906 makes the following observation best stated in terms of λ. When λ is real and either $\lambda > 4$ or $\lambda < -2$, the largest bounded set that the transformation $x \rightarrow f(x) = \lambda x(1-x)$ leaves invariant is a dust contained in [0,1]. These close relatives of the Cantor dust deserve to be called *real Fatou dust.*

Figure 1 illustrates those dusts as the vertical coordinate $-4/\lambda$ varies from -1 to 0. The black intervals mark the end points of the tremas of orders 1 to 5. The end points x_1 and x_2 of the mid-trema are solutions of the equation $\lambda x(1-x) = 1$; they draw a parabola. Second-order tremas end at the points $x_{1,2}$, $x_{1,2}$, $x_{2,1}$, and $x_{2,2}$, such that $\lambda x_{m,n}(1 - x_{m,n}) = x_m$, etc.

This remarkable relation between Cantor-like dusts and one of the most elementary among all functions deserves to be known beyond the circle of specialists.

For the above λs, the real Fatou dust is also the largest bounded self-squared set in the complex plane.

Self-squared Julia curves in the plane (M 1980n{C3})

The simplest self-squared curve is obtained for $\lambda = 2$ and $\mu = 0$. In terms of $\tilde{f}$ it is the circle $|z| = 1$. By the transformation $z \rightarrow z^2$, a belt wound *once* around the circle stretches into a belt wound *twice*, leaving the "buckle" at

FIGURE C4-10. [Top of Plate 188 of M 1982F]. The λ-map—minus two unit discs—after inversion with respect to $\lambda = 1$. It looks like a horizontal strip, providing confirmation of the circles seen in Figure 9. Other perceived circles are confirmed by different inversions.

$z = 1$ fixed. The corresponding largest bounded self-squared domain is the disc $|z| \leq 1$.

However, making the parameter a real but $\mu \neq 0$, and then making it complex, opens successive Pandora's boxes of possibilities that satisfy the eye no less than the mind. The *filled-in Julia set* is defined, for each λ, as the maximal bounded domain under $z \to \lambda z(1-z)$, that is, as the set of z that the quadratic map does not send to infinity. The *Julia set* is then defined as the boundary of the filled-in Julia set.

FIGURE C4-11. [Jacket and Plate C5 of M 1982F; the color negative of the original has been reduced to shades of gray]. Julia dragon for $n = n' = 5$. The jacket of M 1982F, with this Figure's full color original, is on the author's homesite.

Julia sets for real values of the parameters λ or μ

On Figure 2, the bottom panel is straightforward but the top panel is a compact composite: meant to be attractive but also a bit mysterious. Imagine five filled-in Julia sets printed on transparent paper using opaque ink: white for $\lambda = 1.5$ and 2.5 and black for $\lambda = 1, 2,$ and 3. Then superposing those prints by increasing λ yields this figure.

To avoid spoiling a nice picture, the coordinates are not drawn. The horizontal symmetry line is the real axis and it interests those Julia sets at the points $x = 0$ and $x = 1$.

Figure 3 interpolates between the special example of Figure 2 and combines all λ ranging vertically from 1 to 4. Each horizontal section of this chapter is a Julia set. Using subtle graphics, this "sculpture" was drawn within a computer's memory by a process that amounts to whittling away all points in an initial cube, whose iterates by $z \rightarrow \lambda z(1 - z)$ converge to infinity.

The drape's "belt" is the Julia set for $\lambda = 2$. All other Julia sets are fractal curves. One perceives striking "pleats" whose position varies con-

FIGURE C4-12. [Top Plate 190 of M 1982F]. A trim and athletic Julia dragon for $n = 11$ (top) and an anorexic one obtained after many bifurcations (bottom).

tinuously with λ; they are "pressed" *in* below the belt, and "pressed" *out* above the belt.

FIGURE C4-13. [Plate 191 of M 1982F]. Multiarmed Julia Dragon for which the parameter value is close to the value to $\lambda = 1$ or $\mu = -1/4$ yielding the basic scallop shape.

The complicated blobs on the wall holding the top of the drape, to which this sculpture cannot possibly do justice, amplify the Myrberg–Feigenbaum theory of iteration of the real map $x \to \lambda x(1-x)$

(A) For every value of λ, the drape includes, as "backbone," a fractal tree formed by the iterated preimages of the real interval [0,1]. For all small, and some high values of $\lambda < 3$, this tree's branches are completely "covered by flesh." For other high values of λ, however, there is no flesh. The branches along either $x = 1/2$ or $y = 0$ are visible here, but the graphic process unavoidably misses the rest.

(B) Certain horizontal strips of the wall behind the drape are entirely covered with tiny "hills" or "corrugations," of which only a few large ones can be seen. In real iteration theory, these strips and hills concern the non-chaotic intervals of the parameter reported in Myrberg 1962. In terms of the M-set, to which we now come, they concern the "island molecules" intersected by the real axis.

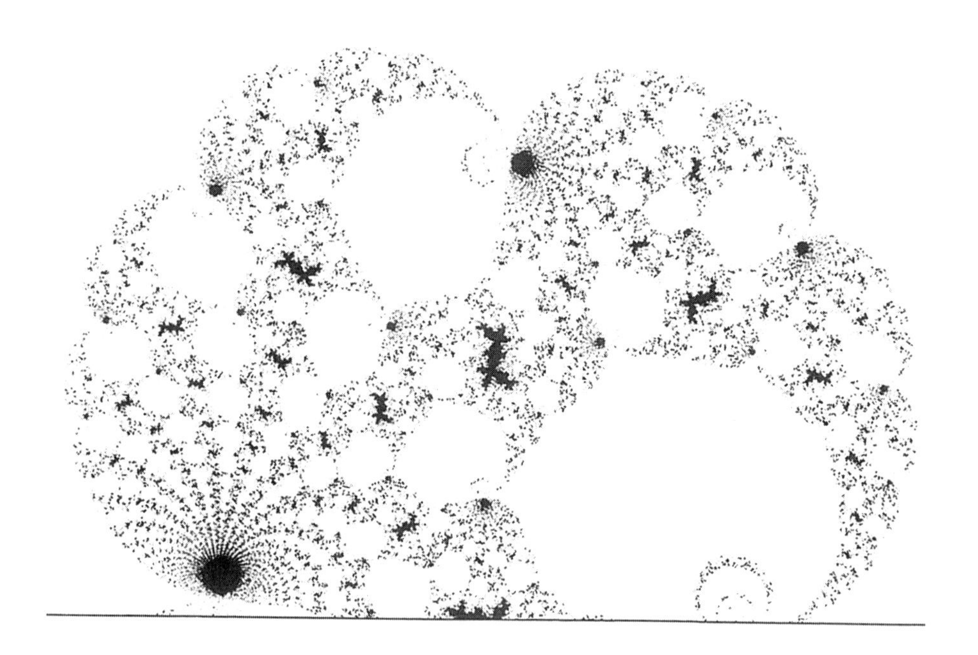

FIGURE C4-14. [Bottom of Plate 190 of M 1982F]. A "dragonfield" or "σ-dragon" (= denumerably infinite assembly of identically shaped dragons) obtained when the parameter lies within the island molecule close to $\lambda = 1$ and $\mu = -1/4$ that is visible on Figure 9.

Julia sets for complex values of the parameters μ or λ and the M-set

Moving on from real to complex μ and λ, Julia sets I drew exhibited an overwhelming variety of "self-squared dragon" shapes that can be sampled by looking a few figures ahead. Each self-squared curve is attractive in its own way. And the most attractive ones to me are the "dragons" exemplified by Figures 11, 12, 13, and 14. Their obvious and extreme variety cried out for an organizing principle. A criterion that imposed itself after a while was the connectedness of the filled-in Julia set. The criterion of correctedness is whether or not the repeated transformation $z \rightarrow z^2 - \mu$ *fails* to move $z_0 = 0$ on to infinity. This criterion defines a set I discovered and—generically— will be called M-set.

Specifically, the parameter μ yields the "μ-map," the black shape in Figure 4. In theory, it is a "limit lemniscate," namely the limit for $n \rightarrow \infty$ of the algebraic curves called "lemniscates," defined by $|\tilde{f}_n(0)| =$ for some large R. By prescribing $m = \max n$ and stopping when $n = m$ one obtains approximate μ-maps illustrated by Figures 4 and 5. The approximation improves by increasing m.

Similarly, the parameter λ yields the "λ-map" shown on Figure 6, with added details on Figures 7, 8, 9, and 10.

The fine structure of the M-set : λ- or μ-atoms and λ- or μ-molecules, continent and islands

Examples of the shapes of Julia sets for non-real λ or μ must be preceded by a discussion of the fine structure of the parameter map. Depending on the case, the description is easier when the parameter is either λ or μ.

Atoms and molecules. The maximal open set within the M-set splits into an infinity of maximal open connected domains I propose to call "atoms." Two atoms' boundaries either fail to overlap, or have in common one point, to be called "bond," that belongs to M-set. A seed's cusp is never a bond.

An infinity of atoms bind either directly or through intermediate atoms to form a "molecule."

The μ-map with a central cardiord and the λ-map with two disks $|\lambda - 2| = 1$ and $|\lambda| = 1$ "punctured" by removal of $\lambda = 0$. The λ values in these domains are such that the iterates of the point $z_0 = 1/2$ converge to a limit point not at infinity. In Figure 7 these discs are left empty to emphasize the remainder as a "corona" whose outside boundary is a fractal curve.

Sprouts. This corona splits into "sprouts" rooted on "receptor bonds" sited on the circles $|\lambda| = 1$ or $|\lambda - 2| = 1$. A root's λ is of the form $\lambda = \exp(2\pi i m/n)$ or $\lambda = 2 - \exp(2\pi i m/n)$, with m/n an irreducible rational number < 1.

The inverse of the λ-map with respect to a circle centered at $\lambda = 1$. Examining the sprouts in the corona of the λ-map of Figure 7, one gains the impression that "corresponding points" lie on a "circumscribing" circle through $\lambda = 1$ and centered to the right of $\lambda = 2$. Figure 10 provides confirmation since the inverted map has a horizontal upper "envelope." Other perceived circles are confirmed by different inversions.

{P.S. 2003. This observation became the origin of a number of significant developments. They include the "n-squared" conjecture and the "normalized" M-set, which are described in Chapters C6 and C7 below.}

Island molecules. Many of the "spots" around the maps are genuine "island molecules," first reported in M 1980n{C3}. They are shaped like the large μ molecule, except for a nonlinear distortion.

Periods. To each atom is attached an integer "period" w. When μ lies in an atom of period w, the iterates $\tilde{f}_n(z)$ converge to ∞ or to a stable limit cycle containing w points.

Within an atom of period w, $|\tilde{f}_w(z_\mu)| < 1$, where z_μ is any point of the limit cycle corresponding to μ. On the atom's boundary, $|\tilde{f}'_w(z_\mu)| = 1$, with $\tilde{f}'_w(z_\mu) = 1$ characterizing a cusp or a "root."

Spines and trees. When the atom's period is w, the curve where $\tilde{f}'_w(z_\mu)$ is real defines its *spine*.

The spines lying on the real axis are known in the theory of self-mapping as [0,1], and their closure is known to be [−2,4].

Nuclei. Each atom contains a point to be called "nucleus," satisfying $\tilde{f}'_w(z_\mu) = 0$, and $\tilde{f}_w(0) = 0$.

The nuclei on the real axis were described in Myrberg 1962 and rediscovered in Metropolis, Stein, & Stein 1973. The corresponding parameters are often called "superstable" (Collet & Eckman 1980).

Viewed as algebraic equation in μ, $|\tilde{f}_w(0)| = 0$ is of order 2^{w-1}. Hence, there could be at most 2^{w-1} atoms of period w, but in fact there are fewer, except for $w = 1$. For $w = 2$, $\tilde{f}_2(0) = 0$ has 2 roots, but one of them is already the nucleus of an atom of period 1. More generally, all the roots of $\tilde{f}_m(0) = 0$ are also roots of $\tilde{f}_{km}(0) = 0$ where k is an integer > 1.

Next, define each rational boundary point on the boundary of an atom of period w, as satisfying $f_w^{*\prime}(z_\mu) = \exp(2\pi i m/n)$, where m/n is an irreducible rational number < 1, and observe that it carries a "receptor bond" ready to connect to an atom of period nw. As a result, some new atoms bind to existing receptor bonds.

But not all new atoms are thereby exhausted, and the remaining ones have no choice but to seed new island molecules. The molecules are therefore infinite in number.

Bifurcation and confluence. When μ varies continuously in a molecule, each outbound traversal of a bond leads to bifurcation: w is multiplied by n. Example: increasing a real-valued μ leads to Myrberg's period doubling sequence. The inverse of bifurcation, which M 1980n{C3} investigates and calls *confluence*, must stop at the period of the molecule's seed. The continent molecule is the region of confluence to $c = 1$. Each island molecule is not a region of confluence to $c = 1$, but to this island molecule's nucleus. The dragon's or sub-dragon's shape is ruled by the values of $f_w^{*\prime}(z_\mu)$ and w/c.

The M-set is fractal; Feigenbaum's δ as a corollary. I conjecture via a "renormalization" argument that atoms increasingly removed from their molecule's seed come increasingly close to being *identical* in shape.

A corollary is that the boundary of each molecule is locally self-similar. Since it is not smooth on small scales, it is a fractal curve.

This local self-similarity generalizes a fact concerning Myrberg bifurcation, due to Grossmann & Thomae and to Feigenbaum. The widths of increasingly small sprouts' intercepts by the real axis of λ or μ, converge to a geometrically decreasing sequence, of ratio $\delta = 4.66920...$ (Collet & Eckman 1980). In its original form, the existence of δ seems a technical analytic result. Now it proves to be an aspect of a broader property of fractal scaling.

Each bifurcation into $m > 2$ introduces an additional basic ratio.

Topological dimension. When μ lies outside M-set, the largest bounded self-squared set is a (Fatou) dust. When μ lies within M-set, or is a bond, the largest such set is a domain bounded by a self-squared curve. At least some μ on M-set yield a tree-like curve. The topological dimension of the set that fails to go to ∞ is 0 for the Fatou dusts, 1 for starved dragons, and 2 for other dragons.

Ramification. When λ lies in one of the open empty discs of top Plate 189, the self-squared curve is a closed simple curve (not ramified, a loop), as in Figures 2 and 3.

When λ lies on the circles $|\lambda| = 1$ or $|\lambda - 2| = 1$, or in the surrounding open connected region, the self-squared curve is a ramified net, with tremas bounded by fractal loops, like the dragons in Figure 13.

When λ lies in the very important island molecules, which will soon prove to be *regions of nonconfluence to* 1, the self-squared curve is either a σ-loop, or a σ-dragon, as in Figure 14. The σ introduces no new loop.

I discovered more generally that the closure of the other atom spines decomposes into a collection of trees, each rooted on a receptor bond. The list of orders of ramification at different points of such a tree is made up of 1 for the branch tips, plus the orders of bifurcation leading to the tree's root. Furthermore, when the tree is rooted on an island atom, one must add the orders of bifurcation leading from $|\lambda - 2| \leq 1$ or $|\lambda| \leq 1$ to this atom.

A sampling of Julia set "dragons" for which μ and λ are not reals

Julia sets being fractal when $\mu \neq 0$ is rumored to have been proven fully in some further cases by Dennis Sullivan and I harbor no doubt it will be proven in all cases. Moving beyond real-valued μ and λ greatly increases the diversity and attractiveness of the Julia sets.

Successive bifurcations and "starved dragon." In a sprout that corresponds to $\theta/2\pi = m/n$, with integers n and m, the bifurcation order is n, and the number of dragon heads or tails (or whatever these domains should be called) around each articulation point is n. A second bifurcation of order m'/n' splits each of these domains into n' "sausage links," and thins them down.

The best known of all dragons is on Figure 11, reproduced from the jacket and Plate C5 of M 1982F. It corresponds to $n = n' = 5$. The top half of Figure 12 corresponds to $n = 11$. The bottom half shows that a dragon subjected to many bifurcations loses all flesh eventually, it collapses into a skeletal branched curve.

Draconic molting. To watch a dragon in the process of self-squaring would be a fascinating sight! A monstrous "molting" detaches the skin of a dragon's belly and back from innumerable folds. Then, it stretches the skin to twice its length, which of course remains infinite all along! Next, it folds each skin around the back as well as the belly. And finally, it re-attaches all the folds neatly in their new positions.

The singular limit $\lambda = 1$. Peano dragons. Dragons with a nice heft, neither obese nor skinny, obtain when λ lies within a sprout, at some distance away from the root at $\theta = 2\pi/n$. Dragons with a nice twisted obtained when λ lies near one of the 2 subsprouts corresponding to an order of bifurcation of 4 to 10: one subsprout yields a leftward, the other a rightward, twist. As $n \to \infty$, $\theta \to 0$, hence λ tends to 1. The corresponding dragon must necessarily converge to the scallop shape that surrounds the top panel in Figure 2 and forms the base of the drape in Figure 3. But this limit is of a kind that is called "singular," meaning that, between $n = \infty$ and n large but finite, enormous qualitative differences are observed. First, consider the difference between the scallop and Figure 13 corresponding to $n = 100$ As $n \to \infty$, the dragon's arms grow in number, the skin crumples, and the skin's fractal dimension increases. The whole really attempts to converge to a "hermit-dragon" that would fill the shell of a $\lambda = 1$ scallop to the brim, i.e., to the dimension $D = 2$. A self-squared Peano curve? Yes, but we know from Chapter 7 of M 1982F that Peano curves are not curves: as it attains $D = 2$, our dragon curve dies as a curve to become a plane domain.

"Dragonfields or "σ-dragons" obtained by choosing λ in an "island molecule." The parameter's position with respect to the M-set allows the Julia set to fall in either of three general categories. Parameters outside the M-set yield generalized Fatou dusts. Parameters within the continental molecule yield "dragons." The remaining possibility is when the parameter lies within an island molecule, it is exemplified by Figure 14.

{P.S. 2003. This illustration cries out for substantial elaboration. It is by design that Figures 13 and 14 use parameter values close to the value $\mu = 1/4$ (or $\lambda = 1$) that yields the scalloped Julia set in Figure 2.

This proximity serves to make two separate points. On the one hand, in terms of "overall shape," the Julia set is not overly sensitive to the exact parameter value: not only Figure 13 but even Figure 14 are perceived as having an "overall envelope" that is close in shape to the scallop.

On the other hand, the detailed structure within that envelope is extremely sensitive to the parameter. Specifically, Figure 14 is made of an infinity of individual dragons. All have the same shape that is determined by very local conditions, namely, the position of the parameter within the "island molecule" to which it belongs. Choosing as parameter the nucleus of that molecule would have yielded disc-shaped individual dragons. Choosing the cusp of the island molecule would have yielded an infinity of scallops. And Figure 14 shows instantly that the parameter does not lie along the spine and lies in an atom obtained by one bifurcation of order 4.

The term "sprout" was never used after M 1982F. But an extension of the chemical terminology of atoms and molecules was proposed in M 1985g and deserves mention. It involves structures of devil causeway, polymer, and glue. The fact that the M-set is connected implies that all the molecules are also linked to each other, but the links between the M-molecules are very different from the links between the M-atoms. Take the nuclei of two molecules A and B, and link them within the M-set by a continuous curve parametrized by $u \in [0, 1]$, in such a way that the us contained within a molecule form an open interval. Two such intervals never touch, and the complement of these intervals' union is a fractal dust. By analogy with the widely used term "devil's staircase," the molecules crossed between A and B will be said to form a "devil's causeway," and the M-set will be said to have a "devil's polymer" structure. In the case of the quadratic map it is branching with no closed loops. The difference between the full closed M-set and the semi-open M-set is made of the scattered points that "glue" the molecules together. These points form a fractal dust.}

Physica D, 7, 1983, 224-239

C5

The complex quadratic map and its $\mathcal{M}$-set

• *Chapter foreword (2003).* Blithely, this text assumes that the $\mathcal{M}$-set is the closure of the $\mathcal{M}_0$-set of parameter values such that quadratic dynamics has a finite stable limit cycle in addition to the point at infinity. Actually, this property was an open mathematical conjecture in 1983 and, despite gigantic efforts, remains open today. It is equivalent to the "MLC conjecture" that the $\mathcal{M}$-set is locally connected. The original title was "On the quadratic mapping $z \rightarrow z^2 - \mu$ for complex μ and z; the fractal structure of its $\mathcal{M}$-set and scaling." The original's subsections were numbered. •

✦ **Abstract.** For each complex μ, denote by $\mathcal{F}(\mu)$ the largest bounded set in the complex plane that is invariant under the action of the map $z \rightarrow f(z) = z^2 - \mu$. M 1980n{C3} and M 1982F{*FGN*}, Chapter 19 {C4} reported various remarkable properties of the $\mathcal{M}_0$ set (the set of those values of the complex μ for which $\mathcal{F}(\mu)$ contains domains) and of the closure $\mathcal{M}$ of $\mathcal{M}_0$. {P.S. 2003: see *Chapter foreword.*} The goals of this work are as follows.

(A) To restate some previously reported properties of $\mathcal{F}(\mu)$, $\mathcal{M}_0$ and $\mathcal{M}$ in new ways and report new observations.

(B) To deduce some known properties of the map f for real μ and z, with $\mu \in -\,]-1/4, 2[$ and $z \in\,]-1/2 - 1/2\sqrt{1+4\mu}\,, 1/2 + 1/2\sqrt{1+4\mu}\,]$ In many ways, the properties of the quadratic map are easier to grasp in the complex plane than in an interval. (This illustrates the saying that, "when one wishes to simplify a theory, one should complexify the variables."{P.S. 2003. This was a favorite saying of Gaston Julia, my teacher in differential geometry in 1945.})

(C) To introduce some recent pure mathematical work triggered by M 1980n{C3}. Further pure mathematical work is strongly urged. ✦

THE ILLUSTRATIONS ARE THE FOCUS OF THIS PAPER, and the text is organized in the form of extended comments. Additional illustrations are found in M 1980n{C3} and Chapter 19 of M 1982F{*FGN*}{C4}.

1. DISCUSSION OF FIGURES 1A TO 1E. ILLUSTRATION OF THE ACTION OF $z \to f(z, \mu) = z^2 - \mu$ ON A LARGE COMPLEX CIRCLE. SEQUENCES OF ALGEBRAIC CURVES APPROXIMATING THE REPELLER (JULIA) SETS $\mathcal{F}*(\mu)$

A transformation becomes easier to study when one has a concrete visual feeling for its action. In the case of $z \to f(z, \mu)$ it is known that the point at infinity is a stable fixed point of f. Hence, it is an attractive point. In order for a circle of sufficiently large radius r and center 0 to be in the domain of attraction of ∞, $r > r_W = 1/2 + (1/2)\sqrt{1 + 4|\mu|}$ is a sufficient condition. The circle of radius r_W and center 0 will be denoted by W^0 and called "whirlpool circle," and r_W will be called the "whirlpool radius," because the orbits of all the points outside W^0 "whirl away" from 0.

On the other hand, some complex z are not attracted to ∞. They include the bounded fixed points $1/2 \pm (1/2)\sqrt{1 + 4\mu}$ and their successive pre-images. Let the maximal bounded set invariant under $f(z, \mu)$ be denoted by $\mathcal{F}(\mu)$. By definition of the whirlpool circles, $\mathcal{F}(\mu)$ is contained within W^0. Also, for every value of k, $\mathcal{F}(\mu)$ is contained in the k-th pre-image of W^0 under $f(z)$, i.e., the pre-image of W under $f_k(z)$. This last set is defined by $|f_k(z, \mu)| = r_w$ denoted by W^{-k}. It is an algebraic curve called "lemniscate" (Walsh 1956). The lemniscates corresponding to increasing values of k are non-overlapping and are monotonically imbedded in sequence. They can be called "parallel under $f(z, \mu)$. The set $\mathcal{F}(\mu)$ is the limit of these curves plus their interiors.

Denote by $\mathcal{F}*$ the boundary of $\mathcal{F}$. The set $\mathcal{F}*$ is the limit of W^{-k} for $k \to \infty$. It is the repeller set of $f(z, \mu)$ and is also called Julia set.

History: the earliest basic facts about global iteration were described in Fatou 1906, and the bulk of the original theory was described nearly simultaneously in Julia 1918 and Fatou 1919. Since the term "Julia set" has become entrenched, I chose to honor Fatou by denoting this set by $\mathcal{F}*$.

Now for some illuminating illustrations. For four selected values of μ, the parts a to d of Figure 1 represent the interiors of W_0 and of several curves W^{-k} in superposition. The goal is to demonstrate intuitively that

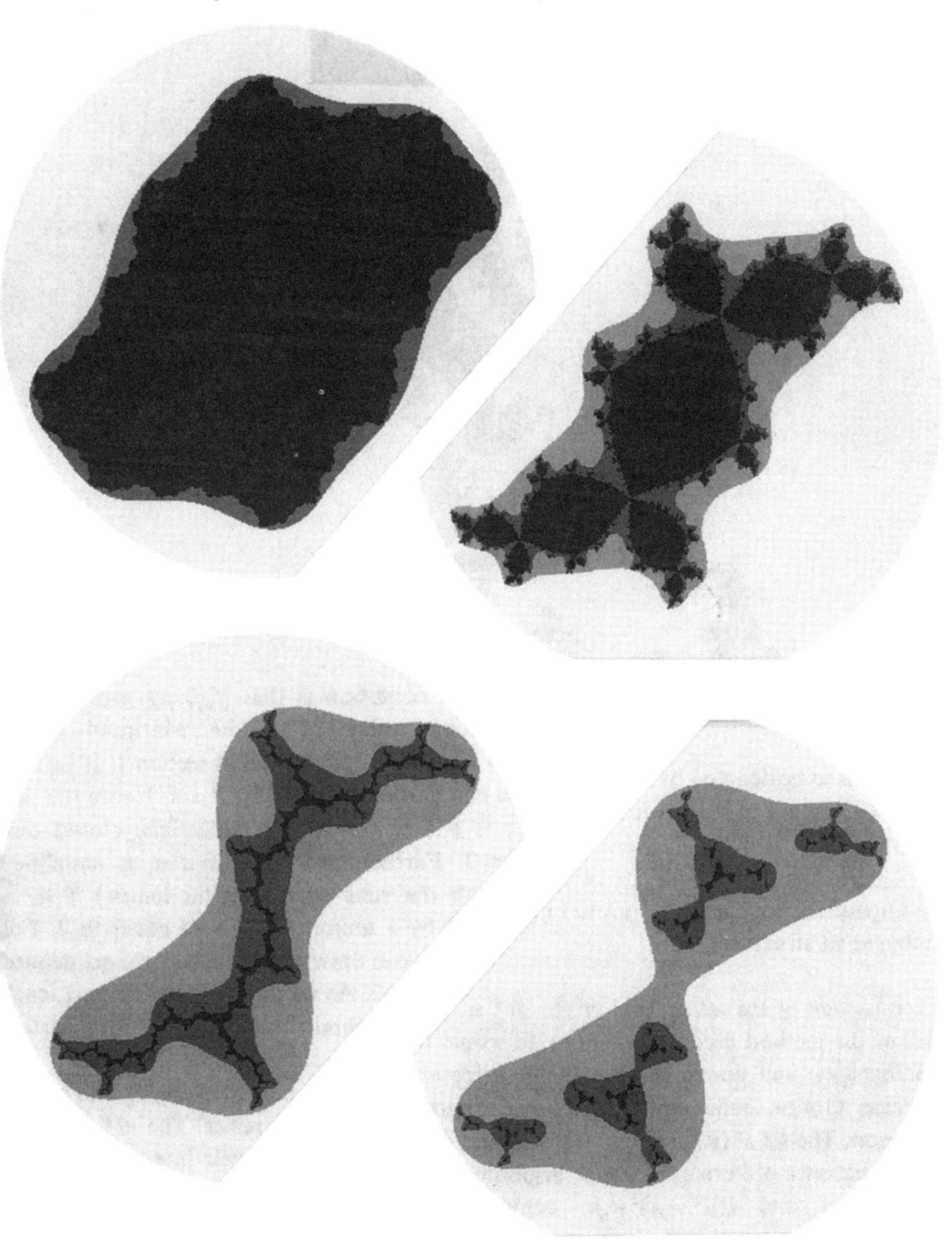

FIGURE C5-1. Action of $z \to z^2 - \mu$ on a gray circle (partly truncated): (a) the Julia set is a loop-free closed fractal curve, (b) the Julia set is a closed fractal curve with loops, (c) the Julia set is a fractal tree, (d) the Julia set is a fractal dust ("Cantor set").

the topology of $\mathcal{F}$ of the Julia set $\mathcal{F}*$ greatly depends on the value of μ. In particular, $\mathcal{F}*$ can be (a) a loop-free ("simple") curve that bounds a domain, (b) a curve with multiple points that bounds an infinite number of domains, (c) a tree ("branching curve without loop," "dendrite") that does not surround a domain, (d) a totally disconnected dust.

Figure 2 represents an attractive example of $\mathcal{F}*$.

2. DISCUSSIONS OF FIGURES 2A TO 2F. CLASSIFICATION OF THE VALUES OF μ BY THE TOPOLOGY OF $\mathcal{F}(\mu)$. THE SETS $\mathcal{M}_0$ AND $\mathcal{M}_0\mathcal{M}$. SEQUENCES OF ALGEBRAIC CURVES APPROXIMATING $\mathcal{M}_0\mathcal{M}$. THE CONTINENT, ISLANDS, STELLATE STRUCTURES, DEVIL'S CAUSEWAYS

This series of figures investigates in detail the set of those values of μ for which $\mathcal{F}(\mu)$ is connected. This set is to be denoted by $\mathcal{M}$ and $\mathcal{M}_0$ will denote the set of values of μ for which $\mathcal{F}(\mu)$ has interior points, that is, includes domains. On a graph, e.g., on Figure 3a, the $\mathcal{M}_0$ and $\mathcal{M}$ sets cannot be distinguished, but they differ significantly in their structure.

2.1. ***Construction of the $\mathcal{M}_0$ and $\mathcal{M}$ sets.*** To follow the method used in Figures 1a to 1d would be cumbersome and unreliable, but is not necessary because Gaston Julia gave the following direct criterion. The set $\mathcal{F}(\mu)$ is disconnected if and only if the sequence of iterates of $z=0$, beginning with $-\mu, \mu^2-\mu$ and $(\mu^2-\mu)^2-\mu^2-\mu$, converges to infinity.

For this to be the case, a necessary and sufficient condition is that $|f_k(0,\mu)|$ must exceed, for some value of k, the whirlpool radius $r_W=1/2+1/2\sqrt{1+4|\mu|}$ derived in Section 1. If $|\mu|>2$, this condition is satisfied for $k=1$. Hence, the $\mathcal{M}_0$ and $\mathcal{M}$ sets are entirely contained within the closed disc $|\mu|\leq 2$. Furthermore, the program is simplified (though the runs become a bit longer) if r_W is replaced by a uniform threshold equal to 2. We saw in Section 1 that the identity $|f_k(0,\mu)|=2$ defines in the $\mu-$*plane* an algebraic curve called "lemniscate." All the lemniscates include the point $\mu=2$, but otherwise they are non-overlapping and monotonically imbedded in sequence. The $\mathcal{M}$ set is the limit of these curves plus their interiors.

As is known, μ is called superstable of minimal period k if $f_k(0,\mu)=0$, but $f_h(0,\mu)\neq 0$ for every $h<k$. It follows that $f_{nk}(0,\mu)=0$ for every integer n, hence all superstable μs fail to iterate to infinity, meaning that they belong to the $\mathcal{M}$ set. Walsh 1956 reports that a lemniscate cannot contain a

loop within a loop: it is necessarily either a single loop, or a finite union of loops with non-overlapping interiors. In the present case, it turns out that the lemniscates are simple loops for all values of k, hence $\mathcal{M}$ is a connected set. But other features of the $\mathcal{M}$ set must first be considered.

2.2. The continental subset of $\mathcal{M}_0$. For reasons to be introduced momentarily, the structure of Figure 2a is clarified by positioning the grid of μ's that real valued μ's are not tested. A first glance reveals that the great bulk of the black points lie in a large and very highly structured "continent." It has a striking "cactus tree" structure, which I propose to

FIGURE C5-2. Interior of a Julia set (repeller set) of $z \rightarrow z^2 - \mu$ after two successive sixfold bifurcations.

describe as a "molecule" made of an infinity of "atoms." At the center is a "seed atom," which has the shape of a cardioid, and contains all the μ's for which $f(z, \mu)$ has a single stable limit point besides ∞. The exactly cir-

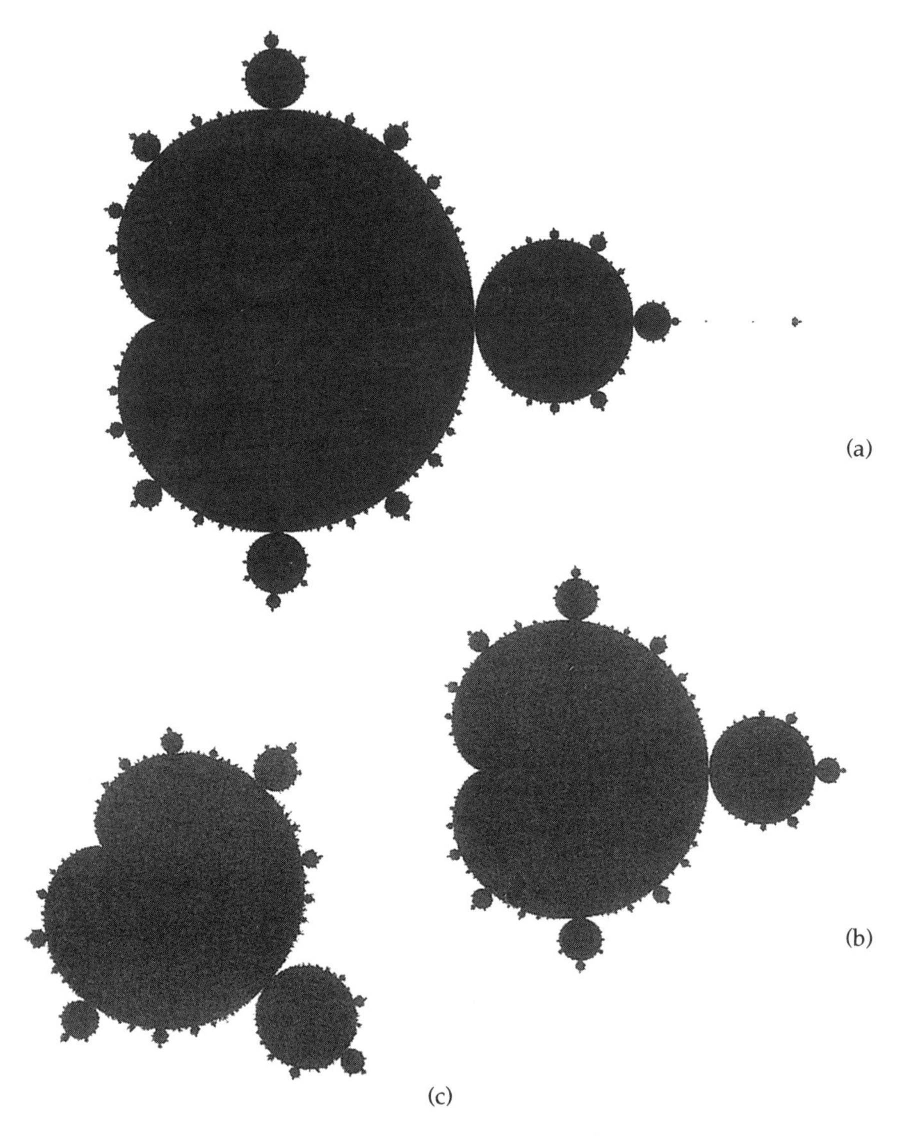

FIGURE C5-3. (a) Overall view of the $\mathcal{M}_0$-set of $z \rightarrow z^2 - \mu$. (b) Detail of an $\mathcal{M}_0$-island: the "speck" to the right of Figure 3a. (c) Detail of an $\mathcal{M}_0$-island : the "speck" at the bottom of Figure 3a.

cular atom to the right of the cardioid contains all the μ's for which $f(z, \mu)$ has a stable cycle of period 2, and the near circular atoms that follow to its right correspond to stable cycles of periods 4, 8, etc. The points where these atoms join are the μ's corresponding to the basic real μ bifurcations. Other near-circular atoms that touch the cardioid correspond to cycles of order $k > 2$.

2.3. The shape of the $\mathcal{M}$-set near the value μ_∞ at the rightmost tip of the continental subset. Scaling property in the plane, and its use to rederive (as corollary) the known scaling property of bifurcations on the real line. Consider the sequence of atoms that converge to the tip of the continent. They seem essentially alike, and seem tangent to two straight half-lines that are symmetric with respect to the real axis and converge to μ_∞ defined as the accumulation point of bifurcations. This is a geometric property of scaling, more precisely, of asymptotic geometric scaling.

An inferred consequence is that these atoms' horizontal intercepts decrease geometrically at each bifurcation. This inference is of course well-known to be true, having been discovered by Grossman & Thomae and by Feigenbaum.

To verify the identity of the atom shapes with a more exacting test, $\mathcal{M}$ was redrawn by replacing the parameter μ by $\nu = \log(\mu - \mu_\infty)$. The cardioid shaped seed atom is thereby made much smaller, and the other atoms become near-identical.

2.4. The big island to the right of the continent. Other islands. In addition to the continent, the $\mathcal{M}_0$ set contains a number of scattered specks. It is hoped that these specks escaped the watchful eye of the editors and the printers of the present Proceedings. A reason for concern is that the printers of M 1980n{C3} erased their counterparts on the firm assumption that they could only be dirt!

They are, in fact, very real, and it may be useful to devote a few lines to telling how I discovered them. Examining my first rough graph of $\mathcal{M}$, I too took most of them to be dirt. But the biggest one, positioned to the right of the continent, looked too big to be spurious, and it was easy to verify that it intersects the axis of real μ's along the interval, discovered by Myrberg and Metropolis, Stein and Stein (see Collet & Eckmann 1980), for which $f(z, \mu)$ has a stable cycle of period 3. I had this speck examined in closeup, Figure 2b and it was revealed to be essentially a downsized version of the continent. Other Myrberg intervals that I examined in closeup were also revealed to intersect very small downsized versions of the continent.

Thus, the rightmost tip of the continent continues along the real axis by a peculiar causeway. Because of analogy with the devil's staircase (M 1982F{*FGN*} page 83), I propose to call it the "devil's stepstones." The metaphor starts with large stones set in a stream to accommodate ordinary super giants, then smaller stones are set to accommodate ordinary giants, small giants, supermen, and so on down to devilishly tiny beasts. Ultimately the stones leave no gap of positive width, however small. The real axis runs along the center of this causeway, in a way that is familiar to students of the real transform $f(z, \mu)$.

At this point, I traced several puzzling observations to the same source. The first observation was that for periods 1, 2 and 3, each superstable μ is the "nucleus" of an atom known to belong to the continent or an island off the real axis. However, two superstable μ's of period 4 remained "unattached," and, for higher periods, the number of "unattached" superstable μs kept increasing rapidly. One may have argued that some atoms contain multiple-root nuclei, or several distinct nuclei, but these atoms should have looked different from atoms of smallest period 1, 2 or 3, while, in fact, all atom shapes fell into either of the two patterns exemplified by the seed cardioid and the circle to the right of it.

The second puzzling observation was that, except for the point μ_∞, the tips of the continent gave no evidence of being followed by the devil's stepstones. The third puzzling observation was already mentioned: when the $\mathcal{M}_0$-set was traced with low precision on a medium-tight lattice, it seemed surrounded by unattached specks of dirt.

A close-up view of the big speck to the right of $\mathcal{M}$ (Figure 3c) settled the three puzzles in one sweep: most specks did not vanish but turned out to be islands identical to the continent in their topology and overall form. It soon became clear that these islands do not scatter around haphazardly, but form "stellate" arrays (see M 1980n{C3}, M 1982F{*FGN*} for details).

An array close to a point of bifurcation of order 11 is seen on Figure 4. Further closeups revealed increasing numbers of increasingly small islands between larger islands along each "ray." This was reminiscent of the above-mentioned fact that increasingly small islands are "pierced through" by the real axis. At this point, the growing analogy with the real axis suggested the following: the islands in every ray are linked together by curves that are counterparts of the real axis, but could not be seen because curves other than the axis nearly always fall between the lattice points used in computation. This implies that the stellate structure reflects an underlying tree structure, and that the $\mathcal{M}$-set is connected. It may be recalled that $\mathcal{M}$ was approximated by a sequence of lemniscates, which (according to general results) might have split into separate loops. The fact that $\mathcal{M}$ remains connected implies that the approximating lemniscates are single loops. This was verified to be the case until high order preimages of the circle $|\mu| = 2$, parts of which started falling between the lattice points.

FIGURE C5-4. Detail of the $\mathcal{M}_0$-web offshore the continent on Figure 3a.

Needless to say, computer-based observations do not provide a substitute for actual proofs. (It is also true that some full proofs bring rigor without insight.) In any event, the mathematical study of my computer-based observations on the $\mathcal{M}$-set turned out to be fruitful and useful, witness Douady & Hubbard 1982, and could not have been undertaken without those observations.

2.5. The inverse of bifurcation: the notion of confluence. The literature of bifurcation never seems to refer to the opposite effect that is observed, say, when μ starts with a value in an atom other than a seed atom, and changes continuously, without leaving the island, until it reaches a seed atom. M 1980n gave to this inverse operation the name *confluence*. The point is that the continent is the domain of confluence to a stable limit point, and each island is a domain of confluence to a periodic cycle, but not of confluence to a limit point.

2.6. The atoms' and the islands' intrinsic coordinates. Homologous points. Given an atom of minimal period k, denote by z_μ at any point in the stable cycle corresponding to the parameter value μ. We know that the complex number $f'_k(z_\mu, \mu)$ is less than 1 in modulus. Its real and imaginary parts form intrinsic coordinates for the point μ within the atom to which it belongs. Two points with identical intrinsic coordinates can be called homologous within their atoms. The set of μs for which $f'_n(z_\mu, \mu)$ is real will be called the atom's "spine." It runs from a point where $f'_k(z_\mu, \mu) = 1$ (which is a cusp in the case of seed atoms), to a point of bifurcation of order 2, where $f'_k(z_\mu, \mu) = -1$.

Furthermore, each atom's position in its island can be identified by an "address," namely the sequence of integers that identify the sequence of bifurcations that lead to this atom starting from the seed cardioid. Each bifurcation is indeed marked by a rational number n_i/m_i, with $m_i \geq 2$ and $0 \geq n_i \geq m_i$. Thus, it suffices to write these n_i and m_i separated by commas. One can agree that the seed cardioid's address is 0 (and other addresses may, but need not, start by 0).

The combination of the address of the atom and of the value of $f'(z_\mu, \mu)$ forms an intrinsic coordinate for the point μ within the island to which μ belongs. Two points having identical intrinsic coordinates can be called "homologous" within their islands.

An island's spine combines its seed cardioid's spine with the spines of atoms corresponding to $m_i = 2$ bifurcation. Every island spine's endpoint is homologous to the tip μ_∞ of the continental subset of the $\mathcal{M}$ -set.

2.7. *"Universality class" argument to explain why the islands are alike.* Assume that μ is a superstable value μ of minimum period k. We wish to determine the shape of the atom nucleated by μ^*.

The lowest order terms in the expansion of $f_k(z, \mu)$ near $z=0$ and $\mu=\mu^*$ can be written as $\beta_k z^2 + \lambda_k(\mu - \mu^*)$. Now let us state and test out a brutal "universality class" argument, then a milder version of it.

The brutal argument claims that the shape of the atom nucleated by μ^* depends only on the lowest terms in the expansion of $f_k(z, \mu)$ near $z = 0$ and $\mu=\mu^*$. If this were the case, μ^* would nucleate a cardioid-shaped seed atom. This atom and the molecule grown upon it would be identical to the continent, except for its size being reduced in the ratio $1/\beta_k\lambda_k$. The milder argument agrees to take account of a few higher order terms near $z=0$ and $\mu=\mu^*$, while continuing to disregard the behavior of $f_k(z, \mu)$ far from $z=0$ and $\mu=\mu^*$. This milder argument suggests the following properties. (A) The atom nucleated by μ^* is the seed atom of a molecule, and its shape resembles the continental cardioid except for some mild nonlinear deformation. (B) Other atoms obtained by bifurcation are arrayed around this seed as on the continent, except again for a mild deformation.

Inspection of the actual $\mathcal{M}_0$ set indicates that, when prediction (A) is correct, (B) is also correct. Moreover, (A) can only fail by (A) and, (B), being replaced by the following properties. (A') The atom is not a seed and its shape is near circular. (B') Other atoms obtained by bifurcation are arrayed around the atom nucleated by $\mu\mathcal{M}^*$, in the same way as their counterparts are arrayed around the continent, except for a transformation that straightens out the cusp.

For $n=2$ the superstable values satisfying $\mu^2-\mu=0$ are $\mu^*=0$ or 1. Near $\mu^*=1, f_z(z, \mu)=(z^2-\mu)^2-\mu=z^4-2\mu z^2+\mu^2-\mu \sim -2z^2+(\mu-1)$. This suggests an atom equal to the basic cardioid downsized in the ratio of $1/2$ and translated to the right by 1. But the actual atom is bigger in every direction and happens to be precisely a disc.

One may expect to find that the condition of validity of the milder universality class predictions (A) and (B) is that $\beta_k\lambda_k$ be large.

A further universality class argument (not as yet well developed) suggests that atoms increasingly removed from the seed of their islands tend to the universal shape.

2.8. *"Universality class" argument to explain the shape of $\mathcal{F}^*(\mu)$ when μ^* lies in an island.* The brutal universality class argument also makes a prediction concerning $\mathcal{F}^*(\mu)$: C) The Julia set $\mathcal{F}^*(\mu)$, call it a "little dragon," obtains by reducing in the ratio β_k, the Julia set which the full $\mathcal{F}^*(z, \mu)$ predicts for the point that lies in the continent and is homologous to μ, namely $\mu' = \beta_k \lambda_k (\mu - \mu^*)$. The milder universality class argument makes the prediction C' The portion of the Julia set $\mathcal{F}^*(\mu)$ obtains, by reducing in the ratio β_k, the Julia set, which the full $\mathcal{F}(z, \mu)$ predicts for the point that lies in the continent and is homologous to μ.

Inspection of actual sets $\mathcal{F}^*(\mu)$ indicates that the portion of $\mathcal{F}^*(\mu)$ near $z = 0$ is indeed the little dragon predicted by the milder (C'). But the brutal (C) provides a quite incorrectly pallid idea of the structure of the whole of $\mathcal{F}^*(\mu)$. This set does *not* reduce to the little dragon near $z = 0$, but is made up of an infinity of mildly deformed replicas of this little dragon.

These replicas form devil's stepstones with the same structure we already encountered in the shape of $\mathcal{M}$. (Figure 5 is relative to a case where μ is very close to the nucleus of the cardioid on Figure 3c.) That is, these replicas are strung along a tree. As to this tree's shape, it brings in something entirely foreign to the universality argument. Indeed, this shape is determined by μ, and not merely by the point in the continent that is homologous to μ. This shape varies fairly slowly with μ, and is approximately determined by μ^*.

To introduce an even rougher but useful approximation, let us begin by bringing in the parameter value μ'' corresponding to the tip of the island containing μ^*. This point is homologous to the classical real point μ_∞ at the tip of the continent. In the next section's discussion of the Julia sets $\mathcal{F}^*$, we shall see that μ_∞ is among the values of μ for which $\mathcal{F}^*(\mu)$ reduces to a tree: it has a real interval as its spine, other ribs as well, but no flesh. When μ'' is the tip of an island other than the continent, $\mathcal{F}(\mu')$ is also a tree (though it contains no straight interval). Now we come back to $\mathcal{F}^*(\mu)$: it is found that the replica dragons belonging to this set string along a tree approximated by $\mathcal{F}^*(\mu')$.

2.9. Rough estimates of the counterparts of the ratio δ for bifurcations of order > 2 . For the purpose of this subsection, it is best to change the coordinates by replacing the parameter μ by the parameter $\lambda = 1 \pm \sqrt{1 + 4\mu}$. This corresponds to the map $z \rightarrow f^*(z, \lambda) = \lambda z(1 - z)$. The corresponding transform of the $\mathcal{M}_0$ set is shown on page 250 of M 1982F{*FGN*}. The continent is no longer the same shape as the islands, since, instead of being seeded by a cardioid, it is now seeded by two discs. But this transformed shape has its own assets. A first advantage, which had been ascertained with pen and paper, is that the bifurcation from a stable fixed point to a cycle of period m recurs at the points where either λ or $2 - \lambda$ is of the form $\exp(2\pi in/m)$, with an integer n smaller than m.

A second advantage only appeared after the $\mathcal{M}_0$ set had been computed and could be examined. It was observed that the tips of the "major" sprouts around the circle $|\lambda - 2| = 1$, defined as the sprouts rooted at the points $\exp(2\pi i/m)$, appear to be placed along a larger circle whose diameter begins at $\lambda = 1$ and ends somewhere beyond $\lambda = 3$. This suggests that one perform an inversion of the $\mathcal{M}_0$ set with respect to $\lambda = 1$, with $\lambda = 3$ remaining fixed. This inversion should yield sprouts placed between parallel lines. Furthermore, the transform of the root of the mth major sprout should lie at a vertical distance from $\lambda = 3$ equal to $2\tan(1/\pi/2 - \pi/m) = 2\cot(\pi/m)$. For large m, this yields $2m/\pi$, i.e., a series of equally spaced points.

The inverted $\mathcal{M}_0$ set shown in Figure 6 is plotted using very different units along the two axes, so that the graph remains legible while covering many values of m. The above hunch is confirmed, except for $m = 2$ and 3. That is, an extrapolation from sprouts with a higher value of m would

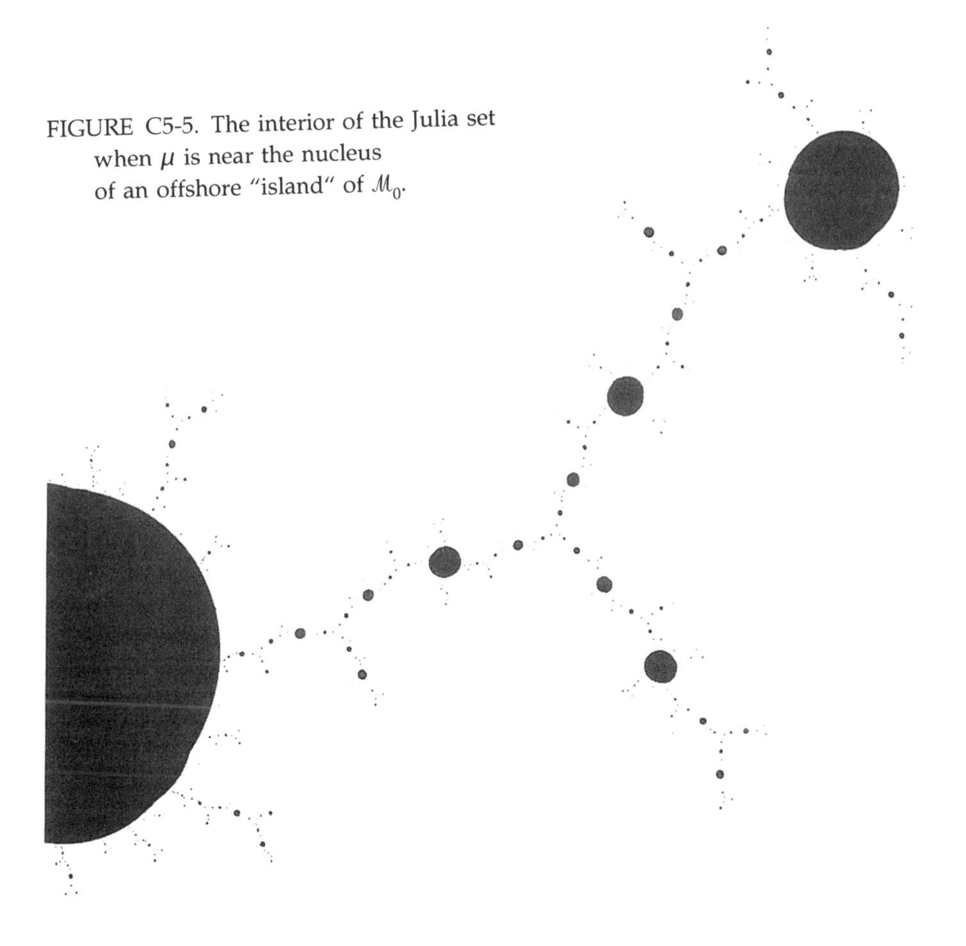

FIGURE C5-5. The interior of the Julia set when μ is near the nucleus of an offshore "island" of $\mathcal{M}_0$.

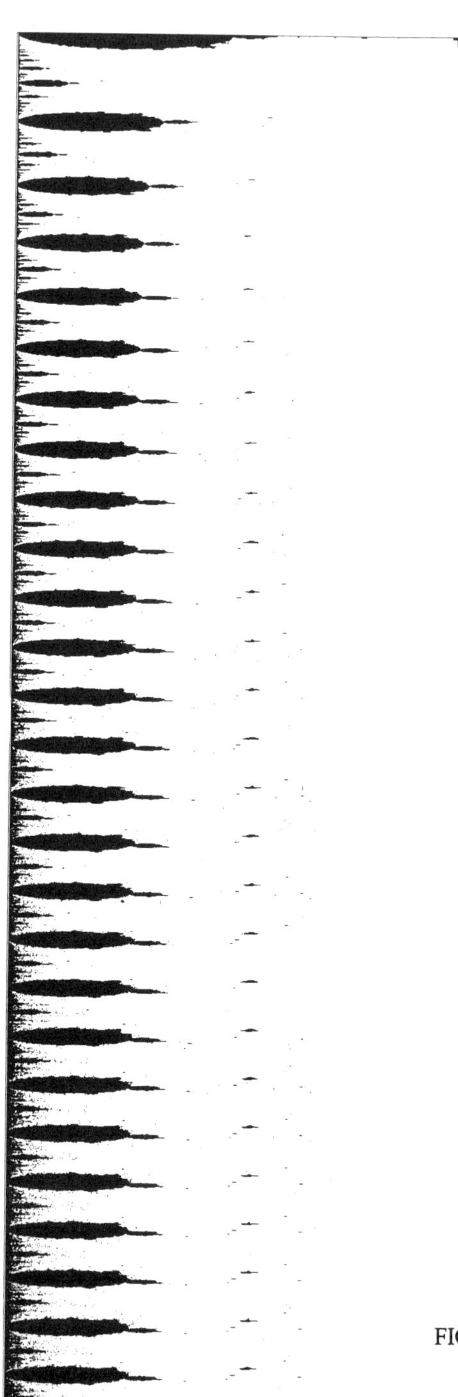

FIGURE C5-6. Detail of the $\mathcal{M}_0$ set of $z \rightarrow \lambda z(1-z)$ after inversion and compression.

yield a smaller sprout for $m = 2$. Denote by A the height of the inverted sprouts for large m. Assuming circular atoms, these properties of inversion yield the result that the relative linear size of the sprout of order m is $A \sin^2(\pi/m)[2 - A \sin^2(\pi/m)]^{-1}$. This is roughly the ratio of successive absolute changes in μ between bifurcations of order m, that is, the m-th counterpart of the $1/\delta$ ratio of Grossman & Thomae and of Feigenbaum. In fact, this ratio is close to m^{-2} for all m.

3. DISCUSSION OF FIGURES 7 to 10. THE REPELLER STACK FOR REAL μ AND COMPLEX z. ILLUSTRATIONS OF THE INFLUENCE OF THE VALUE OF μ ON THE SHAPE AND THE TOPOLOGY OF THE REPELLER (JULIA) SET $\mathcal{F}^*(\mu)$

In Figures 7 to 10, the horizontal coordinates x and y are the real and the imaginary parts of z, and the vertical coordinate is μ. The figures represent a stack of Julia sets $\mathcal{F}^*(\mu)$ ranging from $-1/4$ to 2. The goal is to show that the shape of $\mathcal{F}^*(\mu)$ varies continuously, while the topology of $\mathcal{F}^*(\mu)$ moves around discontinuously. The stack was sliced along the plane $xO\mu$, and the two halves have been separated... .

3.1. *Computation of the stack.* The theory of $\mathcal{F}^*(\mu)$ involves two "proof-of-existence" constructions. The first is used in Figures 1a to 1d. The second consists in tracing the pre-images of the unstable fixed point $z' = 1/2 + 1/2\sqrt{1 + 4\mu}$. This second construction is efficient only if μ is near 0, so that $\mathcal{F}^*(\mu)$ is an uncomplicated loop. In general, either construction requires prohibitively long computer runs to yield an acceptable approximation. For the sake of efficient computation, it was found best to devise several alternative constructions and to use them in combination. After the fact, these programs turned out to help in understanding the facts. Figure 7 combines some "veils" and a "shell," while Figure 8 represents the veils alone (with fewer stages for the sake of clarity), and Figures 9 and 10 represent the shell alone.

3.2. *The ribs and veils.* For each μ, the backbone of the horizontal section of the stack is the real interval from $]-z', z'[$. The other ribs are the pre-images of $]-z', z'[$ under $f(z)$; Fig. 8, shows them up to order 8. The interval $]-z', z'[$ is well-known to fail to converge to infinity under $f(z)$, hence the backbone and the ribs belong to the set $\mathcal{F}(\mu)$ and can be said to form its "skeleton." The ribs corresponding to different μ's merge together to form a series of "veils." They include a square wall in the plane $y = 0$ and a rounded wall in the plane $x = 0$. Moving through a superstable μ, the veils change from hanging on the rounded wall to hanging on the square wall, or from hanging on a high-order veil to hanging on one of lower order. The pre-images of the unstable fixed point z' are the rib tips. The precise relationship between the ribs' closure and the set $\mathcal{F}(\mu)$ depends on the value of μ.

3.3. ***Superstable*** μs. For superstable values of μ, the ribs' closure is a domain identical to $\mathcal{F}(\mu)$. Hence, (by the same anatomical analogy) $\mathcal{F}(\mu)$ can be said to include no proper flesh. The obvious example is $\mu = 0$, when $\mathcal{F}(m)$ is the disc of unit radius, and the kth order ribs are segments joining 0 to the points of the form $\exp(2^{-k}\pi i/n)$, with n an integer and $0 < n < 2^{k+1}$. • Figure 11 • Figure 12

3.4. ***Chaotic*** μs. For the chaotic values of μ, the closure of the ribs is a tree, and is again identical to $\mathcal{F}(\mu)$. The obvious example (though a degenerate one) is $\mu = 2$. Indeed, the set $\mathcal{F}(2)$ and its ribs both reduce to the backbone $[-2, 2]$. To obtain $\mathcal{F}(2)$, it is obviously faster to draw the backbone than to use the proof of existence construction that dots $\mathcal{F}(2)$ the dense pre-images of $z = 2$.

FIGURE C5-7. Perspective view of the top portion of the repeller stack of Julia sets for real μ (vertical) and complex z (horizontal).

Whenever μ is close to either a superstable or a chaotic value, the maximal invariant set $\mathcal{F}(\mu)$ is rapidly approximated by only a few levels of ribs. Since $\mathcal{F}(0)$ is simply a disc and all the other superstable or chaotic μs fall in]1,2[it was found best to draw a few levels of ribs for every μ in]1,2[. "Unnecessary" computation costs less than determining whether or not the computation was worth performing.

3.5. *Stable but not superstable* μs. The remaining real values of $\mu \in]-1/4, 2[$ are the μ's for which a stable fixed point or a finite period exists but is not superstable. For these values of μ, the set of ribs is not dense in a domain and does not remain a tree even in the limit. To describe the resulting structure of $\mathcal{F}^*(\mu)$, let us mix the previous anatomical metaphor with a botanical one: we can say that for these μs, the trees' branches join asymptotically to form a "canopy." Clearly, an inspection of the pre-images of the unstable fixed point z' could not distinguish between the cases when the branch tips are disconnected and form dusts, and cases where the branch tips are connected.

FIGURE C5-8. Enlargement of the top portion of the "veils" within Figure 7.

FIGURE C5-9. Perspective view of the repeller stack minus the web: outside view.

FIGURE C5-10. Perspective view of the repeller stack: inside cut showing bifurcations.

3.6. The shell. As already mentioned, the proof of existence construction of $\mathcal{F}^*(\mu)$ via the pre-images of z' is efficient when $\mathcal{F}^*(\mu)$ is an uncomplicated loop, that is, for μ near $\mu = 0$. Whenever $\mathcal{F}^*(\mu)$ even moderately kinky, the cusp shaped kinks remain unfilled even after other portions of $\mathcal{F}^*(\mu)$ have been covered many times over. (For the cognoscendi: the reason is that this method reconstitutes the invariant measure on the Julia set, and this measure can be extraordinarily uneven.)

FIGURE C5-11. Detail of the top portion of Figure 9: cut along the plane $y = 10^{-3}$, using x and μ as coordinates.

An efficient graphic method is one that spends roughly equal times on each portion of $\mathcal{F}^*(\mu)$. The shell in Figure 1 was drawn by the following shell generator (Norton 1982). Each horizontal plane was covered with a square lattice and the position of a lattice point was saved in computer memory whenever, (a), its kth iterate falls within a circle of radius 2, and, (b), the kth iterate of at least one of its neighbors falls outside of that circle. These points are identified by a search method that starts with the unstable fixed point z', and is very efficient, because the number of wasted points (tested but not saved) is only a small multiple of the number of points that are saved.

Unfortunately, whenever $\mathcal{F}^*(\mu)$ is *very* kinky, as is the case for $\mu \in]1,2[$, the shell generator misses many points in $\mathcal{F}^*(\mu)$. It misses "A-pieces" that (by definition) are so thin that they squeeze between the lattice points. And it misses "B-pieces" that (by definition) are large but connect to the unstable fixed point z' through A-pieces.

When the shell is examined from the inside (Figure 4) these A- and B-pieces above do not matter, because they would be hidden anyhow. Inclusion of the ribs would hide the evidence discussed below. When, to the contrary, the shell is examined from the outside, as on Figures 7, 8 and 9, A- and B-pieces do matter. Luckily, many of the points missed by the

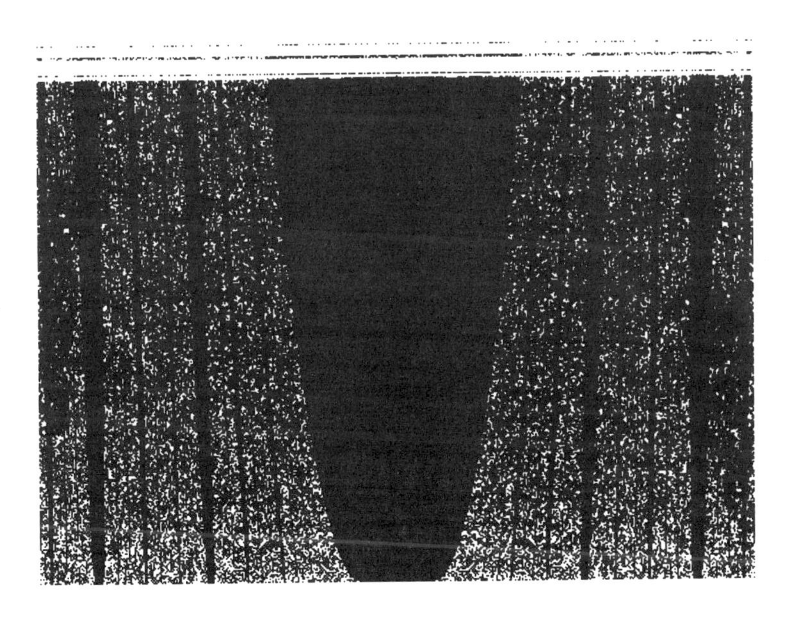

FIGURE C5-12. Even finer detail of the top portion of Figure 12 after it has been cut along the plane $y = 10^{-10}$, using x and μ as coordinates.

shell generator are picked up by the rib generator, and the combination of the two yields a sensible idea of the outside shape of the stack.

I did not intend to split this figure open and the inside view was computed by mistake. *Felix culpa.*

3.7. Basic observations. As μ increases, the repeller set varies continuously, but its topological characteristics change back and forth. The largest invariant set varies continuously within each Myrberg interval of μ. However, let μ decrease through the value $-1/4$, or through a value homologous to $-1/4$ within an island. The result is that a continuous canopy becomes punctured, leaving a dust and allowing the flesh to "evaporate." On opposite sides of a chaotic μ, the tree tips combine into canopies in different fashions.

Near $\mu = 0$, the shell is extremely smooth. More generally, as μ moves away from $\mu = 0$, the unsmoothness of $\mathcal{F}^*(\mu)$ increases very slowly and gradually. This led me to conjecture that the fractal dimension of $\mathcal{F}^*(\mu)$ is a very regular function of μ: infinitely differentiable and perhaps analytic. This hunch was proven true in Ruelle 1982.

3.8. Interpretation of the bottom portion of the inside view of the shell. Bifurcations. Aside from a clearly visible circle for $\mu = 0$, the most striking feature of this view resides in rows upon rows of protuberances. The lowest row lies at height $\mu = 3/4$. For real μ below 3/4, the map $f(z)$, whether in real or complex z, has a stable limit point. For $\mu = 3/4$, this stable limit bifurcates into a stable limit cycle of period 2. At the same time, one sees that $\mathcal{F}^*(\mu)$) changes from being a simple loop to being an infinitely knotted one. (The protuberances are denumerable and have two limit points: the unstable fixed point z' and $-z'$. The next highest row of protuberances marks the second bifurcation, and so on.

The reader is surely acquainted with the tree diagram described in May 1976, which maps the variation with μ of the values of x for all the points in the corresponding stable cycle. Were it superimposed on Figure 4, this tree would be rooted at the sharp tip to the bottom left, and each branch point would hang on a suitable protuberance. One may extend May's diagram to map the variation with μ of the "real preorbit of the cycle," defined as the set of real z that eventually fall exactly in the cycle. The resulting preorbit map diagram would be made of many trees, with a branch hanging on every protuberance of Figure 4.

3.9. Interpretation of the top portions of the inside and outside views of the shell. The Myrberg intervals of μ. The top of the stack is characterized by a nearly blank wall, which we know from the veil generating construction. However, this wall is interrupted by hanging "knobs" forming horizontal strips. Each strip corresponds to a Myrberg interval of μ.

3.10. Interpretation of some horizontal or vertical sections of the stack. When μ lives in a Myrberg interval, a horizontal section of the stack is a tree formed of Devil's Stepstones. Since the stepstones vary continuously with μ, those which intersect the plane $x = 0$ would form a kind of Devil's Corduroy.

Now take vertical sections. The $y = 0$ section of the whole stack is bounded to the side by the half parabola $\mu = x^2 + 2x$ for $x < 1/2$, and the intervals from $(\mu = -1/2)$ to $(\mu = -1/4, x = 1/2)$ and from $(\mu = 2, x = -1)$ to $(\mu = 2, x = 2)$. The bottom of this vase-shaped outline is filled solid and the top is surmounted by Myrberg strips.

Now consider analogous vertical sections of the top of the shell for $y = 10^{-3}$ (Figure 11) and of a detail for $y = 10^{-10}$ (Figure 12). Here, one sees a large number of black shapes, each of them a deformed version of the vase shape of the overall stack. When μ is such that the iterates of $f(z, \mu)$ are not chaotic, the intersection of $\mathcal{F}^*(\mu)$ and the wall $y = 0$ is a denumerable set. Its points of accumulation number 2 when μ lies in the continent, and are themselves denumerable when lies in an island. When μ is chaotic, the intersection of $\mathcal{F}^*(\mu)$ and the wall is an interval.

Acknowledgment. The programs to generate the illustrations in this paper, many of them elaborate, are due to my colleague V. A. Norton.

Chaos, Fractals and Dynamical Systems
Eds. W. Smith & P. Fischer, 1985

C6

Bifurcation points and the "n-squared" approximation and conjecture, illustrated by M.L. Frame and K. Mitchell

• ***Foreword to this chapter and the appended figure (2003).*** The n^2 conjecture advanced in this chapter's Section 2 was first proven in Guckenheimer & McGehee 1984. The two authors and I were participating in a special year on iteration that Lennart Carleson and Peter W. Jones convened during 1983–1984 at the Mittag-Leffler Institute in Djursholm (Sweden). During a seminar that I was giving, two auditors suddenly stopped listening and started writing furiously. After my talk ended, they rushed up with proofs that turned out to be identical and led to a joint report. They explained the n^2 phenomenon in terms of the normal forms of resonant bifurcations with multiplier $\exp(2\pi i/n)$. More extensive results establish that these stability domains have a limiting shape following rescaling. They are corollaries of the theory of analytic normal forms for parabolic points. See, for example, Shishikura 2000.

Section 1 states a "rough n-squared approximation" that automatically defines deviations from the approximation. Those deviations were numerically studied by Michael L. Frame and Kerry Mitchell, and Figure 1 provides an opportunity to make their results public. It is an exception to a rule, since the rest of this book limits itself to my original observations and their immediate "fallout."

For each "atom" rooted on the circle $|\lambda| = 1$ of the Mandelbrot set of $z \to \lambda z(1-z)$, Frame and Mitchell consider the vector from the root to the center defined by $f' = 0$. The top figure plots the deviation of the product n^2 {length of that vector} from its rough approximation equal to 1. The bottom figure plots the angle between that vector and one that joins the atom's root to the origin $\lambda = 0$.

Both figures are strikingly "fractal," reminiscent of the multifractal diagrams in Figure C21.2, relative to the Minkowski measure. •

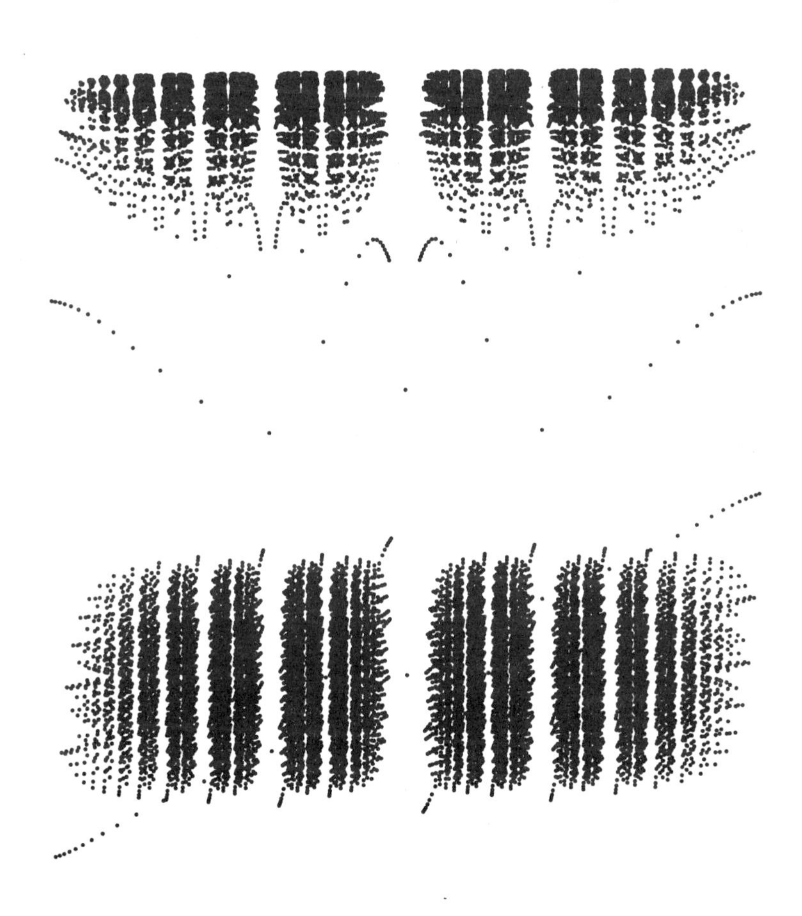

FIGURE C6-1. Deviations from the "n-squared" approximation, as explained in this chapter's foreword.

✦ **Abstract.** It is conjectured that the derivatives of $f(z, \lambda)$ for certain values of z and λ exactly satisfy a certain property called n^2 rule. Numerical calculations confirm this rule very accurately but a mathematical proof is lacking. ✦

1. Empirical observations on the radicals of M_λ

The set $\mathcal{M}_\lambda$ relative to the map $z \to \lambda z(1-2)$ includes a circle surrounded by sprout-shaped aggregates of atoms, "$\mathcal{M}$-radicals" in my chemical terminology whose roots are the points of the form $\lambda_0 = \exp(2\pi i m/n)$. The (open) atom rooted at the bond λ_0 is the locus of the points that satisfy the condition $|f_n'(z_\lambda, \lambda)| < 1$, where z_λ is one of the values in the stable periodic cycle of n points corresponding to λ. The distance between λ_0 and the atom's nucleus will be called the atom's "radius," and denoted by $r_{m,n}$.

First Observation. *The radicals are near-identical in shape to one half of $\mathcal{M}_\lambda$ itself. Hence the same is true of the sub-radicals of any order, and one half of $\mathcal{M}_\lambda$ is nearly self-similar.*

Second Observation. *Taking m/n to be an irreducible fraction, the radius $r_{m,n}$ is nearly independent of m.*

Third Observation in Rough Approximation (Mandelbrot 1983p {C5}). *The radius $r_{m,n}$ is nearly equal to n^{-2}.*

Corollary. Compare the radicals whose roots result from different sequences of bifurcation, but with orders having the same product $n_1, n_2 \ldots n_k$. Combining the above observations predicts that these sprouts have nearly identical radii.

Third Observation in a More Refined Approximation. *The quantity n^{-2} gives a close approximation to the distance from the bond λ_0 to the projection of the nucleus upon the half-line from 0 to λ_0.*

2. A mathematical conjecture whose validity would account for the above observations

In the preceding section, the terms "nearly identical", "nearly independent" and "nearly equal" reflect significant variability, to be explored in a forthcoming work. Nevertheless, these observations suggested the following precise mathematical statement. It has been verified numerically for a large number of values of m/n, but I have no general proof of it.

The "Special n^2 Conjecture". *Consider the half-line that radiates from $\lambda = 0$ to the bond $\lambda = \exp(2\pi i m/n)$, which corresponds to bifurcation from 1 to* *n. Take the derivative of $f_n(z\lambda, \lambda)$ along this half line. At the point λ_0, this derivative is equal to $-n^2$.*

Generalization. The function $g^{(\lambda)}$. The special n^2 conjecture concerns derivatives along a line that is orthogonal to the boundaries of the atoms bonded at λ_0. Furthermore, if $|\lambda| \leq 1$, the function $|f'(z_\lambda, \lambda)|$ is identical to $|\lambda|$; hence, its derivative along the same radial direction is 1.

Now define the function $g(\lambda)$ for λ in the semi-open $\mathcal{M}$-set as equal to $f_n'(z_\lambda, \lambda)$, where n is the period of the limit cycle corresponding to λ and z_λ is a point in that limit cycle. Clearly, $|g(\lambda)| = 1$ when λ is a bond. Restated in terms of $g(\lambda)$, the special n^2 conjecture asserts that when λ crosses certain special points of bifurcation into n, the gradient of $|g(\lambda)|$ is multiplied by $-n^2$.

This restatement is clearly invariant, in the sense that it remains unchanged if the parameter λ is replaced by μ or by any of a wide range of functions of λ that are sufficiently smooth near λ_0 and preserve orthogonality. Furthermore, the invariant form lends itself directly to the following generalization.

The "General n^2 Conjecture". *Let λ_0 be a bond between two atoms, corresponding to bifurcation from order n_0 to order $n_0 n$, and consider $g(\lambda)$ along a curve that crosses λ_0 in a direction orthogonal to the bonded atoms' boundaries. The ratio of the derivatives of $|g(\lambda)|$ to the right and the left of the point λ_0 is $-n^2$.*

Comment. The observations would hold true if $f_n'(z_\lambda, \lambda)$ were a linear function of λ within each atom, and the n^2 conjecture were confirmed. It would follow indeed that every atom in the continent of the $\mathcal{M}_\lambda$ -set would be a disc, and that the function $f_n'(z_\lambda, \lambda)$ would vanish at the nucleus and be equal to 1 in modulus along the disc's circular boundary. The atom's radius would be the inverse of the derivative of $f_n'(z_\lambda, \lambda)$ at the root of this atom, hence would be independent of m and equal to n^{-2}.

Acknowledgments. I acknowledge numerous useful discussions with V. Alan Norton and James A. Given. The computer programs used to evaluate the radii $r_{m,n}$ were written by J. A. G., under the guidance of V. A. N. The assistance of Linda Soloff and of Janice H. Cook was valuable.

Chaos, Fractals and Dynamical Systems,
Eds. W. Smith & P. Fischer, 1985

C7

The "normalized radical" of the $\mathcal{M}$-set

• *Chapter foreword (2003).* The original abstract was expanded. Is the "normalized radical" an idle exercise, or is it interesting in any way? To my disappointment, it has attracted little further attention. •

✦ **Abstract.** A "normalized radical" $\mathcal{R}$ of the $\mathcal{M}$-set is defined as the shape that satisfies exactly all the self-similarity properties that hold approximately for the molecules of the $\mathcal{M}$-set of the quadratic map. Explicit constructions show that the complement of $\mathcal{R}$ is a σ-lune, and prove that the $\mathcal{R}$-set does not self-overlap. The fractal dimension D of the boundary of $\mathcal{R}$ is shown to satisfy $\Sigma_2^\infty \phi(n) n^{-2D} = 1$, where $\phi(n)$ is Euler's number-theoretic function.

A rough first approximation is the solution $D = 1.239375$ of the equation $\Sigma_2^\infty n^{1-2D} = \zeta(2D-1) - 1 = \pi^2/6$, where ζ is the Riemann zeta function. A less elegant but doubtless closer second approximation is $D = 1.234802$. The same D applies to the $\mathcal{M}$-sets of other maps in the same class of universality.

Interesting "rank-size" probability distributions are introduced. ✦

1. INTRODUCTION

A remarkable set in the complex plane, the $\mathcal{M}$-set, plays a central role in the dynamics implicit in the iteration of polynomials, of rational functions and of other analytic maps. This paper focuses on the quadratic map, written as $z \rightarrow f(z, \lambda) = \lambda z(1 - z)$, where z and λ are complex numbers. The molecules of the corresponding $\mathcal{M}$-set are approximately self-similar. Section 2 defines a "normalized radical" $\mathcal{R}$, as being a shape for which these self-similarity rules hold exactly, and asserts that $\mathcal{R}$ fails to self-overlap. Section 3 proves that $\mathcal{R}$ fails to self-overlap via an explicit construction of the complement of $\mathcal{R}$. Section 4 is an aside into number theory, needed in Section 5. Section 5 evaluates the fractal dimension of the boundary of $\mathcal{R}$ while Section 6 introduces some interesting probability distributions.

2. THE NORMALIZED $\mathcal{R}$. DEFINITION, ABSENCE OF SELF-OVERLAP

***Definition of $\mathcal{R}$ as a σ*-disc.** The normalized $\mathcal{M}$-radical $\mathcal{R}$ is seen on Figure 1 to be a σ-disc, that is, a denumerable collection of closed disc-shaped atoms, and is constructed recursively as follows. The 0-th generation atom is the unit disc and the 1-st generation atoms are discs tangent to the unit disc, the points of tangency ("bonds") being of the form $\lambda = \exp(2\pi i m/n)$ and the radii being equal to n^{-2}. Each k-th generation atom of radius n^{-2} is given intrinsic coordinates such that its center is of coordinate 0 and its bond to the $(k-1)$th generation disc is of coordinate-n^{-1}. Then this k-th generation atom is "decorated" by $(k+1)$st generation atoms whose bonds' intrinsic coordinates are of the form $n^{-2}\exp(2\pi i m/\nu)$ and whose radii are equal to $n^{-1}\nu^{-2}$.

Unbounded variant of the radical. A variant of this construction alternates stages of interpolation extrapolation. The process is tedious to write down, but the result is, in effect, the same as the limit of simple interpolation (as above) that has been blown up in the ratio 4^{∞}, while its rightmost point remains fixed. A piece of this extrapolation is seen on Figure 1, using a partly-filled disc.

Definition of the term, "lune". Given a circle C' and a circle C" that is tangent to C' and otherwise contained in C', the set of points inside C' and outside C" is called an open lune.

Representation of the complement of $\mathcal{R}$ as a σ -lune. *The complement of $\mathcal{R}$ is covered by a σ -lune, namely by a denumerable infinite collection of open lunes.* This construction results from various properties of $\mathcal{R}$ embodied in the lemmas of Section 3. An advanced stage of construction is illustrated on Figure 2. The construction is so effective that Figure 2 is not very telling. Therefore, an alternative illustration is given in Figure 3, where

FIGURE C7-1. By definition, the normalized radical $\mathcal{R}$ of the $\mathcal{M}$-set is a σ-disc. Here it is approximated by the union of a fairly large number of closed discs.

each lune is represented by a fan of lines, and it is clearly seen how the various fans overlap. (It is amusing to report that this attractive and illustrative Figure 3 had originally resulted from an error of computer programming.)

Corollary. Property of non-self-overlap. The closed discs constructed as in the above definition of $\mathcal{R}$ do not overlap except at their bonds.

A map, $z \to h(z\eta)$ a complex parameter, such that its $\mathcal{M}$-set is made of normalized $\mathcal{M}$-radicals $\mathcal{R}$; the dependence of η on μ is singular. Denote by η the complex variable in the plane in which $\mathcal{R}$ is embedded. The interiors of $\mathcal{R}$ and of the continental $\mathcal{M}$-molecule of the map written as

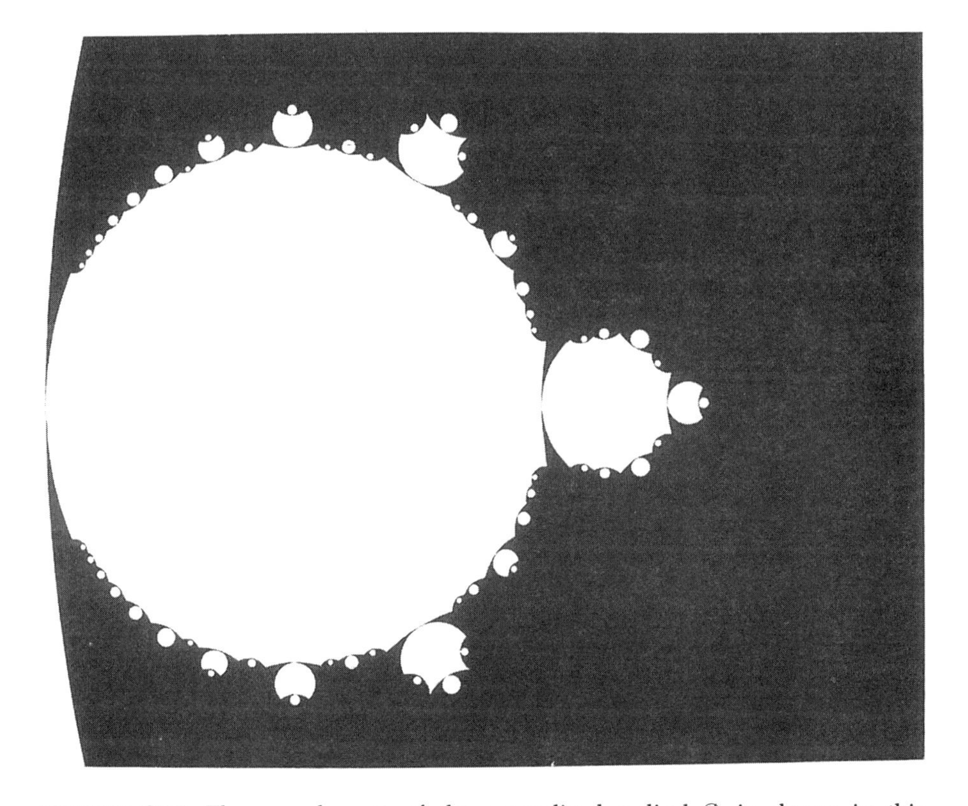

FIGURE C7-2. The complement of the normalized radical $\mathcal{R}$ is shown in this paper to be a σ-lune. Here it is approximated by the union of a not-too-large number of lunes. By definition, the subradicals are reduced-scale versions of the radical itself. Therefore, the examination of subradicals of increasing size shows how an increasing number of lunes defines the complement of $\mathcal{R}$: first as a disc, then in increasing detail.

$f*(z, \mu) = z^2 - \mu$ can be put in a doubly continuous one-to-one correspondence $\mu(\eta$ in such a way that, for $h(z, \eta) = f*[z, \mu(\eta)]$, the $g(\lambda)$ function (M 1985g{C6}, Section 6) is piecewise linear within each $\mathcal{M}$-atom. Thus, the $\mathcal{M}$ -molecule of $h(z, \eta)$ is $\mathcal{R}$ itself. Recall, however, that Feigenbaum "universality," as applied to the map f^* restricted to the real axis, tells us that the ratio of the diameters of successive atoms intersected by the real axis takes on the value 4.66920 for a wide class of maps. For the map, $h(z, \eta)$ its value is 4, therefore $h(z, \eta)$ does not belong to the same class of universality as $f*(z, \mu)$.

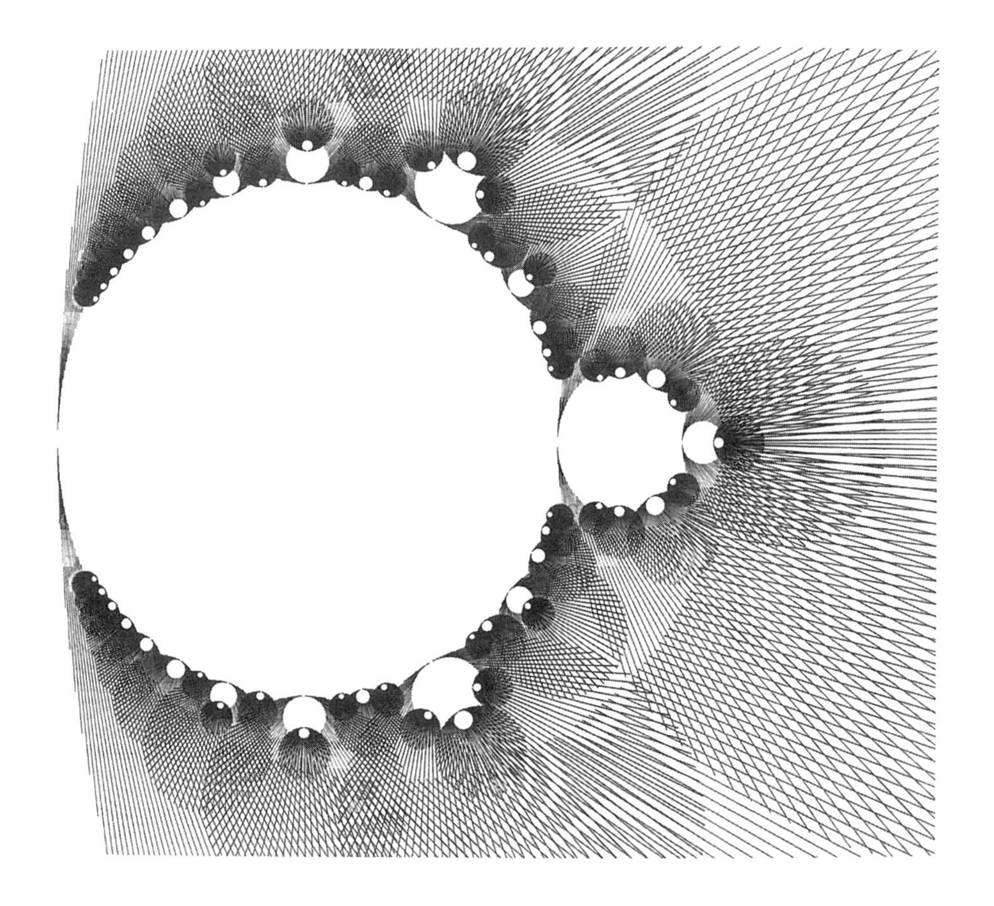

FIGURE C7-3. Identical to Figure 2, except for a programming error. Each lune should have been drawn as the union of filled-in quadrangles, but is drawn here as the union of these quadrangles' outlines. This Figure explains the σ-lune algorithm better, and it has the great virtue of being attractive.

As an example, consider the neighborhood of the limit points μ_∞ and η_∞ corresponding to an infinite sequence of 2-bifurcations. In this neighborhood, $(\mu_\infty - \mu) \propto (\eta_\infty - \eta)^{log4.66/log4}$. Similarly, the counterparts of λ for series of n-bifurcations, with $n > 2$ were evaluated by Cvitanovic & Myrheim 1982; they are also close to, but different from, n^{-2}. Thus, the corresponding $\mu(\eta)$ is very singular on the boundary of $\mathcal{R}$.

Deflated form of $\mathcal{R}$. Figure 4 shows what would happen to $\mathcal{R}$ if each disc of perimeter $2\pi n^{-2}$ flattened into a "doublet": an interval of length πn^{-2} covered in one direction and then in the opposite one. The right side is parameterized by θ in $[0, \pi]$, and the left side by θ in $[\pi, 2\pi]$. Two atoms mutually tangent for $\theta = \pi$ are replaced by doublets colinear to the originators and all other orthogonal atoms are replaced by doublets orthogonal to the original one.

3. CIRCLES THAT OSCULATE $\mathcal{R}$, AND PROOF THAT THE COMPLEMENT OF $\mathcal{R}$ CAN BE COVERED BY A σ -LUNE

Comment. The lemmas in this section have obvious approximate counterparts in the actual $\mathcal{M}$-set of the map $z \to \lambda z(1-z)$.

Lemma A. *An infinite sequence of bifurcations leads to the point* $[-5/3,0]$ *Proof.* The diameters of the discs in this infinite sequence are $2, 2(1/4), 2(1/4)^2$, etc.. These diameters' sum is $8/3$.

Lemma B. *$\mathcal{R}$ is contained in the closed disc of center 0 and radius 5/3. Proof.* By Lemma A, this disc contains all the bonds that obtain by a finite sequence of bifurcations of order 2. It suffices to consider a sequence of bifurcations, such that the orders $n_1, n_2 ... n_g$, are not all equal to 2, and to show that it always leads to a bond that lies within the above disc. Indeed, consider the broken line that goes from 0 to the first bifurcation bond, then to the center of the corresponding atom, then to the next bifurcation bond, etc... Each vector in this broken line is at most equal in modulus to that of the line which corresponds to $n_g \equiv 2$, and at least one of these vectors is smaller in modulus. Therefore, those vectors' sum is less than $5/3$.

Lemma C. *Near its root at* $\exp(2\pi i\theta)$ *with* $\theta = 0$, *$\mathcal{R}$ is locally osculated by a circle of curvature (inverse radius) equal to* $\rho^{-1} = 1 - 4/3\pi^2$. *Proof.* Apply

Lemma B to each subradical of $\mathcal{R}$. A subradical whose bond coordinates are $\exp(i\theta) = \exp(2\pi im/n)$ lies within a disc whose radius is $(5/3)\ n^{-2}$ and whose center is at a distance of $1 + n^{-2}$ from 0. A fortiori, it lies in a disc whose radius is $(5/3)(m/n)^2$ and whose center is at a distance $1 + (m/n)^2$ These enlarged bounding discs badly overlap, but their union's outside boundary is a smooth curve. Near $\theta = 0$, this boundary is the curve of equation $x = -y^2/2 + (8/3)(y/2\pi)^2 = -y^2[1/2 - 2/3\pi^2] = y^2/2\rho$ Hence its local curvature is $\rho - 1 = 1 - 4/3\pi^2$, meaning that the osculating circle has the radius $\rho \sim 1.1562$.

Lemma D. *$\mathcal{R}$ is contained in a disc of center -1/3 and radius 4/3. Proof.* Writing $\theta/2\pi = u$, the distance from $-1/3$ to a point is an enlarged bounding disc is at most $1/3\ \{5u^2 + \sqrt{[1 + 9\ (1 + u^2)^2 + 6(1 + u^2)\cos(1\pi u)]}\}$. This quantity is readily seen to be 4/3 for $u = -1{,}0$ or 1, and to be smaller than 4/3 for all other u in $[-1{,}1]$.

Lemma E. *Near the bond point* $\exp(2\pi im/n)$, *divide $\mathcal{R}$ into a subradical, whose large atom is a circle of radius n^{-2}, and a remainder. The subradical is locally osculated by a circle of curvature $n^2(1 - 4/3\pi^2)$, and the remainder is locally osculated by a circle of curvature* $4n^2/3\pi^2 - 1$.

Corollary of Lemma E. Osculating lunes. *The subradical and the remainder do not overlap locally; in fact, they are locally separated by the "osculating lune" contained between the osculating circles.*

Comment on Lemma E. Globally, each osculating lune intersects $\mathcal{R}$. Therefore, covering the complement of $\mathcal{R}$ requires lunes that are smaller than the osculating lunes.

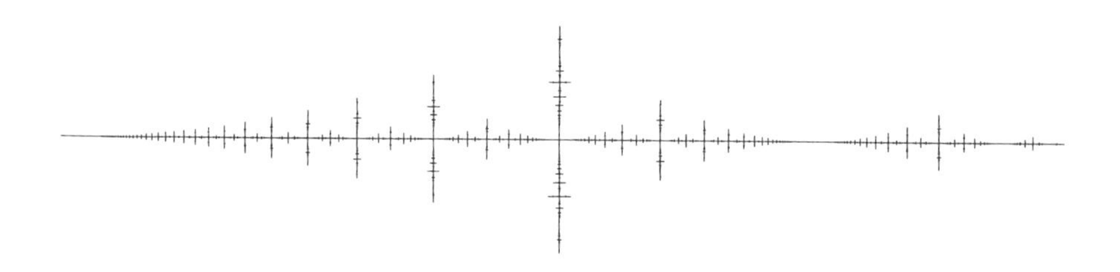

FIGURE C7-4. The boundary of $\mathcal{R}$, after each of its discs has been "deflated" into two superposed segments. The segments corresponding to 2-bifurcations are aligned, and those corresponding to higher bifurcations are othogonal.

Lemma F. Complementary lunes. *To each subradical of* $\mathcal{R}$ *whose big atom is a disc of radius r, attach the "complementary lune" that is contained between the following two circles: the atom's circumscribed circle, whose radius is (4/3)r, and the circle of radius (6.6)(4/3)r that is tangent to the circumscribed circle at the bond of the subradical. This lune is called complementary because it lies outside* $\mathcal{R}$. The intersection of the exterior of $\mathcal{R}$ with the whole radical's circumscribed circle is covered by $\sigma-$ lune defined as the union of the subradicals' complementary lunes.

Comment on Lemma F. One of the circles that includes this lune is obvious, since, in order for a lune to lie outside of $\mathcal{R}$, it is necessary and sufficient that one of the bounding circles be circumscribed to the atom. The numerical factor 6.6 was obtained by trial and error. However, it can easily be proven that the exterior of $\mathcal{R}$ can be covered by slightly more involved sets.

As seen on Figure 3, the structure of the exterior of $\mathcal{R}$ is well outlined by a small number of lunes. The same structure is also clearly visible on one half of the $\mathcal{M}$-set. It has motivated the present investigations, and is explained by them.

4. EULER'S NUMBER-THEORETICAL FUNCTION $\phi(n)$ AND KIN

The number of irreducible fractions of denominator n is Euler's number theoretical function $\phi(n)$. Define $\phi*(n) = \Sigma_{u=1}^{n}\phi(u)$. By a theorem of Mertens, $\phi*(n) \sim (3/\pi^2)n^2 + 0(n\log n)$ as $n \to \infty$ (Hardy & Wright 1960, Theorem 330) yielding in a rough approximation $\phi(n) \sim (6/\pi^2)n \sim 0.6079271n$.

A more detailed investigation suggested the more precise representation $\phi* = (3/\pi^2)\{n^2 + n[1 + \beta(n)] + \alpha\}$. This formula defines a number α and a function $\beta(n)$. This β looks very much like a stationary random function of n that oscillates in a bounded interval centered at zero, and the "random term" $n[1 + \beta(n)]$ representation is of the order of magnitude of the square root of the "drift term" $n^2 + n$. The function $\beta(n)$ has very interesting properties that will be reported in greater detail elsewhere. Observe that $(6/\pi^2)n$ is the finite difference of the above $\phi*(n)$ when $\beta(n)$ is set to 0.

5. VALUE OF THE FRACTAL DIMENSION D OF THE BOUNDARY OF $\mathcal{R}$

By the self-similarity property of $\mathcal{R}$, D is the solution of the D-generating equation $\sum_{n=2}^{\infty}\phi(n)n^{-2D}=1$. A numerical solution obtained by Newton's method is $D=1.239375$. Using the leading term of the Mertens formula for $\phi(n)$, one obtains a simple approximate D-generating function, namely $(6/\pi^2)\sum_2^{\infty}n^{1-2D}=1$, that is $\zeta(2D-1)=1+\pi^2/6$. The numerical solution is $D=1.245947$. The smallness of this value of D explains why Figure 4 is so "skinny".

6. RANK-SIZE PROBABILITY DISTRIBUTIONS SUGGESTED BY THE SUBRADICALS OF $\mathcal{R}$

Rank the subradicals directly attached to $\mathcal{R}$ by increasing values of n and, for each n, by increasing values of m. The rank of a subradical in this list will be denoted by ρ. The number of subradicals whose base circle is of radius n^{-1} Euler's $\phi(n)$ (Section 3), hence $\sum_{n=1}^{n}\phi(u)=\phi*(n)$ is the number of ρs such that base circle radius is $\geq n^2$. In the Mertens approximation, $\phi*(n)\sim(3/\pi^2)n^2$; hence we find the rules that "a subradical's linear size is $\sim 1/\mathit{rank}$" and "a subradical's area is $\sim 1/(\mathit{rank})^2$"

These are examples of the so called "statistical rank-size" rule, which is followed by many natural phenomena; a few are mentioned in M 1982{*FGN*} (e.g., Chapter 38) {P.S. 2001. See also M 1997E, Chapter E7.} Unfortunately, examples where this rule is obtained by a theoretical argument are comparatively rare, which adds value to the present example. Because $\Sigma(1/n)=\infty$, linear size cannot be weighted to yield a probability but area can be weighted in this way: it suffices to divide it by the sum of all the subradicals' areas.

Does there exist an exponent D such that N^{-D} is the probability of a subradical, without the need for a weighting prefactor. Such a D must satisfy $\sum_{n=2}^{\infty}\phi(n)n^{-2D}=1$, therefore is the fractal dimension of the boundary of $\mathcal{R}$. Analogy with other nonrandom fractals suggests that the Hausdorff measure of the boundary of $\mathcal{R}$ is positive and finite and can be taken to be 1. If so, n^{-2D} (without any prefactor) is the Hausdorff measure of the boundary of the subradical rooted at $\exp(2\pi im/n)$. This measure gives mathematical meaning to the intuitive notion of point distributed uniformly on the boundary of $\mathcal{R}$.

7. THE NUMBER-THEORETICAL FUNCTION $\nu(n)$

Denote by $\nu(n)$ the number of circles of radius $\geq n^{-2}$ contained in $\mathcal{R}$. For many fractal curves whose complement is constructed as a union of open "gaps" (for example, the Apollonian gasket; M 1982{*FGN*}, p. 170 {C16}) the dimension is known to be the value of D that rules the distribution of the linear sizes of the gaps. That is, F being a numerical prefactor, $\nu(n) \sim F(n^{-2})^{-D} = Fn^{2D}$. It is reasonable to assume (but remains to be proven) that this relation also holds for $\mathcal{R}$. This conjecture suggests that $\nu(n) \infty n^{2D}$.

The function ν(k) has a direct arithmetical interpretation. It is the number of distinct products of irreducible fractions, such that the product's denominator is n, granted that each permutation of distinct multiplicands is counted separately. This definition would look contrived, were it not for the application that motivated it. One wonders, inevitably, whether or not this $\nu(n)$ plays any other role in "non-abelian" arithmetic.

Acknowledgments. I acknowledge numerous useful discussions with V. Alan Norton, James A. Given and Janice H. Cook. The computer programs used to draw Figures 1, 2 and 3, and to estimate D, were written by J.H.C.

Chaos, Fractals and Dynamical Systems
Eds. W. Smith & P. Fischer, 1985

C8

The boundary of the $\mathcal{M}$-set is of dimension 2

• *Chapter foreword (2003).* To great acclaim — and belatedly — the conjecture advanced in this text was eventually confirmed in Shishikura 1984. An obvious next question concerns the boundary's Hausdorff measure in the dimension 2. Whether this measure is 0 or positive is an unsolved problem. •

✦ **Abstract.** It is conjectured that the boundary of the $\mathcal{M}$-set of the complex map has a Hausdorff-Besicovich fractal dimension equal to 2. ✦

THE AMBITION OF THIS PRESENT PAPER is to help understand the "devil's polymer" structure that binds together the molecules of the $\mathcal{M}$-set. It may be a branching polymer (tree), as for the map $z \to \lambda z(1-z)$, in which case the complement is connected. It may be a net whose complement is a collection of open sets. And it may be a net whose complement's structure is not yet well-understood, as for the map $z \to \lambda(z+1/z)$. The present text is devoted to $z \to \lambda z(1-z)$, but the conjecture it describes is surely of broad applicability.

Conjecture. *The boundary of the* $\mathcal{M}$-set *is a curve of fractal dimension equal to* $D = 2$.

Background. Figure 1 — a negative of Plate 164 of M 1982{*FGN*} — results from a recursive creation of (white) branches. The ratio of branch lengths before and after each branch point increases slowly to $\sqrt{2}/2$ as one moves towards the branch tips, and the ratio of width to length decreases to zero. On the left side of the illustration, this width/length ratio decreases even faster than on the right side. If one zooms in, the picture remains unchanged in overall shape, but the relative thicknesses of similarly positioned branches decrease to zero. This means that — as intended — both diagrams *fail* to be self-similar. This is how both achieve the desired result: the fractal dimension of the branch sides and the branch tips is D=2. (The rate at which the ratio width/length decreases to zero does not affect the value of D.)

Evidence. Figures 2 and 3 describe the local structure of the boundary of the $\mathcal{M}$-set by "zooming" onto the neighborhood of a typical point of the boundary of the continental molecule. More precisely, if σ denotes the "golden ratio" $\sigma = (\sqrt{5} - 1)/2$, Figures 2 and 3 inspect in increasing detail a *small* piece of the $\mathcal{M}$-set near $\lambda = \exp(2\pi i\sigma)$, and a smaller sub-piece. Each piece is roughly centered on a parameter value of the form $\lambda_m = \exp(2\pi i\sigma_m)$, where σ_m is the ratio of the m-th and the $(m+1)$st Fibonacci numbers. These σ_m are successive approximations to the golden ratio σ. And the size of each piece contains the radical rooted at λ_m and the stellate structure of rays and island molecules beyond its tip.

The striking overall resemblance between Figures 2 and 3 and Figure 1 provides the evidence for the above conjecture.

FIGURE C8-1. Construction of two curves of fractal dimension $D = 2$; from M1982F{FGN} p. 164.

Discussion. What causes the behavior shown in Figures 2 and 3? To find out, let us compare the stellate structures of rays and island molecules that exist beyond the trip of earth radical in the $\mathcal{M}$-set. The orders of branching of these structures are the orders of bifurcation of the sequence that leads to the tip in question. When the radical is large, Paper III shows that the product of the orders of bifurcation is small, hence the largest among them is also small. Therefore, the large radicals' stellate structures have the appearance of sparse "trees". For example, the tip corresponding to an infinity of 2-bifurcations continues by a spear without branches. Similarly, M 1982F{*FGN*} p. 189 shows the tip corresponding to one 3-bifurcation and an infinity of 2-bifurcations: there is little branching. The same p. 189 of M 1982{*FGN*} shows, however, that one 100-bifurcation followed by an infinity of 2-bifurcations yields a much thicker structure. Increasingly small radicals have a rapidly increasing probability of

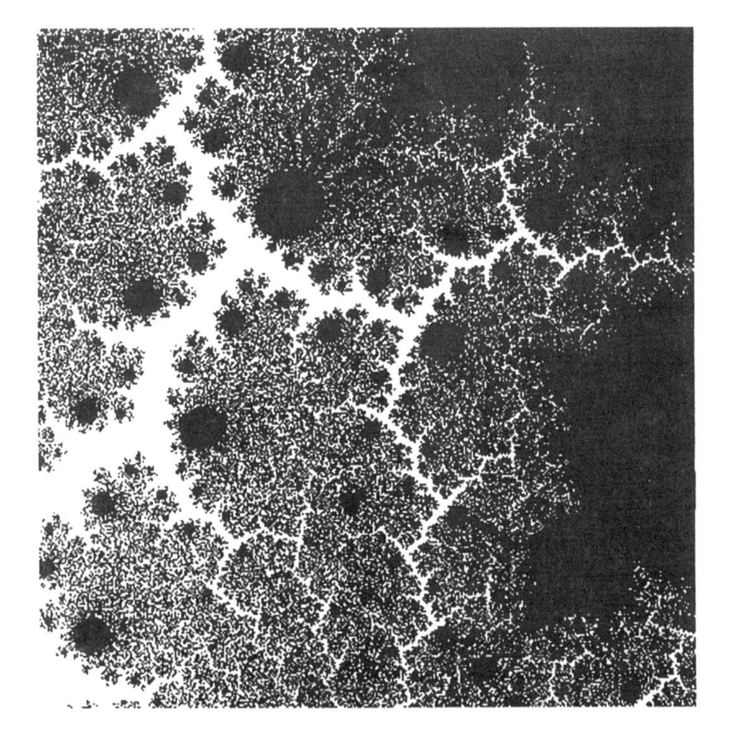

FIGURE C8-2. In parameter space, a detail of the $\mathcal{M}$-set of $z \to \lambda z(1-z)$, when $-\log \lambda / 2\pi i$ lies in a neighborhood of the golden ratio $(\sqrt{5}-1)/2$. The disc $|\lambda| < 1$ is visible to the top right.

involving very high bifurcations; as a result, their stellate structures become increasingly thick "bushes".

It is clear that the local shape of the boundary of the $\mathcal{M}$-set is dominated by an infinity of infinitely small and "bushy" structures. They crowd each other, hence the above conjecture.

Comment. Figure 1 was drawn 7 years before Figures 2-3. The latter exhibit small branches throughout, while the small branches in Figure 1 only arise by division of slightly larger branches. Figure 1 is a variant of a fractal model of the lung; a more suitable background for the present analogy would have been provided by a fractal model of the vasculature!

Acknowledgments. Figure 1 was prepared by Sigmund W. Handelman. Figures 2 and 3, computed with programs written by V. Alan Norton, were prepared by James A. Given.

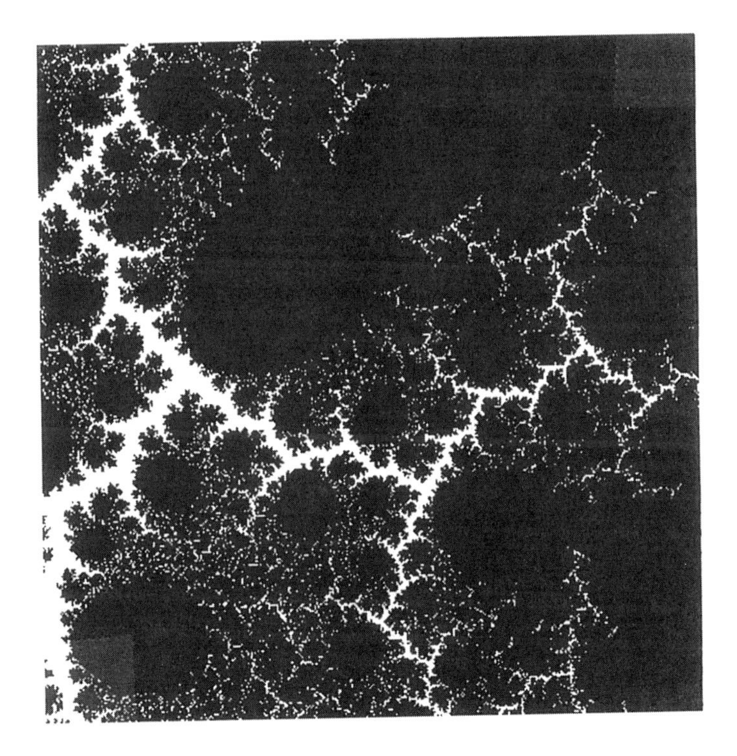

FIGURE C8-3. In parameter space, an expanded detail of Figure 2, in the region of its top right corner. The disc $|\lambda| < 1$ is again visible at the top right. This Figure is very analogous to Figure 2, but the white strips are narrower. This is a symptom of $D = 2$.

Chaos, Fractals and Dynamical Systems
Eds. W. Smith & P. Fischer, 1985

Certain Julia sets include smooth components

The Julia set F^* of the map $z \to \tilde{f}(z) = z^2 - \mu$ may be the boundary of an atom, of a molecule, or of a "devil's polymer" in the z-plane. Denote the boundary of one of the atoms of F^* by H. When $\mu \neq 0$ is the nucleus of a cardioid-shaped atom of the M-set, it is conjectured that the fractal dimension D of H is 1. Thus, H may be a be a rectifiable curve (of well defined length) or perhaps only a borderline fractal curve (of logarithmically diverging length). This paper comments on a clearer version of Figure 5 of M1983l{C5} and develops a remark made there, but not very explicitly.

The map $z \to \tilde{f}(z) = z^2 - \mu$ and its Julia sets $\tilde{F}$

For the quadratic map, the Julia set $\tilde{F}$ is the boundary of the set F of points that do not iterate to infinity under $\tilde{f}$. When $\mu = 0$, this $\tilde{F}$ is the unit circle and when $\mu = 2$, it is an interval. For all other μs, $\tilde{F}$ is unsmooth.

Of special interest are the μs such that the set F has interior points, namely those in the interior of the M-set, plus the bonds between atoms of the M-set. The structure of F depends on μ and several possibilities will be explored.

When μ lies in the large atom of the continental molecule of the M-set, F is an atom; that is, $\tilde{F}$ is a Jordan curve (simple loop).

When μ lies in the continental molecule of the M-set, but not in its large atom, F is a molecule, the union of a (denumerably infinite) number of joined atoms.

When μ lies in an island molecule of the M-set, F has a "devil's polymer" structure.

In the second and third of the above possibilities, the complement of $\tilde{F}$ includes a denumerable infinity of maximal open components that do not include the point at infinity. The boundary of one of these components is to be called H, and others are the preimages of H under $\tilde{f}(z)$. The set and its preimages being smooth images of each other, they have the same "partial" fractal dimension D, at most equal to the dimension of $\tilde{F}$. When a fractal set is the union of curves that are individually smoother than the whole, then (M 1982F, p. 119) the overall fractal dimension does not measure the roughness of the whole but its fragmentation.

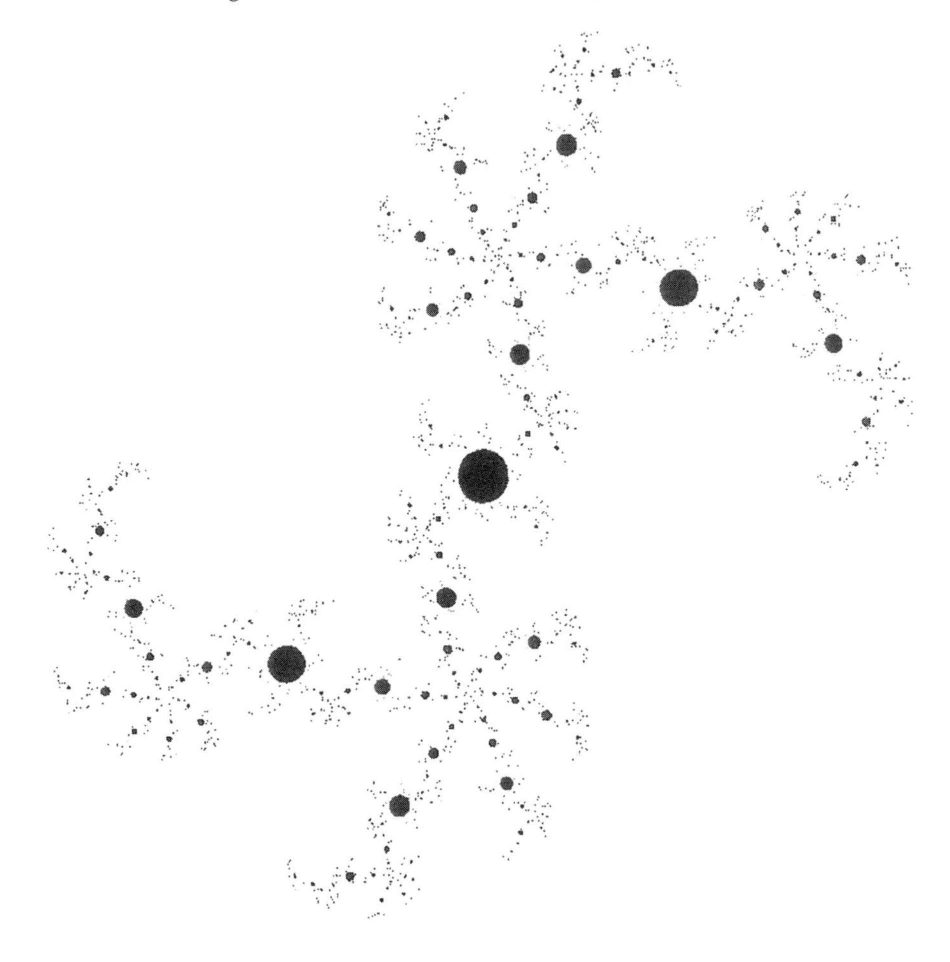

FIGURE C9-1. A Julia set made up of near circular mono-atomic molecules. The map is $z \to z^2 - \mu$, where μ is the nucleus of the largest atom of the largest island off a radical corresponding to 5-bifurcation. The cycle's order is 6.

Fatou 1919 had shown that in general it is impossible for a subset of $\tilde{F}$ to be an isolated analytic arc. Under wider conditions, $\tilde{F}$ may include non-isolated analytic arcs.

Well-known example

When μ is real and the map $\tilde{f}(z)$ is chaotic, $\tilde{F}$ includes a real interval and its preimages, which are smooth algebraic curves. In this case, the "branched polymer" structure of $\tilde{F}$ is of special simplicity. Indeed, the real interval $[-z', z']$, with $z = 1/2 + 1/2(1+4\mu)^{1/2}$, is contained in F and can be said to form its "spine", in the sense that H and a collection of its preimages appear as "beads" strung along $[-z', z']$. Also, the preimages of $[-z', z']$ are smooth curves contained in F, and they form "ribs" along which other preimages of H are strung as "beads". At the origin O and at each preimage of O, an infinity of ribs intersect, and near the origin these ribs are nearly straight and radiate from O.

Observations concerning the Julia sets when μ is the nucleus of a cardiod-shaped large atom of the M-set and $\mu \neq 0$

In this case, H can be defined as the boundary of the domain of immediate attraction of $z = 0$. The origin may be called the nucleus of H, and the preimages of H have as nuclei the corresponding preimages of O. When $\mu = 0$, the set H is identical to $\tilde{F}$ and is a circle. When $\mu \neq 0$, the observation is that H is nearly a disc, hence $\tilde{F}$ is the union of near circles. They look so smooth as to seem to have tangents and therefore to be rectifiable. It has been mentioned that they are linked in a "branched devil's polymer" structure. An example is shown on Figure 1.

Acknowledgements. Figure 1, improving on Figure 5 of M1983l{C5}, used programs written by V. Alan Norton and was prepared by James A. Given.

Chaos, Fractals and Dynamical Systems
Eds. W. Smith & P. Fischer, 1985

C10

Domain-filling sequences of Julia sets, and intuitive rationale for the Siegel discs

• *Chapter foreword (2003).* No claim is made in this piece to *prove* that quadratic dynamics involves Siegel discs. The only goal is to make their occurrence seem "natural" and even "obvious." •

✦ **Abstract.** Within the $\mathcal{M}$-set of the map $z \to \lambda z(1-z)$, consider a sequence of points λ_m having a limit point λ. Denote the corresponding $\mathcal{F}*$ -sets by $\mathcal{F}*(\lambda_m)$ and $\mathcal{F}*(\lambda)$. In general, $\lim \mathcal{F}*(\lambda_m) = \mathcal{F}*(\lim \lambda_m)$, but there is a very important exception. In some cases, the sets $\mathcal{F}*(\lambda_m)$ do not converge to either a curve or a dust, but converge to a domain of the λ -plane, part of which is called the Siegel disc $\mathcal{S}$, while the rest is made of the preimages of $\mathcal{S}$. In such cases, $\mathcal{F}*(\lim \lambda_m)$ is not the set $\lim \mathcal{F}*\lambda_m$ but only that set's boundary. The intuitive meaning of this behavior is discussed and illustrated in terms of the so-called Peano curves, and a mathematical question is raised concerning the nonrational and non-Siegel λ. ✦

1. INTRODUCTION

Special cases of otherwise intuitive mathematical theories often seem to exhibit a "pathological behavior" that requires a technically difficult special treatment. For example, physicists find it hard to believe that there can be very concrete significance to functions F for which $\lim F(X_n)$ is completely different from $F(\lim X_n)$. The discs whose existence is shown in Siegel 1942 are such a special case of the theory of iteration of rational functions (and of other analytic functions). The argument and the illustrations in the present paper relate "Siegel discs" to Peano curves, hence ought to show that the Siegel disc behavior is in fact eminently reasonable. One could say it "should have been expected on physical grounds".

We consider the dynamics of certain extremely non-linear maps $z \rightarrow f(x)$ of the complex plane on itself, and separate the linear and non-linear terms at the origin by writing the map in the form $z \rightarrow \lambda z + g(z)$, where $g(z)$ satisfies $g(0) = g'(0) = 0$. In order to strip down inessential complications, we stay with the quadratic case $g(z) = -\lambda z^2$.

2. BACKGROUND: PEANO CURVES

The key to the argument resides in the concept of Peano "plane-filling curve," which the fractal geometry of nature retrieved from among the monster sets and showed to have great concrete importance. A "plane-filling curve" is really a *sequence* of ordinary curves, $\mathcal{C}_n$, having an unexpected limit. In this instance, we need a slight variant: we assume that the $\mathcal{C}_n$ are loops with no double point (Jordan curves). The criterion of Peano behavior is that there exists a plane domain, $\mathcal{D}$, such that every point P of $\mathcal{D}$ can be written as $P = \lim_{n \rightarrow \infty} P_n$, where P_n belongs to $\mathcal{C}_n$. Thus, the limit of the $\mathcal{C}_n$ is not at all a curve, but is the whole domain $\mathcal{D}$. However, in order to have a curve as limit, one may say that the limit of the $\mathcal{C}_n$ is the boundary of $\mathcal{D}$, which is a curve to be denoted by $\mathcal{C}$. In the classic examples of Peano, Hilbert and Sierpinski, $\mathcal{C}$ is a square. But in some more recent constructions $\mathcal{D}$ is a very irregular fractal curve; M 1982F{*FGN*} p. 68-69 illustrates an example of mine, where the boundary $\mathcal{C}$ of a Peano curve is the Koch snowflake curve.

The point of the present paper is that the same "Peano" behavior is characteristic of certain sequences of Julia sets. The fact that the $\mathcal{C}_n$ are Jordan curves is inessential. We shall also encounter Peano sequences of curves that have infinitely many double points.

3. SIEGEL DISCS

The map $\rightarrow f(z)$ is said to have a Siegel disc, when there exists a bounded domain of the plane, open and connected, within which $z \rightarrow \lambda z + g(z)$ is essentially reduced to its linear term, that is, the map is equivalent to a rotation of angle θ. This means that a suitable deformation of z (a holomorphic function obtained as the solution of Schröder's equation) yields a quantity ω in terms of which the map becomes $(\omega - \omega_0) \rightarrow (\omega - \omega_0)e^z 2\pi i\theta$. The circles centered on ω_0 transform into nested Jordan curves whose union covers the inside of the Siegel disc.

When a Siegel disc exists, λ is of the form $\lambda = \exp(2\pi i\theta)$ with θ a "Siegel number," meaning an irrational number satisfying certain conditions that are rather complicated but sufficiently mild for such numbers to be of unit measure on [0,1]. The rough idea is that a Siegel θ *cannot* be

FIGURE C10-1. The Julia sets of $z \to \lambda z(1-z)$ in the space of the variable z drawn for four values of the parameter λ. To demonstrate the nature of convergence to a Siegel disc, the $\log \lambda_m / 2\pi i$ are the rational numbers 2/3, 8/13, 34/55, and 144/233. They belong to the Fibonacci sequence that converges to the golden ratio $\sqrt{5} - 1)/2$.. The numbers of "petals" in these "flowers" are known to be the denominators of the $\log \lambda_m / 2\pi i$; unfortunately, these petals are not very distinct on this Figure. See the Figures that follow for a method of separating the petals.

well approximated by a sequence of rational numbers. The angle of the rotation $z \rightarrow \lambda z$ being irrational, the iterates are ergodic.

How should one call the dynamics corresponding to f(z) when θ is a Siegel number? Depending on one's feelings, one may call it either highly ordered or highly disordered.

In order to understand this behavior, I think the key step is to investigate a Siegel map $f = \lambda z - \lambda z^2$ as the limit of suitable sequence of quadratic maps ${}_m f$. Iteration associates to each ${}_m f$ a $\mathcal{F}*$ Julia set: the boundary of the domain that fails to iterate to infinity. The Julia set of the limit map f is in general the limit of the Julia sets of the maps ${}_m f$. But when the limit f is a Siegel map, the Julia set of the limit map f is *completely different* from the limit of the Julia set of the approximating maps ${}_m f$. Indeed, the Julia set of the limit is a curve. But the limit of the Julia sets is not a curve, but a bounded domain $\mathcal{D}$ of the complex plane. This $\mathcal{D}$ includes a Siegel disc component and is the union of a Siegel disc and its preimages under f. The Julia set, $\mathcal{F}*$, of the limit f of the ${}_m f$ is the boundary of $\mathcal{D}$.

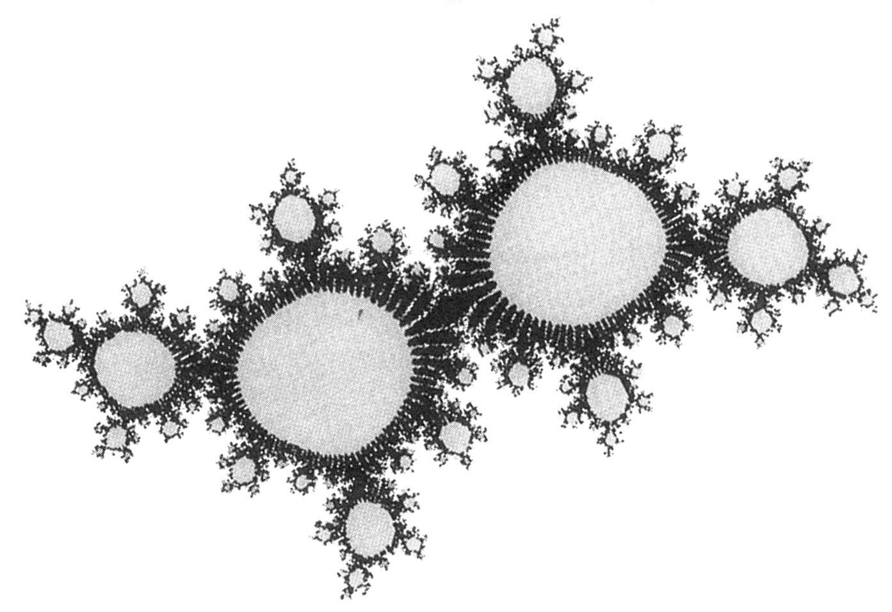

FIGURE C10-2. In order to separate the petals in the space of the variable z, the root $\lambda_m = \exp([2\pi i(55/89)]$ whose exponent belongs to the Fibonacci sequence was replaced by the nucleus of the radical rooted at λ_m. This Figure gives an overall view of the corresponding Julia set. The white, black and gray zones correspond, respectively to values of z_0 that converge to infinity, converge to a cycle, or have failed to converge when the calculation was stopped. Each petal extends to an unstable fixed point in the middle of the gray zone.

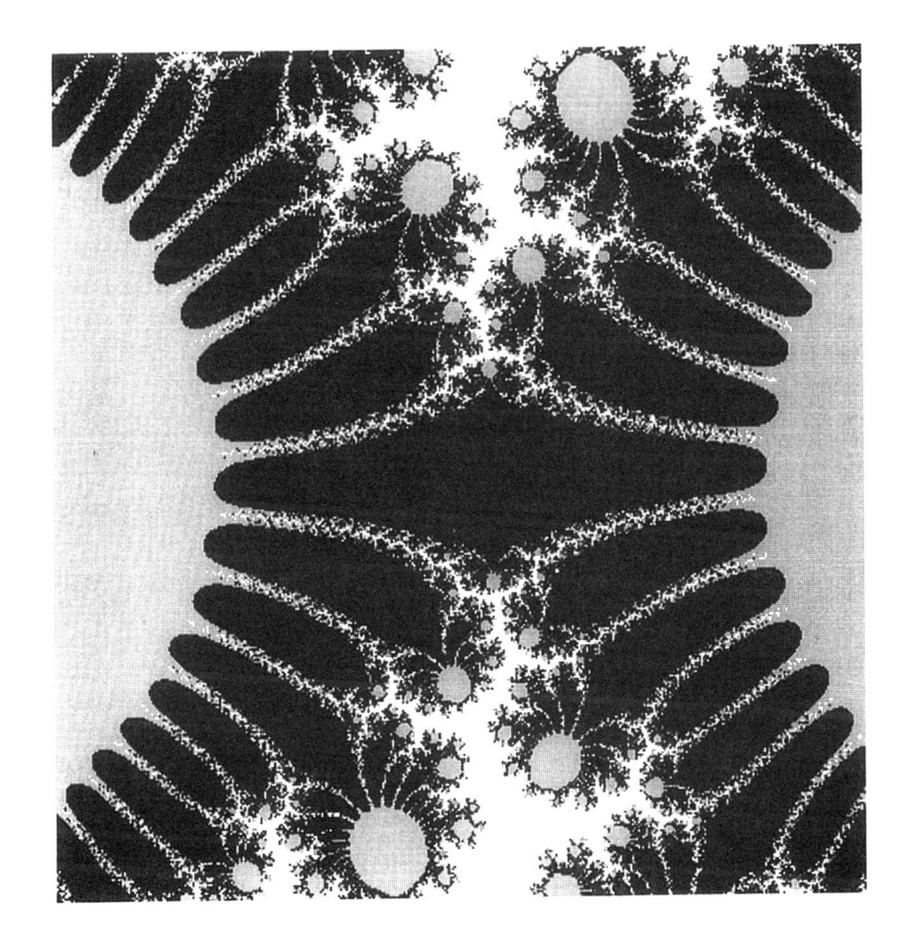

FIGURE C10-3. Figures 3 and 4 are detailed views of the neighborhood of the origin in Figure 2 and the analogous figure for the next Fibonacci ratio, $\log \lambda / 2\pi i = 89/144$. Moving from Figure 3 to Figure 4, the golden ratio is approached, the petals become increasingly thin, and their outline converges to the interior of a domain of the plane.

4. APPROXIMATION OF THE GOLDEN PARAMETER $\lambda = \exp(2\pi i \sigma)$ BY A SEQUENCE OF "BONDS" OR FINITE BIFURCATION VALUES OF λ

Siegel 1942 implies that the Siegel members include the golden ratio σ, which is best defined as the continued fraction exclusively made of 1's, namely $\sigma = 1/(1+1/(1+1/(1+1/...)))$. Clearly, σ, satisfies the equation $\sigma^2 + \sigma - 1$, whose root less than 1 is $1/2(-1+\sqrt{5}) = 0.6180339$.

To gain an intuitive feeling of the action of the map $z \to e^{2\pi i \sigma} z(1-z)$, this action was obtained as the limit of the actions of two approximating sequences of maps $z \to \lambda_m z(1-z)$.

In a first study, λ_m was kept satisfying $|\lambda_m| = 1$, but $z \to e^{2\pi i \sigma} z(1-z)$ was approximated by $z \to e^{2\pi i \sigma m} z(1-z)$, where σ_m the rational number whose continued fraction expansion is made of m repetitions of 1. This is the mth Fibonacci number. By the general theory of the $\mathcal{M}$-set $\mathcal{M}_\lambda$ of $z \to \lambda z(1-z)$, each parameter value $\lambda_m = e^{2\pi i \sigma m}$ is a bifurcation point: the point of contact between the atom $|\lambda| < 1$ and a smaller atom, the order of bifurcation being n(m) = denominator of σ_m.

Figures 1.1 to 1.4 show one half of the interior of the approximating Julia sets for selected values of m. (The other half is obtained by symmetry). A series of flower-like shapes soon start to appear, their petal numbers belonging to the Fibonacci series. At the same time, the Julia set narrows down at some other points.

5. APPROXIMATION OF THE GOLDEN PARAMETER $\lambda = \exp(2\pi i \sigma)$ BY A SEQUENCE OF SUPERSTABLE λ

The trouble with the preceding approximating sequence is that the petals are not visibly separated from each other; to separate them, an alternative approximating sequence was studied. It approximates $z \to e^{2\pi i \sigma z(1-z)}$ by $z \to \lambda_m * z(1-z)$, where $\lambda_m *$ denotes the "nucleus" of the atom of period $n(m)$ whose root lies at $\exp(2\pi i \sigma)$. This $\lambda_m *$ is a superstable value, meaning that the $n(m)th$ iterate of ${}_m f$ satisfies ${}_m f_{n(m)}(0) = {}_m f_{n(m)}{}'(0) = 0$. Figures 2 to 4 show that the same petals that were squeezed together in the case of the sequence λ_m have become well-separated.

6. A MATHEMATICAL QUESTION

The irrational numbers θ split into two classes: the Siegel numbers, which are the great majority, and all the other numbers. Supposing that θ_m converge to a non-Siegel number, let us attempt to visualize the behavior of the Julia sets, $\mathcal{F}*_m$. This attempt is simplified by concentrating on the set Q_m defined as the intersection of $\mathcal{F}*_m$ with a suitable circle. In the Siegel case, the fact that $\mathcal{F}*_m$ converges to a domain filling curve, implies that the set Q_m converges to the whole circle. When $\mathcal{F}*_m$ is not a domain-filling

FIGURE C10-4. An explanation is found in the caption of Figure 3.

sequence because θ converges to a rational number, q_m converges to a Cantor dust of zero measure.

The preceding observations lead to the following query: when θ is neither rational nor a Siegel number, what is the behavior of Q_m ? Does it perhaps converge to a Cantor dust of dimension 1, or even of positive linear measure? If the latter is the case, $\mathcal{F}*$ would be of positive planar measure.

7. SIEGEL DISCS AS LIMIT OF FRACTAL DUSTS

In section 4 (resp. 5), a λ-value on the boundary of the $\mathcal{M}$-set of f(z) is approximated by λ_ms that lie on the boundary of the $\mathcal{M}$-set (resp., inside it). Now, let the approximate values be chosen outside of the $\mathcal{M}$-set The corresponding Julia sets are dusts. When the approximants converge to a rational $\lambda = \exp(2\pi im/n)$, these dusts coalesce into curves. However, convergence of λ_m to a Siegel number λ results in a dust coalescing into a domain.

8. HISTORY

"Siegel" discs were originally described in Julia 1918; see Julia 1968 I pp. 311-317. Next, Julia claimed to prove they could *not* occur (Julia 1968 I p. 321); later he recognized his proof was flawed, and finally dropped the topic (Julia 1968 I, starting p. 21 bottom). Siegel 1942 proved their existence for a certain set of irrational values of θ, having a measure equal to 1. This set has since been broadened.

Acknowledgements. I am in debt to Gregory and David Chudnovsky for interesting comments concerning Siegel discs, and to V. Alan Norton and James A. Given for many interesting discussions as well as for the computer programs used to draw the illustrations in this paper.

Physica Scripta, T9, 1985

C11

Continuous interpolation of the quadratic map and intrinsic tiling of the interiors of Julia sets

• *Chapter foreword (2003).* Granted the immediate and extraordinary popularity of quadratic iteration, the near-total neglect of this paper's subtopic continues to astonish. A contributing factor was "underpublishing" in a volume of proceedings that appeared in an inappropriate journal. It would have helped to republish this work with added detail and splendid color graphics.

Be that as it may, did anybody rediscover and develop the same idea independently? This issue, as many other relative to iteration, must be left to a careful sociological study by a neutral observer.

The original title was "Continuous interpolation of the complex discrete map $z \rightarrow \lambda z(1-z)$ and related topics. On the dynamics of iterated maps, IX." Received September 5, 1984; accepted September 9, 1984. •

✦ **Abstract.** This work reports several observations concerning the dynamics of a continuous interpolate, forward and backward, of the quadratic map of the complex plane. In the difficult limit case $|\lambda| = 1$, the dynamics is known to have rich structures that depend on whether $\arg \lambda / 2\pi$ is rational or a Siegel number. This paper establishes that these structures, a counterpart for $|\lambda| < 1$, are an intrinsic tiling that covers the interior of a $\mathcal{J}$-set and rules the Schröder interpolation of the forward dynamics, its intrinsic inverse, and the periodic or chaotic limit properties of the intrinsic inverse. ✦

A STRIKING FEATURE OF THE ORDERLINESS OF CHAOS is that many of its geometric aspects are governed by fractals, and many of its physical aspects are governed by fractal geometry.

Introduction.

In the geometry of the dynamics of iterated maps, an essential role is played by two fractal sets: the $\mathcal{J}$-set (Julia 1918, Fatou 1919) in the z-plane, and the boundary of the $\mathcal{M}$-set (M 1980n{C3}, in the parameter space. A map's $\mathcal{J}$-set is the closure of the unstable fixed points and fixed cycles, and the $\mathcal{M}$-set is the set of parameter values for which the $\mathcal{J}$-set is connected. To simplify the statements and credit Fatou, the complement of the $\mathcal{J}$-set is called $\mathcal{F}$-set in Blanchard 1984. Here, the maximal open components of the $\mathcal{F}$-set will be called $\mathcal{F}$-components.

This paper concerns the inexhaustible quadratic map, written in the form $z \to f(z) = \lambda z - \lambda z^2$ or $z \to f*(z) = z^2 - \mu$. The $\mathcal{F}$-components are defined for parameter values in a subset of $\mathcal{M}$, namely, the $\mathcal{M}^0$-set of parameter values such that the map $f(z)$ has a finite limit cycle (plus the usual limit point at infinity). The advantage of the form $f(z)$ is that the disc $|\lambda| < 1$ belongs to $\mathcal{M}^0$, and that for $|\lambda| < 1$, the limit point is $z = 0$ and the multiplier at the limit point is λ itself.

This paper reports on new structures that apply for $|\lambda| < 1$, and are dominated by the argument $\theta = \arg \lambda$. It is known that when $|\lambda| = 1$ the arithmetic properties of $\theta/2\mu$ determine such structures as bifurcation the "petal" (Blanchard 1984, p.101), and Siegel discs. This chapter shows that generalizations to the above structures are already present when $|\lambda| < 1$.

These structures were observed thanks to the fractal geometric intuition attained while illustrating (apparently, for the first time) the behavior of the solutions of the Schröder equation, and deducing an intrinsic tiling of the $\mathcal{F}$-components.

The main observation is a "universality" result: The topological structure of the intrinsic tiling depends solely on θ and not on $|\lambda|$. It can be directly inferred from the bifurcation/petal or Siegel disc structure that corresponds to the same θ in the limit case $|\lambda| = 1$.

A corollary is that the fractal dimension of the $\mathcal{J}$-set varies smoothly as $|\lambda| \to 1$ while θ is fixed. This may account intuitively for certain theorems I have heard sketched by N. Sibony.

These and other related observations touch upon a topic that has aroused a little interest for a long time, and is mentioned in the title because of its attractiveness. The iterates $z_0, z_1 = f(z_0)..., z_k = f(z_k - 1) = f_k(z_0)$, which form a sequence with an integer index k, can be embedded intrinsically into a sequence $z_t = f_t(z_0)$ where t is real, by solving the "Schröder equation." When z_0 is in the fundamental tile of the intrinsic tiling, the continuous time can, moreover, be inverted, so that $z_t = f_t(z_0)$ can be defined for positive *and* negative reals. The forward motion is very plain when $|\lambda| < 1$. since there is a stable attractor point, but the arithmetic properties of $\theta/2\pi$ are essential to the dynamics of the backward motion.

It cannot be excluded that some observations reported here are known but obscure.

2. The Schröder function

General values of $\lambda \in \mathcal{M}^0$ will be considered in Section 8, but elsewhere we suppose for simplicity that $\lambda \neq 0$ but $|\lambda| < 1$. The $\mathcal{J}$-set is then a Jordan curve whose interior is the bounded $\mathcal{F}$-component and the domain of attraction of $z_f = 0$. To each point in this interior, Schröder 1870 attached the function $\sigma(z) = \lim_{k \to \infty} \lambda^{-k} f_k(z)$ This function satisfies the Schröder functional equation $\sigma[f(z)] = \lambda\sigma(z)$.

For various values of λ, the Figures give the approximate maps of the function $\sigma(z)$, in the form of a fan of curves of constant argument, and also, in some cases, of curves of constant modulus. To construct them, cut slices of pie fanning from z_f alternatingly colored in black and in white,

FIGURE C11-1. Intrinsic tiling for λ real $\in]0, 1[$.

and separated by equidistant half-lines. Then draw an annulus around $z_f = 0$, whose outer radius is some small $\rho * > 0$. When $f_k(z_0)$ first falls within the disc $|f_k(z_0)| < \rho *$, check whether $\lambda^{-k} f_k(z_0)$ falls on a black or a white point, and color z_0 accordingly. This yields black strips whose boundaries approximate the isolines of the argument of $\theta(z)$. To cut these black strips by white pieces of curves that outline approximate isolines of $|\alpha(z)|$, do not color z_0 if the first $f_k(z_0)$ of modulus $> \rho *$ is of modulus $> \rho *(1 - \varepsilon)$, with a suitably small $\varepsilon > 0$. As $\rho * \to 0$, the approximation improves, and the spurious pattern near $z = 0$ ebbs away.

3. The intrinsic tiling

As the inspection of the Figures clearly show, the Schröder function defines an intrinsic tiling of the bounded F-component. The tiles' boundaries are smooth curves, except for a countable number of 90° kinks. The tile that contains z_f is to be called the *fundamental tile,* and every other tile is a preimage of this fundamental tile.

There is a resemblance with the hyperbolic tiling developed by Henri Poincaré and Felix Klein for Kleinian groups of maps of the complex plane of the fractional linear form $z \to (az + b)/(cz + d)$.

FIGURE C11-2. Intrinsic tiling for a λ near the interval]0, 1[. A small change in λ, compared to Figure 1, changes the tiling drastically.

4. Schröder interpolation of $z^2 - \mu$ to continuous time, in the fundamental tile

In terms of the Schröder function $\sigma(z)$, the interpolation of $f(z)$ in the fundamental tile simply amounts to using a logarithmic spiral to interpolate a sequence of points known to lie along this spiral. In terms of σ, this intrinsic interpolation is continuous and invertible, both forward, between z_0 and the forward attractor $z_f = 0$, and backwards, between z_0 and a backward attractor $\mathcal{B}$, which is a subset of the T-set to be discussed in Section 6. In terms of z, however, the interpolation is not continuous, except when λ is real and positive. Otherwise, this interpolation is continuous in the disc around $z_f = 0$, in which the leaves of the fundamental domain are attached to each other.

(In addition to the above "fundamental", one can define "harmonic" and "subharmonic" interpolations. Given an integer h, the harmonic interpolation is the map of the curve $\sigma(z_f) = (\lambda^{1/h})^t \sigma(z_0)$, and the subharmonic interpolation is the map of the curve $\sigma(z_t) = (\lambda^h)^t \sigma(z_0)$. The harmonics contain all the points $\sigma(Z_k) = \lambda^k \sigma(z_0)$, subharmonics do not.)

(An interpolation in terms of the "Poincaré function" was advanced by S. Lattes in 1917; Dubuc 1982 compares the two interpolations.)

FIGURE C11-3. Intrinsic tiling for λ real $\in]-1,0[$.

5. Dependence of the intrinsic tiling on λ

The intrinsic tiling topology only depends on $\theta = \arg \lambda$ not on $|\lambda|$, and is a sharply discontinuous function of θ (Figures 1 and 2.) This is a noteworthy result, because the $\mathcal{J}$-set that wraps up the tiling varies continuously with $|\lambda|$ and θ, as long as $\lambda \in \mathcal{M}^0$.

Interesting consequences follow, concerning the intrinsic interpolation of $f(z)$ as a function of both time and of λ. The forward motion from a given z_0 is discontinuous in θ and $|\lambda|$. The backward motion is continuous in $|\lambda|$. Its dependence on θ becomes discontinuous after a long enough time, but, for a short time it may be continuous.

6. The backward attractor $\mathcal{B}$, intersection of the $\mathcal{J}$-set with the boundary of the fundamental tile, and the leaves of the fundamental tile

The set $\mathcal{B}$ is, of course, invariant but unstable under the action of f(z). It will be called the fundamental invariant set of $\mathcal{J}$.

The rational case. If $\theta/2\pi = m/n$, the set $\mathcal{B}$ contains n points, each of them being invariant under the action of $f_n(z)$. The direct map f(z) has a fixed point as attractor, hence a cycle of size 1. However, the intrinsic

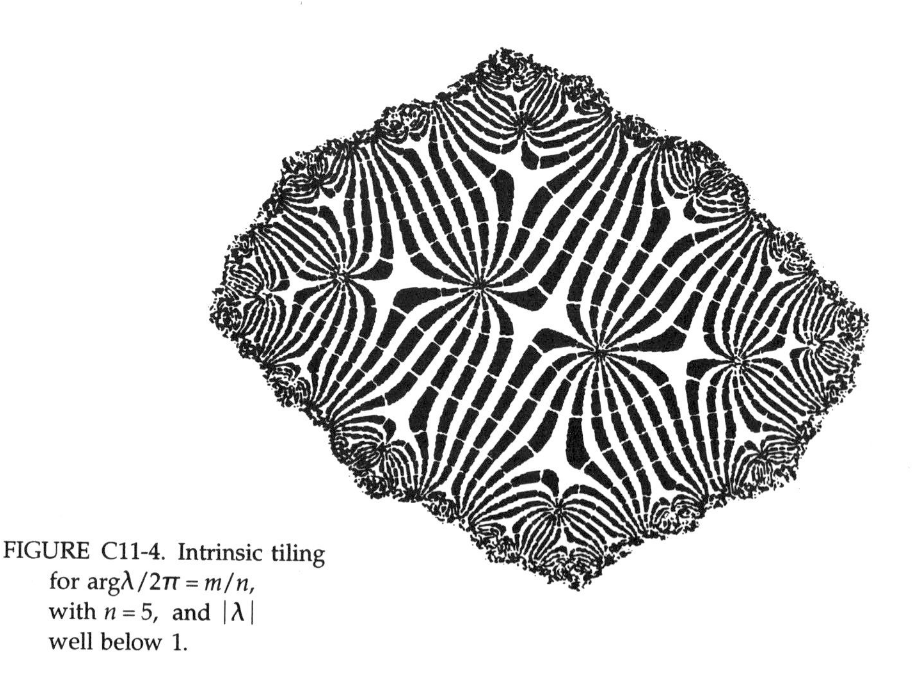

FIGURE C11-4. Intrinsic tiling for $\arg\lambda/2\pi = m/n$, with $n = 5$, and $|\lambda|$ well below 1.

determination of the multivalued inverse map $f(z)$, identified as lying in the fundamental tile, has a limit cycle of size n, namely $\mathcal{B}$.

To each point of $\mathcal{B}$ corresponds an interval of values of $\arg[\sigma(z)]$ of width $2\pi/n$, which reduces to an interval of values of Arg z near $z=0$. Each interval defines a "leaf" in the fundamental tile. The leaves are attached to each other in the neighborhood of $z_f=\{0,0\}$ but they split as

FIGURE C11-5. Intrinsic tiling for the same arg λ as in Figure 4, and $|\lambda|$ approaching 1. The leaves split in an overall radial direction. {P.S. A rotation minimized the reduction.}

one moves away from z_{ρ}, and are separate in the neighborhood of the $\mathcal{J}$-set. Each leaf should be viewed as being made of two half-leaves, separated by a "rib" that bisects the leaf in terms of $arg[\sigma(z)]$

It is known that the $\mathcal{J}$-set can be mapped intrinsically upon a circle, hence, each point of the $\mathcal{J}$-set can be given an intrinsic argument ϕ. Each point of the $\mathcal{B}$-set has an argument ϕ and an interval of arguments $\arg[\sigma(z)]$. The action of $f(z)$ replaces ϕ by 2ϕ, and replaces $\arg[\sigma(z)]$ by $\arg[\sigma(z)] + \arg \lambda$. The mismatch between these two operations is the reason why the study of iteration is rife with fractals.

Observation. The ϕ arguments of the points of the $\mathcal{B}$-set are identical to the ϕ arguments of the point $z_f = 0$ in the limit case where λ is replaced by the parameter $\lambda' = \lambda / |\lambda|$ and bifurcation occurs. In other words, the ϕ arguments of the n points of $\mathcal{B}$ depend only upon λ via the value of θ.

On the circle, when the point of argument ϕ is invariant under the map $\theta \to 2^k\theta$, the ratio $\phi / 2\pi$ is represented by a sequence of binary digits

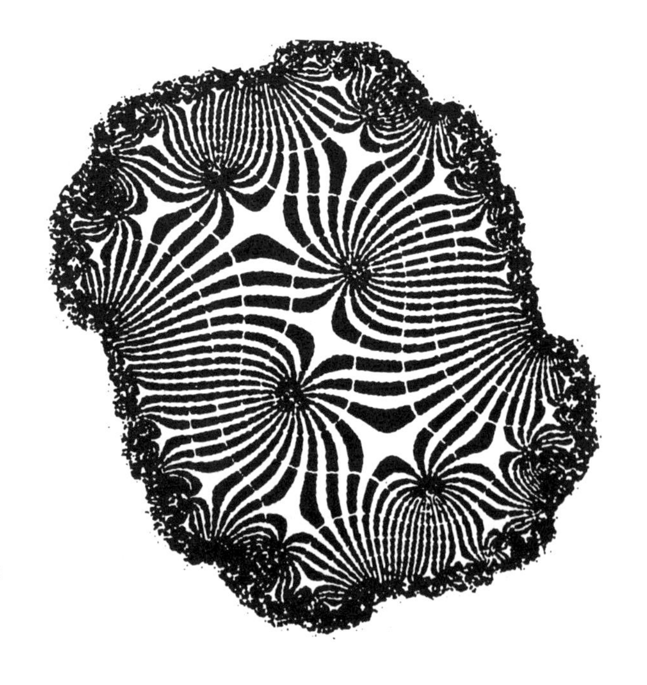

FIGURE C11-6. Intrinsic tiling for arg $\lambda / 1\pi$ the golden mear (a Siegel number) and $|\lambda|$ below 1.

FIGURE C11-7. Intrinsic tiling for the same arg λ as in Figure 6, and $|\lambda|$ approaching 1. The leaves do not split, rather are on the way to coalescing into a Siegel disc. {PS 2003: A rotation minimized the reduction.}

that is periodic with the period 2^k. There are $2^k/k$ cycles, many of them reducible to one another. There is a rule for determining the special invariant unstable set that maps upon $\mathcal{B}$.

The irrational case. If $\theta/2\pi$ is irrational, B is a Cantor set (fractal dust). In order to become convinced that such invariant Cantor dusts exist, it suffices to take on the circle a point for which $\phi/2\pi$ is an infinite sequence of binary digits such that 0 is never followed by 0. The points of the form 2ϕ obviously share this property and D satisfies $2^D = 1.618$.

When the $\mathcal{B}$-set is a Cantor dust, each half-leaf is associated with a trema of $\mathcal{B}$ (an open interval in the complement of I). However, the leaves of the fundamental tile no longer split into two half-leaves. Stated alternatively, one of the half-leaves is degenerate and the rib runs along the side of the leaf that is not pointed towards the midpoint of a trema.

7. Observations relative to the limit $|\lambda| \to 1$, when approached radially

In the limit $|\lambda| = 1$, Julia and Siegel had shown it is important whether the number, $\theta/2\pi$ is (a) rational or (b) irrational and a Siegel number, or (c) irrational but not a Siegel number. The rational cases, $\theta/2\pi = m/n$, correspond to $\lambda \in \mathcal{M}^0$. If so, the topology of the $\mathcal{J}$-set is determined mostly by the value of n. When $|\lambda| = 1$ and $\theta/2\pi$ is a "Siegel number," the $\mathcal{J}$-set is made up of the boundary of a "Siegel disc" and of this disc's preimages under $f(z)$. The novelty reported in the present paper is that the arithmetic nature of $\theta/2\pi$ is already important for $|\lambda| = 1$. To link the properties relative to $|\lambda| = 1$ and $|\lambda| \to 1$ radially, that is, with invariant θ. The $\mathcal{J}$-set acquires double points, as the points of the B-set move either toward or around $z_f = 0$.

When $\theta/2\pi$ is the rational m/n, (Figures 4 and 5) the n points in the $\mathcal{B}$-set all converge to $Z_f = 0$. Each leaf splits, along its rib, into its two halves. Thus, each tile becomes split into n pieces, each of them made of two half-leaves, and all having equal widths in terms of $\arg[\sigma(z)]$. The fundamental tile becomes identified with the petal (Blanchard 1984, p. 101).

When $\theta/2\pi$ is a Siegel irrational (Figures 6 and 7), the endpoints of each trema in the $\mathcal{B}$-set converge circumferentially to the same point - which depends on the trema. As a result, the tiles keep their identity as $|\lambda| \to 1$. The fundamental domain tends to become increasingly separated from the remainder of the interior of the $\mathcal{J}$-set. At the limit $|\lambda| = 1$, it becomes a Siegel disc bounded by a fractal Jordan curve, a subset of the $\mathcal{J}$-set.

The case when $\theta/2\pi$ is a non-Siegel irrational is known to be difficult. A conjectural scenario is that every point in the $\mathcal{B}$-set again converges to $z_f = 0$, and the fundamental domain becomes split into a denumerable infinity of leaves (≡ half-leaves) of unequal widths in terms of $\arg[\sigma(z)]$

Thus, attainment of $|\lambda| = 1$ acts differently on the intrinsic tiling and its $\mathcal{J}$-set wrapping. The tiling depends discontinuously upon θ both when $|\lambda| < 1$ and when $|\lambda| = 1$. The wrapping's dependence upon θ is continuous when $|\lambda| < 1$, but discontinuous when $|\lambda| = 1$.

The shape of the $\mathcal{J}$-set changing smoothly as $|\lambda| \to 1$ radically, the fractal dimension $D(\lambda)$ of the $\mathcal{J}$-set converges smoothly to a limit. But for $|\lambda| = 1$, the shape of J varies discontinuously with θ. Therefore, the radial limit of $D(\lambda)$ seems to be an extremely unsmooth function of θ.

8. Case when $\lambda \in \mathcal{M}^0$ but $|\lambda| > 1$ and $|\lambda - 2| > 1$

For such λ 's there is a limit cycle of size $N(\lambda)$ and the multiplier is some $\Lambda(\lambda)$ satisfying $|\Lambda| < 1$. To each λ with $|\Lambda| \neq 0$, one can associate a parameter value λ' which lies in the same atom of $\mathcal{M}^0$ and satisfies $|\Lambda(\lambda')| = 1$ and $\arg[\Lambda(\lambda')] = \arg[\Lambda(\lambda)]$. The arithmetic properties of $\arg[\Lambda(\lambda')]1/2\pi$ the tiling structure for the parameter value λ, and the limit behavior of the intrinsic inverse of $f_N(z)$.

9. The superstable case

For $\lambda = 0$, the argument θ is not defined. This is a superstable parameter value, the nucleus of an atom in the $\mathcal{M}$-set. In that case, the Schröder equation is replaced by the Boettscher equation, and each $\mathcal{F}$-component — including the exterior of the $\mathcal{J}$-set — is a single tile. The discussion is reserved for a later occasion.

Acknowledgement. The illustrations were prepared by Eriko Hironaka, using computer programs she wrote for this purpose.

Alternative rendering of the bottom of Figure C4.3, in terms of μ. Composite of self-squared Julia fractal curves for real μ ranging from $-1/4$ (bottom) up. *Source*: Alan Norton, "Generation and display of geometric fractals in 3-D." *Computer Graphics*: **16(3)** (1982) 61-67. For permission to reproduce this work of art, I am grateful to Dr. Norton, who also produced Figure C4.3.

Dr. Norton's remarkable extension of quadratic iteration from complex numbers to quaternions is exemplified in the paper quoted above, as well as in his paper "Julia sets in the quaternions," *Computers and Graphics*: **13(2)** (1989) 267-278. Samples are found throughout M1982F.

PART II: NONQUADRATIC RATIONAL DYNAMICS

&&&&&&&&&&&&&&&&&&&&&&&&&&&&&&&&

First publication

C12

Introduction to papers on chaos in nonquadratic dynamics: rational functions devised from doubling formulas

THE PATH OF SCIENTIFIC INVESTIGATION AND DISCOVERY is not necessarily logical, as history never tires of reminding both laymen and scientists. From many viewpoints, the complex quadratic map, reducible to either $z \to z^2 + c$, $z \to \lambda(z^2 - 2)$, or $z \to \lambda z(1-z)$, is the simplest of all nonlinear maps. Its global action was therefore the first to be studied carefully, in Fatou 1906. Later, the 1960s and 1970s brought many studies of the restriction of $z^2 - \mu$ to the real quadratic map $x^2 - \mu$, and everyone became well aware that the dependence of this real map's orbits on the parameter μ involves exquisite complications. It has its Myrberg sequence of bifurcations, its May tree, and its Feigenbaum number. Shortly thereafter, my papers reproduced in Part I made the complex quadratic map widely popular.

Due to the inevitable temptation to "interpolate" between known events, many nonwitnesses take it for granted that my computer-assisted study of iteration began and also ended with the quadratic map. Actual developments followed a very different path. Both before and after the fateful winter of 1980 discussed in Chapter C1, I dealt with many maps far more complex than the quadratic but published only one paper, M1984k{C13}, whose contents were already sketched in M1982F, page 465 of the 1983 update. The extensive unpublished work usefully splits into three parts.

Many of my post-1980 observations are subtle but have already been well documented by others. Any hint that I claim any credit for the resulting discoveries will be avoided by leaving them as they are: unpublished. But it may be worth pointing to the endpaper that adorns M 1982f without being explained. Those "butterflies" arose when I tried to apply the Newton method to the exponential function. The task seemed daunting, and I had the good sense to move on.

A second part of my work is straightforward and is not mentioned in this chapter's title. In that winter of 1980, I also examined the maps $z \to \lambda z(1-z^k)$ for many integers $k > 1$. The results have been endlessly duplicated, not only by experts but also by amateurs. As a footnote to history, a brief Section 1 indulges in illustrating these observations by heroically crude true antiques.

A third part deserved to be sampled in print at long last for these reasons: for the sake of history, to assist a possible future revival of the topic, and because of continuing pedagogical interest: The student of iteration who has become comfortable with the quadratic map is well-inspired also to look sideways. Nonquadratic maps greatly help us to understand the passage from order to chaos, and their variety is literally overwhelming.

1. EXAMPLES OF MAPS OF THE FORM $z \to \lambda z(1-z^k)$

After $z^2 - \mu$ and/or $\lambda z(1-z)$, why not $z^k - \mu$ and/or $\lambda z(1-z^k)$, with k an integer greater than 2? The plots of the M set I drew for $\lambda z(1-z^k)$ have only one virtue, obvious in the primitive Figure 1: They were drawn very early in the game.

2. THREE CHAOTIC MAPS STUDIED BY LATTÈS BASED ON FUNCTIONS FOR WHICH THE DOUBLING FORMULA IS A RATIONAL FUNCTION.

2.1 Background

Early on, I had underestimated the complex quadratic map and expected it to be too "underpowered" to be worth investigating. But experience in experimental mathematics made me proceed cautiously. My uncle had told me about a Frenchman contemporary of Fatou and Julia, ***Samuel Lattès*** (1873-1918), who had singled out several maps that today are called chaotic. Looking for a restrictive guiding principle, I began with families that contain the chaotic examples of Lattès as special cases. Some results I obtained on those maps are reported in this part.

My priorities moved on to writing M 1982F, and I did not seriously pursue the study of iteration. But the quadratic map continued to benefit from an extraordinary level of popularity. Also, nonquadratic maps were briefly mentioned in the widely available M 1986p and were often presented in my "slide shows." It seemed reasonable to expect them to be picked up by many scientists and the general public. Astonishingly, they were not. The very belated first publication of a subset of my illustrations in Chapter C14 may therefore serve as a test. I hope that enough readers will notice them here and that some will proceed beyond the stage where I stopped nearly twenty-five years ago.

Such predictions are tricky. I must also confess to having expected a swing of attention to the quaternionic Julia sets, of which several examples illustrate M 1982F. However, the neglect of quaternionic iteration may be explainable by reluctance to tackle rough surfaces that are hard to render, and quaternions themselves may seem forbidding.

2.1. Lattès

As Fatou and Julia already knew, the complex quadratic map $z \to z^2 - \mu$ for $\mu = 2$ can be rewritten as $(z/2) \to 2(z/2)^2 - 1$. Then the change of variable $z/2 = \cos\theta$ reformulates $z \to z^2 - 2$ into the doubling map $\theta \to 2\theta$. This map is known to be chaotic on the interval $[0, 2\pi]$ of θ, that is, the interval $[-2, 2]$ of z.

Lattès 1918 has pointed out that two other collections of functions of a complex variable share the property that the doubling formula yielding $f(2z)$ is a rational function of $f(z)$.

One such collection centers on cotan θ; the functions tan θ, tanhθ, and coshθ involve only minor modifications.

The second collection of functions includes all the $\wp(z)$ elliptic functions of Weierstrass, which have two parameters g_2 and g_3. Lattès examined one set of those parameters, but I also examined another.

Once again, the example of the real map $x^2 - \mu$ should have led me straight to $\lambda(x^2 - 2)$ and reminded me of Gaston Julia telling Polytechnique students that when things get tough in mathematics, "a good way to simplify is often to complexify." But naiveté and excessive zeal first led me away from $\lambda(z^2 - 2)$. This part reports on work triggered by Lattès.

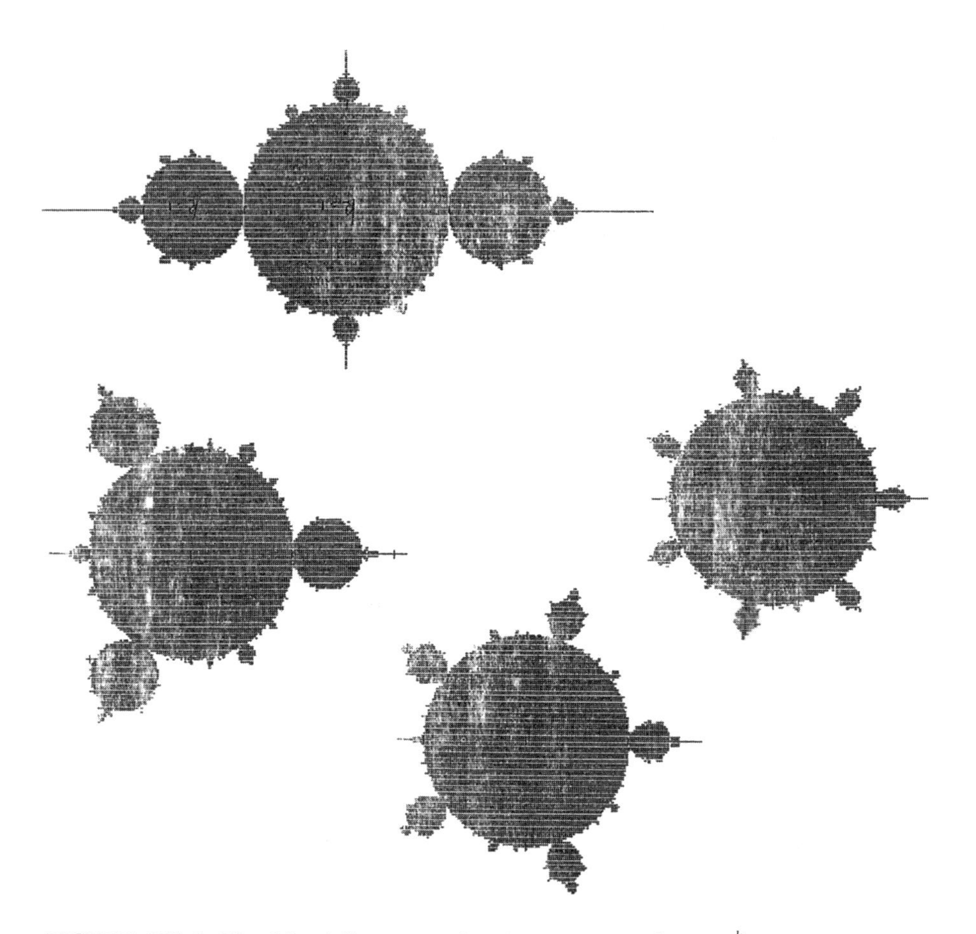

FIGURE C12-1. The Mandelbrot sets for the maps $z \rightarrow \lambda z(1 - z^k)$, as drawn with the very primitive graphic equipment Harvard provided in the winter of 1980. Counterclockwise from top left, $k = 2, 3, 5$, and 7.

2.2 The map $g(z) = (1/2)(z + 1/z)$

2.2.1 Motivation following Lattès, via the doubling formula for the function $z =$ cotan θ. Elementary calculus yields

$$\text{cotan } 2\theta = \frac{\cos(2\theta)}{\sin(2\theta)} = \frac{(\cos^2\theta - \sin^2\theta)}{2\cos\theta\sin\theta} = \frac{1}{2}\left(\text{cotan }\theta - \frac{1}{\text{cotan }\theta}\right).$$

Hence, writing $z =$ cotan θ changes the map $\theta \rightarrow 2\theta$ into

$$z \rightarrow g(z) = \frac{1}{2}\left(z + \frac{1}{z}\right).$$

Apply the Möbius transformation defined by $z = (u+1)/(u-1)$ or $u = (z+1)/(z-1)$. It transforms the map $z \rightarrow g(z)$ into the map $u \rightarrow u^2$. Thus, all starting points z_0 in the half-plane $\text{Re}(z) > 0$ iterate to $z = 1$. In other words, all starting points u_0 (i.e., in the disk $|u| > 1$) iterate to $u = \infty$. All starting points in the half-plane $\text{Re}(z) < 0$ (i.e., in the disk $|u| < 1$) iterate to $z = -1$ (i.e., $u = 0$). As a result, the repeller, that is, the Julia set F*, is the imaginary axis $\text{Re}(z) = 0$, corresponding to the circle $|u| = 1$.

Like all rational maps, $z \rightarrow g(z) = (1/2)(z + 1/z)$ is chaotic on its Julia set, which is a line. In contrast, it is almost perfectly orderly in the plane. We already showed that the same remark applies to the map $z \rightarrow z^2 - 2$ after it is rewritten in terms of θ.

2.2.2 Original motivation of the map $g(z)$, following Cayley, via the solution of the equation $z^2 - 1 = 0$ by the Newton–Raphson method. Historically, the study of the global properties of iteration began when Cayley 1879 used the Newton–Raphson method to solve the equation $f(z) = z^2 - 1 = 0$. That method consists in iterating the map $z \rightarrow z - f(z)/f'(z)$, and hence for $z^2 - 1$, it led to the iteration of the rational function $g(z) = (z^2 - 1)/2z = \lambda(z + 1/z)$. As already mentioned, the repeller is simply the imaginary axis $\text{Re}(z) = 0$. As a result, the application of Newton's method to $z^2 - 1 = 0$ involved a very orderly behavior.

Aside on the solution of $z^3 + 1 = 0$ by the Newton–Raphson method. Sir Arthur Cayley tried to duplicate his success with $z^2 - 1 = 0$ by studying $z^3 - 1 = 0$, but failed. His investigation stalled after unearthing "grey zones" of values of z_0 in which he *could not* disprove that the outcome of Newton's method might depend in complicated fashion on the initial conditions. The desire to clarify Cayley's failure spurred Gaston Julia's general theory of complex iteration and—after a long period of neglect—attracted the attention of John H. Hubbard, who became closely associated with its study.

2.3 The doubling formula for the Weierstrass elliptic function $\wp(z)$ and the resulting rational maps

2.3.1 The map $W_1(z) = (1/4)(1+z^2)^2/[z(z^2-1)]$ and the original Lattès chaos on the plane. Julia credited Lattès repeatedly for singling out the rational map

$$z \to W_1(z) = \frac{1}{4}\frac{(1+z^2)^2}{z(z^2-1)}.$$

As always, this map is chaotic on its Julia set. A special feature of this special map is that its Julia set is the whole complex plane.

2.3.2 Weierstrass elliptic functions $\wp(z)$ and their doubling formula. Lattès devised the map $W_1(z)$ by observing that it follows from the elementary map $\theta \to 2\theta$ by the change of variable $z = \wp(\theta)$, where $\wp(\theta)$ is a special case of the Weierstrass elliptic function. My (probably obsolete) references on those functions are Tannery & Molk 1898 and Whittaker & Watson 1902–1927. In two words, the point of departure is the elliptic functional equation

$$\left(\frac{dy}{dz}\right)^2 = 4y^3 - g_2 y - g_3.$$

A third parameter g_1 has been eliminated beforehand by adding a suitable constant to y in order to eliminate the term in y^2. The solution

$$z = \int_\zeta^\infty [4t^3 - g_2 t - g_3]^{-1/2} dt$$

yields z as a function $z(\zeta)$, and the Weierstrass function is the inverse of the function $z(\zeta)$.

From the gigantic theory of elliptic functions all that will be needed is the doubling formula

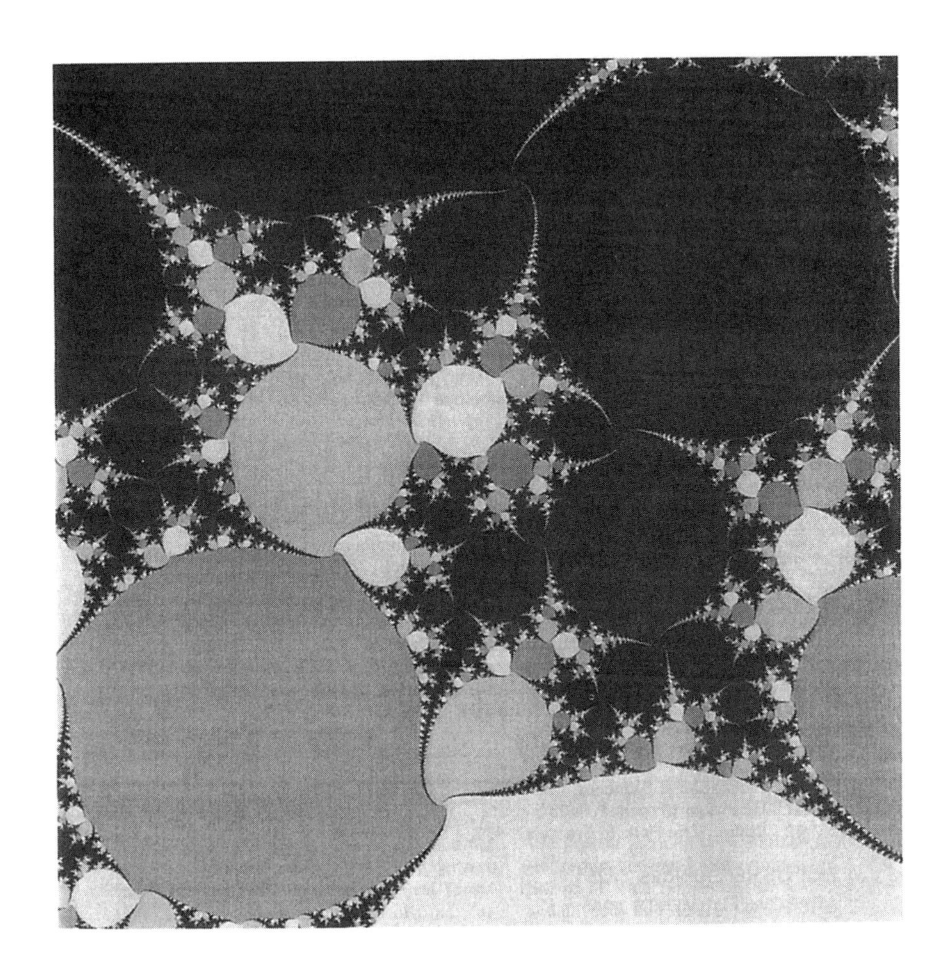

FIGURE C12-2. [Plate viii of M 1982F]. Complementing Chapter 13, the black region illustrates the domain of attraction of one of the two limit cycles of the map $z \rightarrow \lambda(z+1/z)$ with a carefully chosen λ. The domain of attraction of the other limit cycle is symmetric to the black domain with respect to the center of this figure. Here it is split into various shades of gray, each corresponding to one point in this other cycle. The black and nonblack domains recall quadratic iteration. They interlace in a complex fashion, an extreme version of the interlacing to be observed in Chapter C13 for a different nonquadratic map.

$$\wp(2u) = \frac{[\wp^2(u) - g_2/4]^2 + 2g_3\wp(u)}{4\wp^3 u - g_2\wp(u) - g_3}.$$

In Whittaker & Watson 1927, I could not find this formula explicitly, but in Tannery & Molk 1898 it sufficed to plug the elliptic functional equation into the formula $CIII_7$ on p. 93.

The parameters $g_2 = 4$ and $g_3 = 0$ greatly simplify the doubling formula, which is presumably why Lattès chose them and was led to $W_1(z)$.

2.3.3 *Variant chaotic map.* First, as seen in Section 4, I investigated non-chaotic "extensions" of the Lattès chaotic map. Then I chose the next obvious simplifying parameters, which are $g_2 = 0$ and $g_3 = 4$, and investigated "extensions" of the variant chaotic map

$$z \to W_2(z) = \frac{1}{4}\frac{z(z^3 + 8)}{z^3 - 1}.$$

In addition to being simple, the Lattès map and its variant corresponding to $g_2 = 0$ or $g_3 = 0$ exhibit different forms of symmetry to be described and illustrated in Chapter C14.

3. A MULTIPLYING FACTOR λ CREATES THE ONE-PARAMETER FAMILIES OF MAPS $\lambda g(z)$, $\lambda W_1(z)$, AND $\lambda W_2(z)$, EACH OF WHICH INCLUDES AS SPECIAL CASE A MAP KNOWN TO BE CHAOTIC

Now return to my search, before 1980, for worthwhile one-parameter families of maps to study. I saw good reason to restrict this search to families that include as a special case a chaos point that is already well understood but may reward additional study.

Expecting the richest structure to be brought up by complicated maps and being aware of the map investigated by Lattès, I began in 1979 by investigating iteration for the maps $z \to \lambda W_1(z)$, then $z \to \lambda W_2(z)$. In particular, I counted on their behavior for λ close to 1 to help illustrate the nature of the approach to planar chaos. Newly recovered illustrations I made in 1979 are published in Chapter C14 for the first time.

Later, I investigated the map $g(z, \lambda) = \lambda(z + 1/z)$. The results are sketched in M 1983k{C13}, and starting—with the second printing—page 465 and Plate viii of M 1982F concern this map. The least-objectionable spot to reproduce it in this book is in this chapter, as Figure 2. Figure x of M 1982F is reproduced in M 1983k{C13}.

Being eager to share the fun, I described this paper to Adrien Douady on a day when I drove him from New York City to catch a plane at JFK Airport. Our discussion continued in the airport coffee shop, and Douady immediately started referring to it as "Mandelbrot's Kennedy map." When searching for a good name for those other complex maps, I wonder whether I should not name them after La Guardia or Newark.

Be that as it may, Douady went on to describe the interlacing that characterizes Figure 2 as a "marriage." Choosing a different λ, he mated a "rabbit" and a "cathedral," the latter referring to the San Marco dragon illustrated in Figure 3 of M1980n{C3} and in many other places. "Cathedral" is a misnomer, since the Church calls San Marco a basilica, and so do I.

Chaos and Statistical Mechanics,
Ed. Y. Kuramoto, 1984

C13

The map $z \rightarrow \lambda\ (z+1/z)$ and roughening of chaos, from linear to planar (computer-assisted homage to K. Hokusai)

• *Chapter foreword (2003).* The map investigated in this chapter was selected for reasons described in Chapter C12.

The homage to Hokusai Katsushika (1760-1849) consists in the illustration on this page and an elaboration found on next page.

This homage was not an empty gesture towards gracious Japanese hosts but was grounded in strong reasons that deserve to be amplified. Of course, he was a celebrated Japanese artist but to my mind deserves even higher fame for his extraordinarily refined "eye for fractals." The painter Turner "had an eye" for light through the fog. The "cubist" painters around 1900 can be said to have had "an eye" for simple shapes that have been fully formalized long before their time. Presumably, those painters favored scenes that their eye favored. Hokusai was in a way more adventurous because he let his eye be attracted by shapes that were not formalized until long after his death.

Specifically, the illustration on the preceding page is taken from a book-form collection of black-and-white drawings, *One Hundred Views of Mount Fuji,* Hokusai 18.. A closely related "ukiyo-e" variant in color is found in his *Thirty-Six Views of Mount Fuji.* It is far more famous and was the only one I knew in 1983 when I reproduced it in M1982F as Plate C 16. But as an illustration of the theme of "art anticipating mathematics and science," not only is the present version better when printed in black and white, but it is actually the more striking. Hokusai is arguably the best single witness to the fact that since time immemorial, fractal structures were familiar to humanity — but only through art. Mathematics joined them in the late nineteenth century and the sciences around 1960.

Given my chronic habit of publishing only part of my work, I finished this paper solely because of the preparation of a very interesting meeting in Kyoto. Great concern was generated at that time by the property of dynamical systems that basins of attraction often have fractal boundaries. This paper was meant to demonstrate this fact in a context requiring little special mathematics.

Being a new concept at that time, fractal dimension created much confusion and the widespread (often unspoken) belief that it was the only feature of the boundary that mattered. It may be that M1982F contributed to that belief in many readers' minds. But it should not have, since it also introduced the notion of lacunarity. I developed this notion further in M 1993n, M 1995z, and M1998e. One cannot hide, however, that — in 1993 as well as today — lacunarity remains an incompletely developed notion. •

✦ **Abstract.** The terms “chaos” and “order in chaos” prove extremely valuable but elude definition. It remains important to single out instances when the progress to planar chaos can be followed in a detailed and objective fashion. This paper proposes to show that an excellent such example is provided by the iterates of a map for which z and λ are both complex. The subject of this map is touched upon in M 1982F{*FGN*} but only on page 465, which was added in 1983, in the second printing. Therefore, the present paper is self-contained. ✦

THE MAP $z \to g(z) = \lambda(z + 1/z)$ WAS SINGLED OUT FOR STUDY because it has several valuable properties. (A) Within a broad domain of λs, there are two distinct limit cycles, symmetric of each other with respect to $z = 0$. (B) Suitable changes in λ cause both cycles to bifurcate simultaneously into $n > 2$ times larger cycles. (C) The chaos which prevails for certain λ extends over the whole z plane. The features (A), (B) and (C) all fail to hold for the complex quadratic map $z \to \tilde{f}(z) = z^2 - \mu$. Indeed, for every μ, one of the limit cycles of $\tilde{f}$ reduces to the point at infinity, which never bifurcates; and chaos, when it occurs, consists in motion over a small subset of the z-plane.

1. Summary. Relativity of the notion of chaos

The bulk of this paper consists in explanations for a series of figures that illustrate, for diverse λ, the shape of the Julia set $\mathcal{F}^*$, that is, of boundary of the open domains of attraction of the stable limit points and cycles. Different sequences of figures follow different "scenarios" of variation λ, and yield maps that transform gradually from linear chaos and planar order, to either questionable or unquestioned planar chaos.

In order to put these illustrations in perspective, the paper includes comparisons with the polynomial maps. To begin with, the special map $z \to z^2 - 2$ restricted to the real interval $[-2, 2]$ is called thoroughly chaotic. However, the very same map generalized to the complex plane should be called almost completely orderly, since all z_0 except those in the real interval $[-2, 2]$ iterate to ∞. As is well-known (Collet & Eckmann 1980,) there are many other μs for which the maps $z \to z^2 - \mu$ are chaotic on a suitable real interval. But the very same maps are least chaotic in the plane, in the sense that the domain of exceptional z_0 that fail to iterate to ∞ is smaller for a chaotic μ_0 than for any of the nonchaotic μ that can be found arbitrarily close to μ_0.

Thus, there is a clear need for an objective measure of the progress towards chaos. An obvious candidate for measuring orderliness is the fractal dimension D of the Julia set $\mathcal{F}^*$. This paper finds that D is indeed appropriate for some scenarios, but raises very interesting complications for other scenarios, when $\mathcal{F}^*$ involves more than one shape, hence more than one dimension.

The best-known scenario was pioneered by J. Myrberg and is very well explored in many contexts (Collet & Eckmann 1980). It proceeds from linear to planar chaos by an infinite series of finite bifurcations of arbitrary order. When this scenario is applied to $g(z)$, as in Section 5, $\mathcal{F}^*$ remains a fractal curve whose D grows from 1 in to 2, hence its codimension $2 - D$ is indeed an acceptable measure of orderliness.

In an alternative scenario credited to Siegel 1942, D also varies steadily from 1 to 2, but intuition tells us that the limit is very incompletely chaotic in the plane. The key of this paradox is that the corresponding $\mathcal{F}^*$ involves two different shapes, hence two distinct dimensions.

In the third scenario to be examined, planar chaos is approached without bifurcation, and D tends to 2.

2. When λ is real, and $|\lambda| > 1$, iteration is orderly except on $\mathcal{F}^*$

For $\lambda = 0$ all points other than 0 and ∞ move in one step to 0, henceforth the motion is indeterminate. For $|\lambda| > 1$, there is an attractive fixed point at ∞, which contradicts our requirement A).

For real $\lambda > 0$, the map $g(z)$ preserves the sign of Re(z), and, for real $\lambda \neq 0$, the iterated map $g_2(z)$ preserves Re(z). More generally the Julia set is the imaginary axis for all real $\lambda \neq 0$.

3. Non-real λs that satisfy $|\lambda| < 1$

For these λ, the Julia set $\mathcal{F}^*$ is either the whole complex plane or a fractal curve. In the latter case, $\mathcal{F}^*$ has the following properties.

$\mathcal{F}^*$ is (obviously) symmetric with respect to $z = 0$, and is self-inverse with respect to the circle $|z| = 1$..

$\mathcal{F}^*$ includes $z = 0$ and is unbounded. This is obvious when the fixed points $z = \pm\sqrt{1 - \lambda}$ are stable: if z_0 iterates to one of the fixed points, $-z_0$ iterates to the other fixed point, hence the circle of radius mod (z_0) must intersect $\mathcal{F}^*$. (The origin $z = 0$ must be added because $\mathcal{F}^*$ is a closed set.)

$\mathcal{F}^*$ is asympototically self-similar for $z \to \infty$. Indeed, if $|z| < 1$ and z_0 iterates into some cycle, $\lambda(z + 1/z) \sim \lambda z$ into the same cycle. Being self-inverse, $\mathcal{F}^*$ is also asympototically self-similar for $z \to 0$. When $\mathcal{F}^*$ is topologically a line, it winds around a logarithmic spiral for $z \to \infty$ and for

$z \to 0$. These spirals wind in the same direction but *do not* line up, because scale invariance fails near $|z| = 1$.

We wish to start with a real λ for which $\mathcal{F}^*$ is a straight line, and then to change λ and follow $\mathcal{F}^*$ as it changes from a straight line to an increasingly wiggly curve. This requires drawing the semi-open variant of the "Mandelbrot set" defined in M 1980n{C3}.

The semi-open $\mathcal{M}$-set is the maximal set of λs, such that the iteration of the map has a finite limit cycle. Its closure is the ordinary $\mathcal{M}$-set. The semi-open $\mathcal{M}$-set of $z \to \lambda(z + 1/z)$ is shown on Figure 1. (Except for a 90° rotation, Figure 1 reproduces Plate x in the second and later printings of M1982F.)

Inspection reveals that the semi-open $\mathcal{M}$-set is made of $\mathcal{M}$-molecules made of $\mathcal{M}$-atoms, both shapes being common to the maps $\lambda(z + 1/z)$ and $z^2 - \mu$.

The present study is concerned with three different scenarios that start from the extreme order represented by λs in the real interval [0,1] — hence a Julia set identified with the imaginary axis — and end in planar chaos. We focus on the $\mathcal{M}$-molecule that includes the disc-shaped $\mathcal{M}$-atom $1/2| < 1/2$. It is easy to see that this atom collects all λs for which the iteration of $f(z)$ has 2 limit points, $z = \pm\sqrt{\lambda} / \sqrt{1 - \lambda}$.

4. From a flat sea to a Great Wave: computer-assisted homage to Katsushika Hokusal (1760-1849) (Figure 2)

For all λ in the disc $|\lambda - 1/2| < 1/2$, the Julia $\mathcal{F}^*$ is topologically a straight line that winds for $z \to 0$ or $z \to \infty$ around logarithmic spirals symmetric to each other with respect to 0. The spirals are both nicest and most educational when they are neither too loose not too tight. Let us therefore scatter a few parameter values, well within the $\mathcal{M}$-atom $|\lambda - 1/2| < 1/2$ between $\lambda = 1/2$ and the neighborhood of $1/2 + i/2$. To deemphasize the non-spiral complications near $|z| = 1$, the window (portion of the complex plane that is shown) is 200 units wide, and the $\mathcal{F}^*$-sets are rotated to become easier to compare. The $\mathcal{F}^*$-sets show as the boundaries between black "water" and white "air", which are the domains of attraction of two limit points. As intended, the first part of Figure 2 evokes a completely flat black sea, hence planar order. And the figures that follow counterclockwise evoke increasingly threatening black waves.

In parallel, the fractal dimension D of $\mathcal{F}^*$ increases. In this context, D tells how many decimals of z_0 in the counting base b, are needed to know whether z_0 is attracted to $\sqrt{\lambda} / \sqrt{1 - \lambda}$, or to $-\sqrt{\lambda} / \sqrt{1 - \lambda}$. To establish this fact, draw a collection of boxes of relative side $r_1 = 1/b$ on our window. Roughly b^D of these boxes intersect $\mathcal{F}^*$. Write $\beta = b^{D-2}$, and choose $= x_0 + iy_)$ at random in the box. With probability $1-\beta$, the first b-decimals of x_0 and y_0 suffice to determine where z_0 is attracted. More generally, the first k b-decimals of z_0 and y_0 suffice with the probability $(1-\beta)\beta^{k-1} \infty\ b^{k(D-2)}$).

On the average, the number of b-decimals needed to determine the limit is $1/(1-\beta)$. When $2-D$ is small, the expected number of base e-"decimals" needed to determine the limit is $\log_e b/(1-\beta) \sim 1/(2-D)$.

5. First path beyond the Great Wave. The Myrberg scenario of bifurcations

Figure 3 represents the $\mathcal{F}^*$-sets for two values of λ. In the top graph, λ lies past a bifurcation into 4, close to (but short of) a second bifurcation into 3. In the bottom graph, λ is reached by two successive bifurcations into 4, followed by a bifurcation into 3. Thus, the first λ lies off the center of an atom off the atom $|\lambda - 1/2| < 1/2$. And the second λ lies near the nucleus of a small $\mathcal{M}$-atom off a small $\mathcal{M}$-atom attached to the bottom

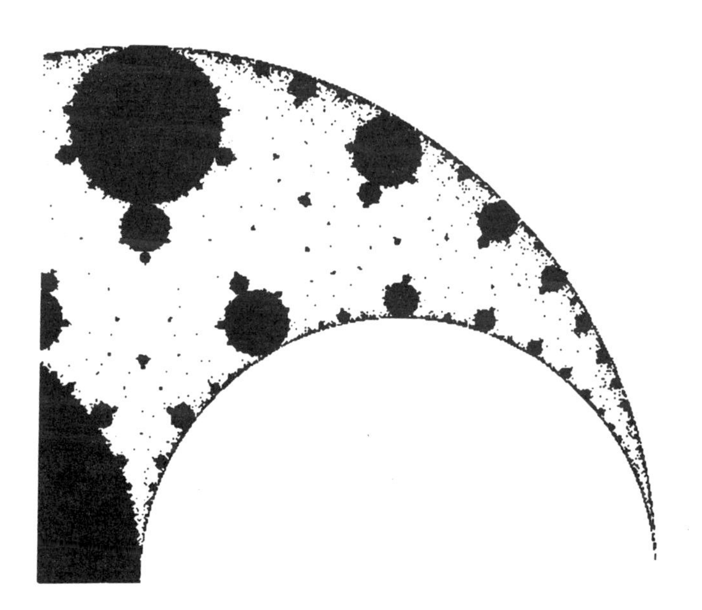

FIGURE C13-1. Largest $\mathcal{M}$-molecules in the upper right quarter of the semi-open $\mathcal{M}$-set of the map $z \to \lambda(z + 1/z)$. For clarity, the atoms $|\lambda| < 1$ and $|\lambda - 1/2| > 1/2$ are not made black but emptied to yield two white discs.

$|\lambda - 1/2| < 1/2$. A fourth λ is seen on p. viii of the second and later printings of M1982F {PS2003: and reproduced in Chapter C12 as Figure 2.

The first bifurcation forms "white water" through the breakdown of connected water and connected air into larger drops, some of them quite large. The bifurcations that follow break these drops into smaller ones, without end. It is clear that one watches a gradual progression towards the ultimate replacement of separate black water and white air by something that is neither water nor air. One cannot help evoking the critical temperature of physics.

The fractal dimension D of $\mathcal{F}^*$ tends toward 2 as planar chaos is approached, and the factor β tends to 1.

6. Second path beyond the Great Wave. A scenario of spiraling towards chaos (Figure 4)

Now select λ to be within the atom $|\lambda - ½| < 1/2$ but extremely close to $\lambda = 1$. The $\mathcal{F}^*$ set is illustrated by Figure 4. It is clear that, as $\lambda \to 1$, hence $\beta \to 1$, and that *chaos is approached without bifurcation.* The facts are perhaps easier to visualize in terms of the parameter $\mu = 1/\lambda$ and the variable $u = 1/z$. This change of variable does not change $\mathcal{F}^*$. For

FIGURE C13-2. From center, counterclockwise; Julia sets of $z \to \lambda(z + 1/z)$ for several λs, "from Flat Sea to Great Wave". Homage to K. Hokusai.

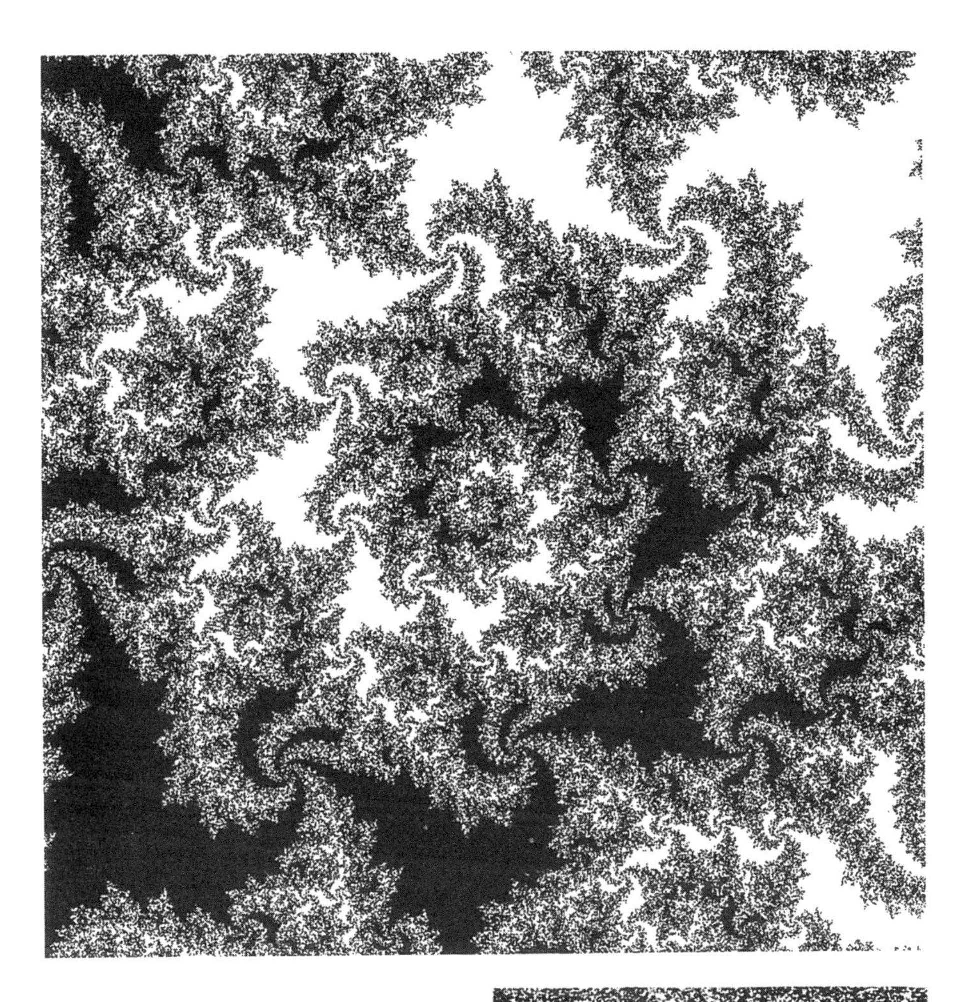

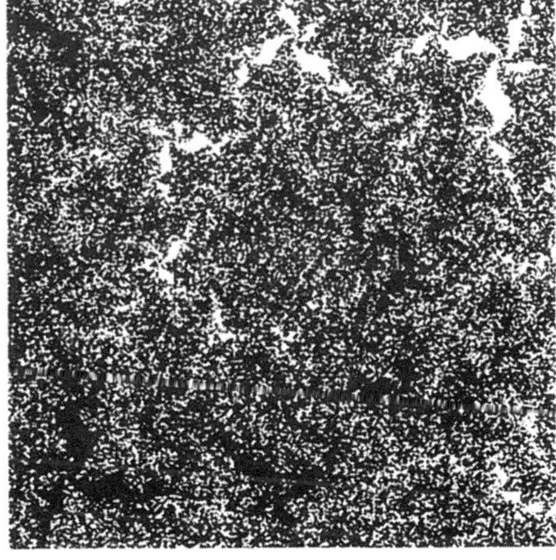

FIGURE C13-3 Julia sets for two λ that yield near totally chaotic maps $g(z)$ in the Myrberg scenario of repeated bifurcations.

discs, and imagine that these cracks converge and join. It follows that water — and air also, by symmetry — become separated into discs attached to each other by single punctual bonds. Two of the discs include the points $\pm + \sqrt{1-\lambda}$ and are called Siegel discs; let the remaining discs be called Siegel pre-discs. At each inter-disc bond, air and water cross each other but over most of the plane they are clearly separated by $\mathcal{F}^*$. Not unexpectedly, the fractal dimension of $\mathcal{F}^*$ takes a value of D_S that is unquestionably less than 2. On the average, one needs about $1/(2-D_S)$ decimals to determine whether a point z_0 is black or white. Incidentally, there is no limit cycle. The iterate z_0 is in the white (black), Siegel pre-discs end up in the white (black) Siegel disc. On the scale of the 200-wide window of Figure 3, the Siegel discs are so small that the Siegel regime looks like convergence.

Next, in order to achieve an idea of how $\mathcal{F}^*$ looks for λ just short of λ_x, it is necessary to know that Siegel discs are created when a curve $\mathcal{F}^*$, topologically a line, folds up and becomes domain- or plane-filling, as described and illustrated in M 1985g{C10}. When λ is just short of λ_S, the cracks invoked in the preceding paragraph have not converged and joined. Instead, the interior of each of the black discs is partly split by many (here, 157) very narrow "fjords", that penetrate deep into the white domains, without quite meeting, but coming close to meeting near the

FIGURE C13-4. Julia set for a λ that yields a near totally chaotic map $g(z)$, and is attained by yet another scenario. Topologically, this curve is a straight line.

center of a spurious white spot. In symmetric fashion, one must visualize black fjords thrusting into the white domain.

Since the boundaries of both white and black fjords are part of $\mathcal{F}^*$ the curve $\mathcal{F}^*$ is very close to filling the whole plane. Its dimension's closeness to 2 tempts us to conclude that the corresponding map $g(z)$ is completely chaotic. But it is not. In fact, the Siegel scenario reveals an important and subtle point: we need a close look at the factor β. For very tiny values of the cell side r_k, we find $\beta \propto r_k^{2-D}$. However, as long as $r_k < \xi$ with ξ a function of $2-D$, we find $\beta = 1$. Thus, the smallness of $2-D$ expresses that *every* cell of side $< \xi$ will be intersected by $\mathcal{F}^*$. But this does not say anything about the relative proportions of black and white in the cells $< \xi$. in the present case, the fjords are so narrow that a cell $< \xi$ is mostly black or mostly white, depending on whether it is in a domain that Figure 5 shows as solid black or solid white. The expected number of decimals of base b depends upon whether one wants to know the color of z_0 precisely, or with high probability. Absolute precision requires $1/[1-b^{D-2}]$ decimals, high probability requires only $1/[1-b^{D_3^2}]$.

On other words, the overall appearance that computer limitations give to Figure 5 is not misleading at all. In fact, it helps reveal a basic truth. When λ is very near λ_s, the shape of $\mathcal{F}^*$ is ruled by *two* distinct dimen-

FIGURE C13-5. Julia set for a λ that yields an unquestionably chaotic map $g(z)$, and is attained by the Siegel scenario.

sions: its own and that of the $\mathcal{F}^*$ corresponding to the nearest Siegel value of λ.

We must agree that near-complete planar chaos should require that all small cells (a) intersect $\mathcal{F}^*$ and (b) be about half black and half white. Under these conditions, a nearly space-filling $\mathcal{F}^*$-set of dimension nearly 2 is not sufficient for complete chaos. The presentation of further results on this topic must be postponed until a later occasion.

Acknowledgement. The illustrations were prepared by James A. Given, using computer programs by V. Alan Norton.

First publication; figures dating to 1974

C14

Two nonquadratic rational maps devised from Weierstrass doubling formulas

THIS CHAPTER DRAWS FROM A CACHE OF LONG "LOST" illustrations that were prepared made in 1979 with the assistance of Mark R. Laff. They are published here for the first time, at long last, together with explanations and comments, describe in documented detail the story told in Chapter C12. Some are attractive, and most affect the basic distinction made in Chapter C1 between seeing and discovering.

Using primitive graphics done at Harvard in 1980, Chapter C1 has established that high graphic quality was *not necessary* for discovering the Mandelbrot set. A fortiori, color was not necessary, but the skills of an experimentalist were. This chapter establishes the converse: the high graphic quality I had in 1979 was *not sufficient*, either.

For evidence, please scan the fine figures of this chapter. With hindsight, the M set is identified throughout. Also, it is immediately "obvious" that many of my earliest Julia sets are complex "composites" or "alloys" built from the "elements" that I went on to discover in 1980 by studying quadratic iteration. I did see those elements in 1979 but could not organize and describe them. Therefore, they remained undiscovered. Could discovery have followed a different path, with the 1980 elements being inferred from the 1979 composites? Dwelling on what might have been is a waste of time.

Those illustrations began as state-of-the-art, then the absence of color made them into used furniture, and today they are valuable antiques. At some points in time, I hoped to clean up their theory before publishing them, but this is no longer my plan. Since those figures' significance materially depends on their age, each is accompanied by the date automatically printed on the original.

1. THE LATTÈS FUNCTION $W_1(z)$ AND THE MAP $z \to \lambda W_1(z)$

Chapter C12 defined the Lattès doubling map $W_1(z)$ as

$$z \to W_1(z) = \frac{1}{4} \frac{(1+z^2)^2}{z(z^2-1)}.$$

This report on the map $z \to \lambda W_1(z)$ begins with Julia sets and continues with less advanced but perhaps more significant early results concerning this map's Mandelbrot set.

1.1 Real parameter λ Julia sets whose "elements" are "San Marco dragons"

Figures 1 and 2 represent part of the z-plane. The value of λ is such that the map $z \to \lambda W_1(z)$ has two cycles, each of order 2. Altogether, the map $z \to \lambda W_1[\lambda W_1(z)]$ has four limit points; hence 4 figures were drawn, meant to be used as direct color separations for a planned four-color display.

In Figure 1, the black points are the starting points z_0 of orbits that converge to one of the cycles, and the white points are those of orbits that converge to the other cycle.

With hindsight, each of the black or white tiles (apart from size) is a nonlinearly deformed San Marco dragon as defined in M 1980n{C3}. That is, the λ I hit upon corresponds to the quadratic map $z \to 2\,z(1-z)$ that marks the first bifurcation of $z \to \lambda z(1-z)$. Too bad that no record indicates whether or not I knew what I was doing.

In Figure 2, each of the black tiles in Figure 1 is split into a collection of black and white tiles, put together shish-kebab style. The color now depends on the position of the orbit after a prescribed number of steps, that is, on the limit point of the squared map $z \to \lambda W_1[\lambda W_1(z)]$.

1.2 Julia set element bounded by self-avoiding "Jordan curves"

As suggested by the shape of the elements in Figures 3 and 4, the underlying values of λ correspond to two values $\lambda < 2$ in the quadratic case. Two colors suffice in each case, because the limit set consists of two nonbifurcated limit points.

1.3 Julia sets for real values λ close to the chaos value $\lambda = 1$

The value $\lambda = 1$ brings us back to the Lattès map, for which the "Julia set" fills the whole complex plane. I wanted to know how this "chaos" is approximated for λ approaching 1. In the case of the real quadratic map $x \to x^2 + c$, this approach was well understood analytically. But, as is always the case with fractal dusts on the line, it was hard to visualize. Phenomena in the plane are harder to study but easier to visualize.

04/24/79

FIGURE C14-1. [04.24.1979] The colors white and black mark the domains of attraction of two cycles, each made of two points, for a map of the form $z \to \lambda_1 W_1(z)$.

Studying $z \to \lambda W_1(z)$ as $\lambda \to 1$ showed that each of the period-two cycles in Figures 1 and 2 is subjected to the sequence of successive bifurcations that the real quadratic map had made familiar. Soon, the number of limit points in a cycle becomes so large that one color separation per limit point would be intractable. Instead, white (resp. black) was used for all the points obtained by bifurcation from the white (resp. black) cycle in Figure 1.

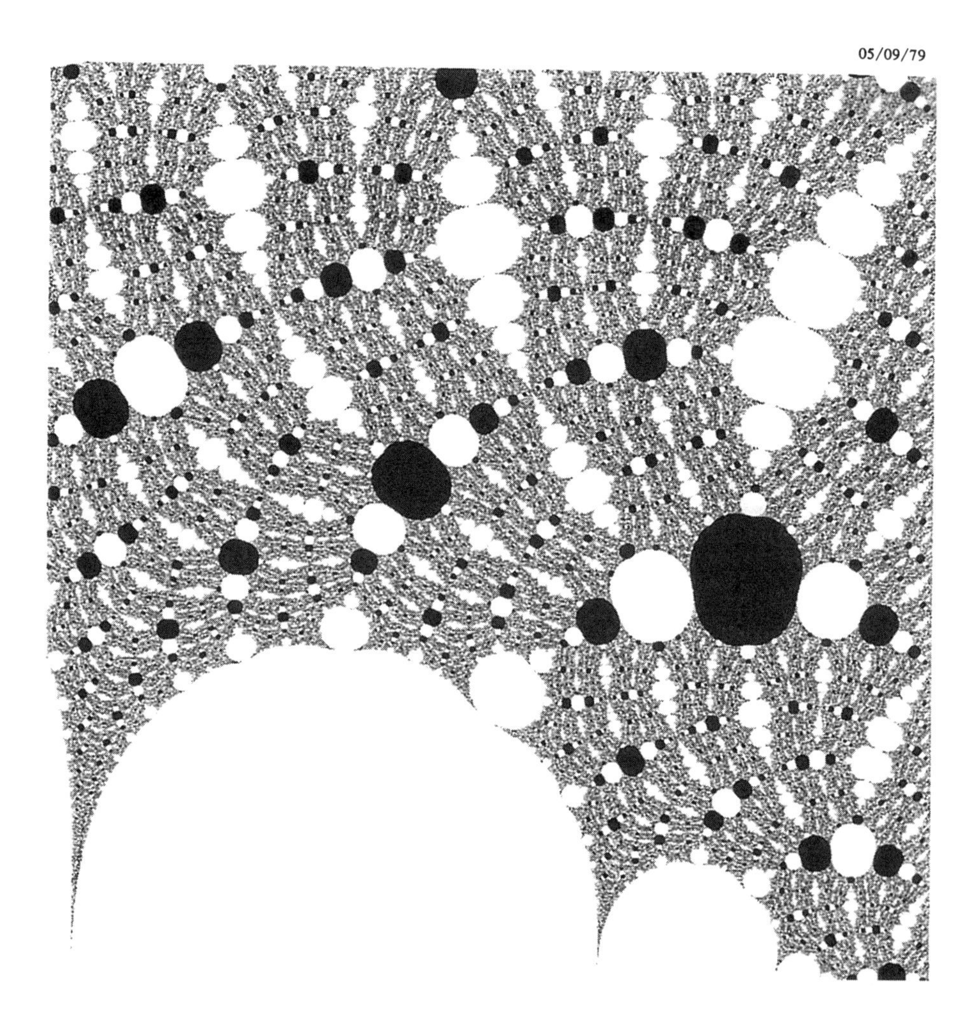

FIGURE C14-2. [05.09.1979] Decomposition of one of the domains of attraction of $\lambda_1 W_1(z)$ in Figure 1. It splits into the domains of attraction of two of the limit points of the map $z \to \lambda_1 W_1(z)[\lambda W_1(z)]$.

Implicitly, Lattès has predicted that choosing λ too close to 1 would yield a meaningless mess. Several values of λ were tried before Figures 5 and 6 were prepared. Throughout, I felt like someone trying to photograph the Cheshire cat in *Alice in Wonderland* at the very moment it is about to disappear. Had I failed and this investigation led to nonsense, the result would have been discarded.

1.4 "Advance shadows" of the Mandelbrot-set-to-be

Figures 1 to 4 were followed by a large number of variants for other values of λ. Increasing confusion motivated the next step. I tried to iden-

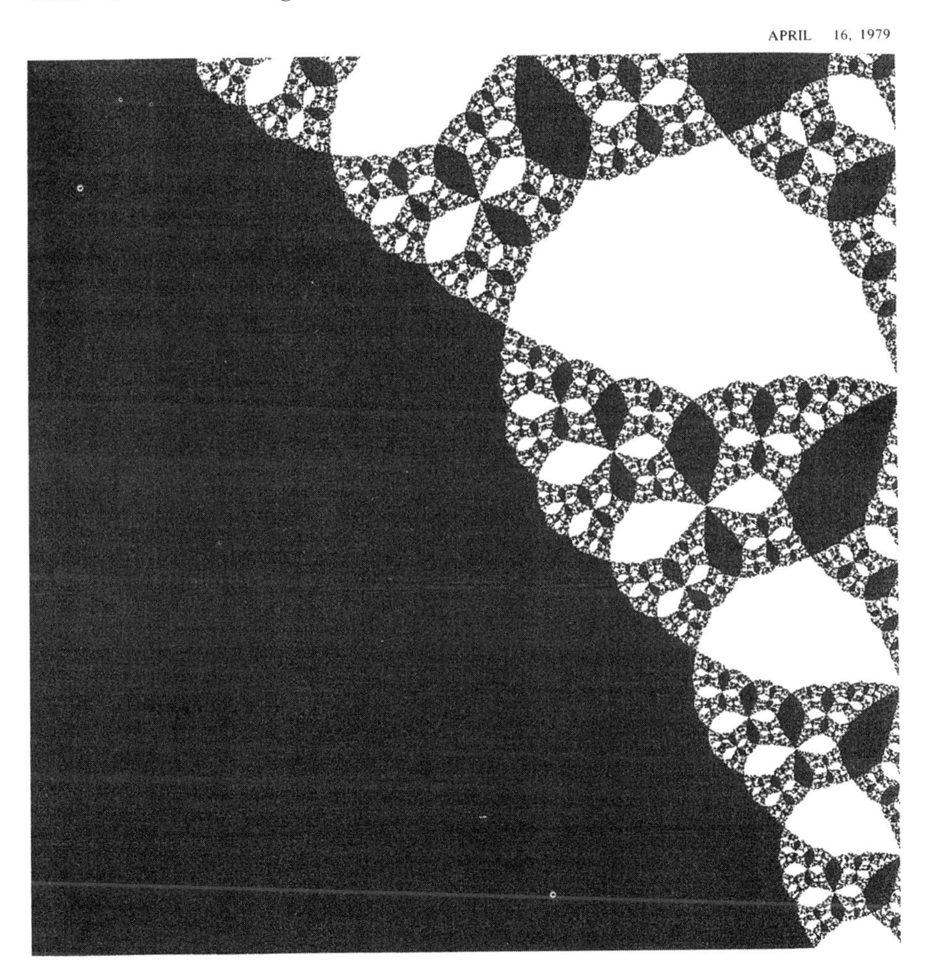

FIGURE C14-3. [04.16.1979] Same as Figure 1, but with a parameter value $\lambda_2 \neq \lambda_1$.

tify the set of complex λ for which the Julia set of $z \rightarrow \lambda W_1(z)$ has a finite number of finite limit cycles. That is, I looked for the counterpart of what became the M^0 set of the quadratic map. If λ belongs to this set, the position of the orbit after a large number of iterations reduces to a finite size set S.

Every early exploratory search must be designed to move fast. Therefore, I began with a shortcut. I reasoned that, if limit cycles exist, they do not lie very far from $z = 0$. The resulting sets of λ were worth examining first, leaving a rigorous justification for later study.

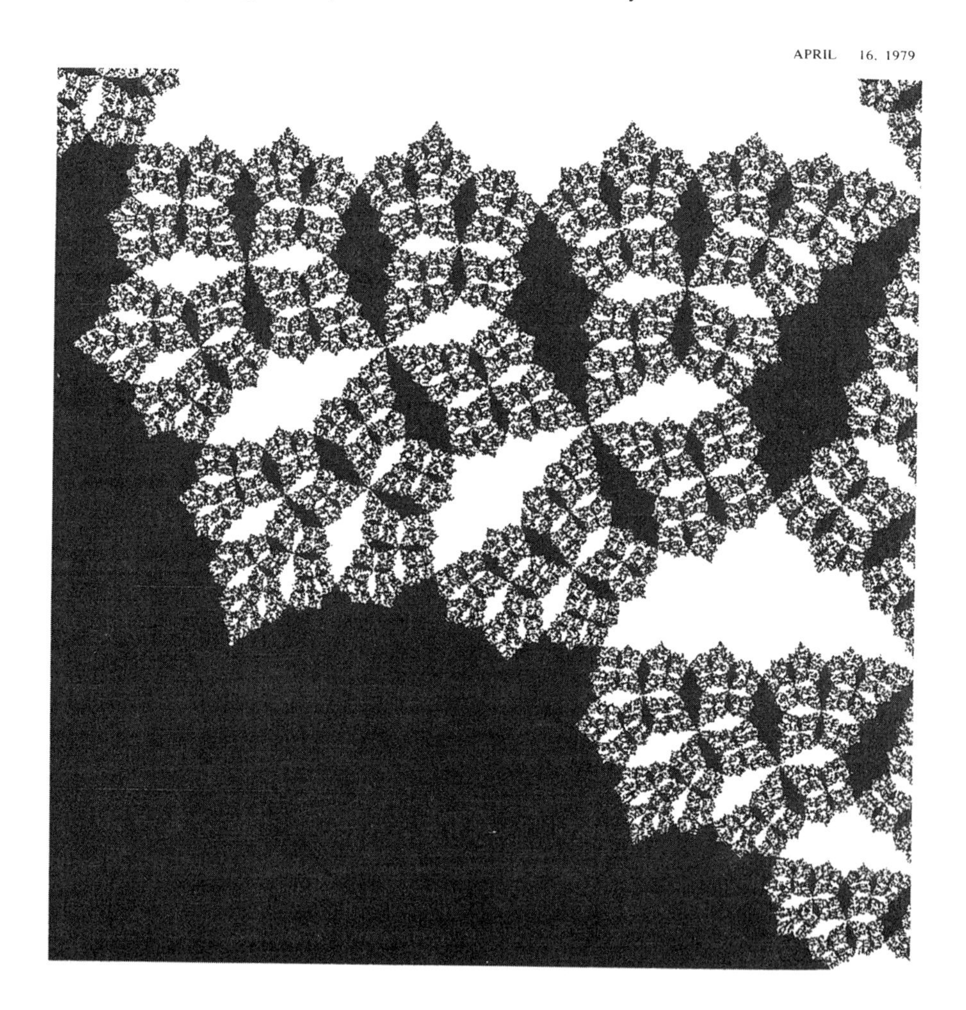

FIGURE C14-4. [04.16.1979] Same as Figure 1, but with a parameter value $\lambda_3 \neq \lambda_2 \neq \lambda_1$.

An illustration from 1979 showing "M set islands." With hindsight from the quadratic case, Figure 7.1 needs no elaboration. Against a black background, everybody instantly recognizes several small versions of the quadratic M^0-set-to-be.

With a second dose of hindsight, those little "things" correspond to island molecules of the quadratic Mandelbrot set. Recall that in the quadratic case those islands are relatively very small. This is why early graphics shown in Chapter C1 showed them as elusive specks of dirt. Their counterparts here are relatively larger and quite obvious. They are even more conspicuous for other complex maps I examined in later years.

04/25/79

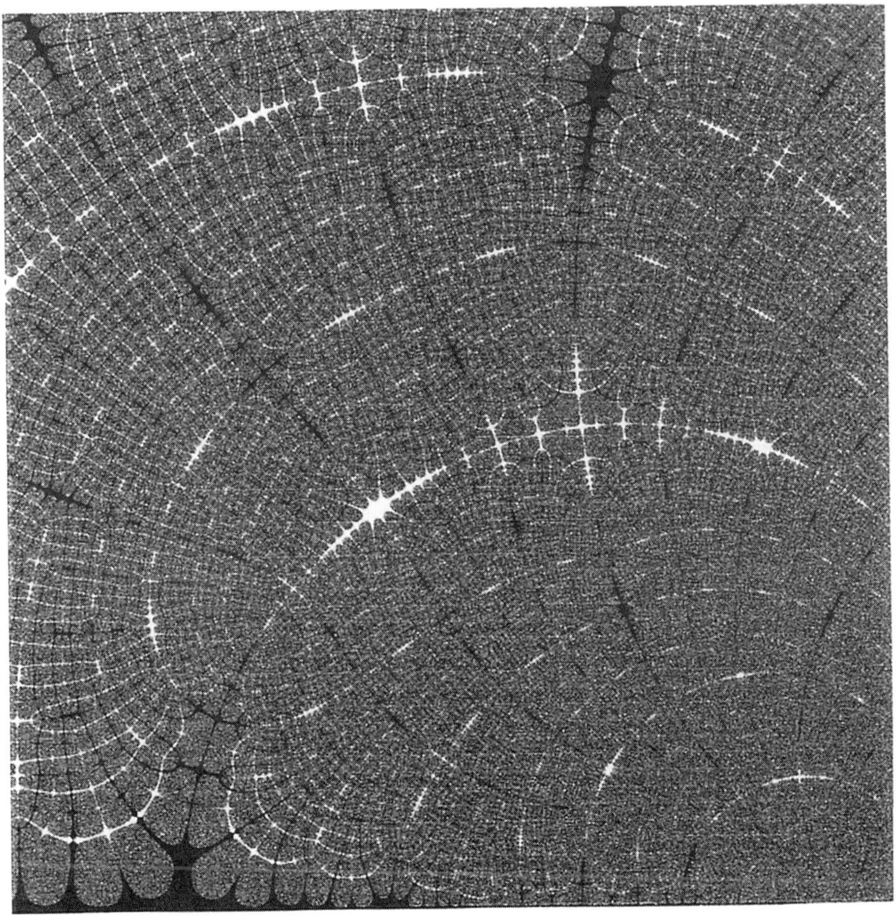

FIGURE C14-5. [04.25.1979] Same as Figure 1, but with a parameter λ_4 near the value $\lambda = 1$ that corresponds to Lattès chaos.

A third dose of hindsight reminds us that the construction of the quadratic M^0 must proceed by successive approximations of the boundary of M *from the inside.* Each disc-like atom of M^0 is approximated by a sequence of smaller discs and only in the limit do atoms acquire a common tangent at the point where they bind. The corresponding effect is clearly seen on Figure 7.1.

The date of Figure 7.1 is significant. Physically I was already at Harvard, but graphics continued to be produced—slowly—at IBM. I shall not speculate on whether or not a faint memory of Figure 7.1 might have

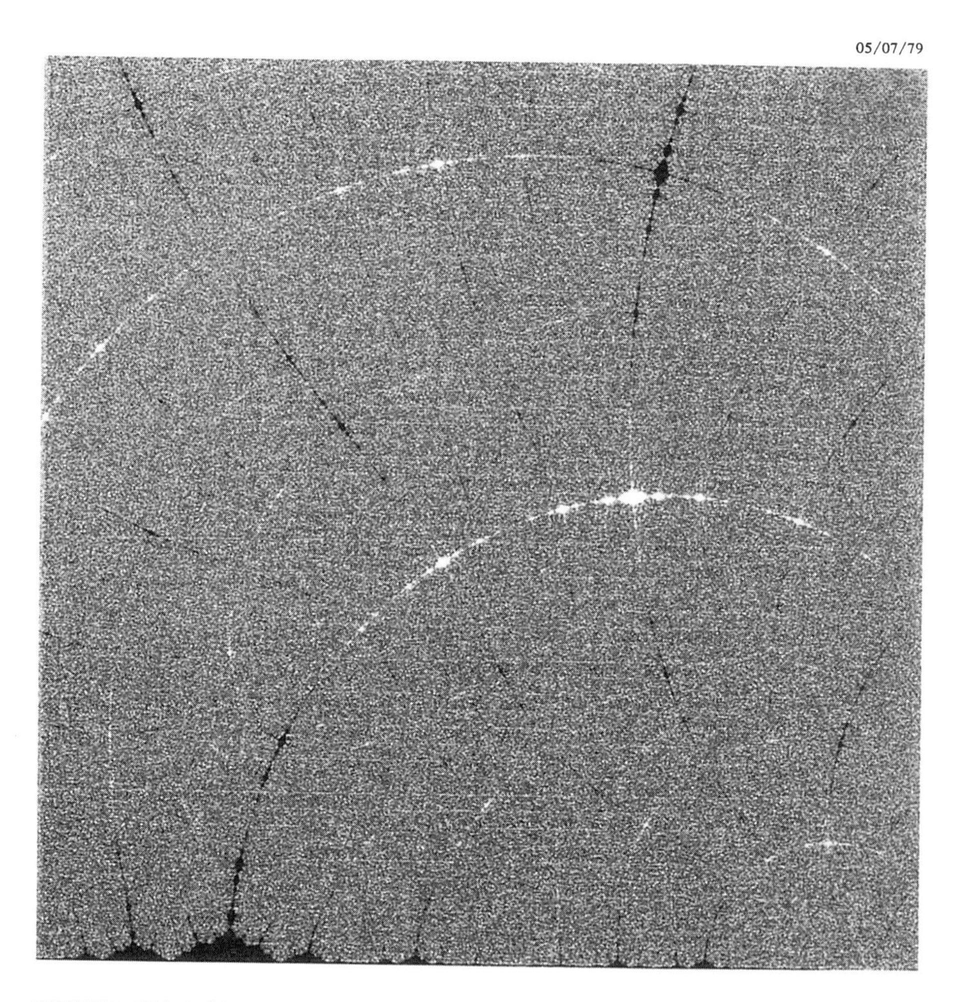

FIGURE C14-6. [05.07.1979] Same as Figure 5, but with a parameter λ_5 even closer to the value $\lambda = 1$ that corresponds to Lattès chaos.

helped me decide a few months later to zoom on the little specks of dirt on the crude graphs of 1980 vintage reproduced in Chapter C1 and C3

With a fourth dose of hindsight, the discovery of the Mandelbrot set and the discovery of its ubiquity were nearly simultaneous.

But hindsight is a waste of time. Together, those objects' smallness and the little I remember of the method of construction implied that Figure 7.1, when new, was a preliminary picture not worth publishing but well worth improving upon under more favorable conditions.

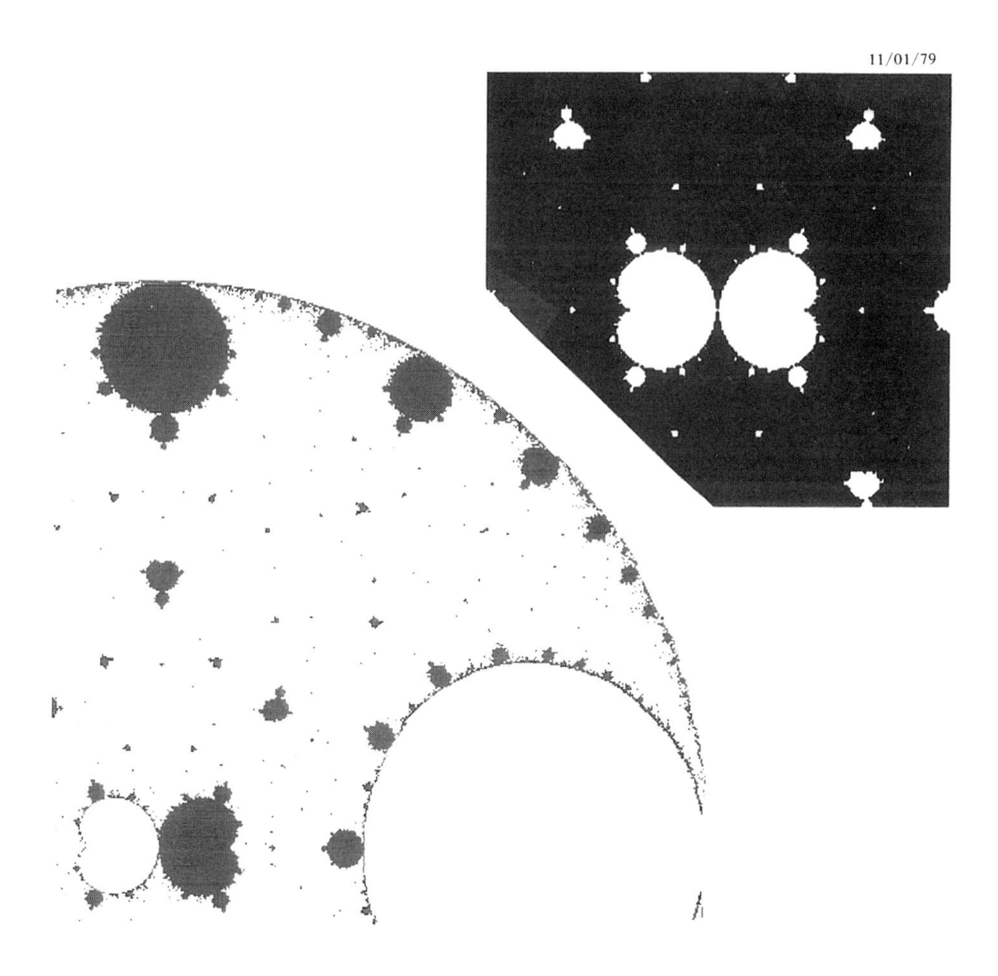

FIGURE C14-7. On the top right, Figure 7.1 reproduces an attempt [11.01.1979] at drawing the M^0 set of the map $z \rightarrow \lambda W_1(z)$. As confirmation, Figure 7.2 reproduces a more sophisticated attempt made at IBM after 1980.

A more recent rendering of the Mandelbrot set of $\lambda W_1(z)$. To help the reader locate Figure 7.1 in a broader context, the much later Figure 7.2 reproduces an M^0 for $z \rightarrow \lambda W_1(z)$ prepared after I had returned to IBM. When first reproduced in M 1986p, this M^0 set did not—to the best of my knowledge—generate attention; once again, I am surprised.

Composite illustration from 1979 of the M-set atoms and their mutual relations. The parts of Figure 8 came to be drawn because to construct M^0, the notion of a "limit set of finite size" was decomposed into the union of the notions of "limit sets of size k, " where computer limitations imposed a

FIGURE C14-8. Composite of domains of the M^0 set of $z \rightarrow \lambda W_1(z)$ yielding cycles of different sizes.

finite k_{max}. Thinking forward to color separations of a proposed color version of M^0, it was natural to output successive values of k separately. This process was overambitious, and for reasons I did not record and no longer recall, each separation-to-be was too fuzzy for the intended use.

With hindsight, however, everyone aware of the quadratic Mandelbrot set recognizes the familiar shape of its "atoms" and some features of their mutual arrangement.

06/02/79

FIGURE C14-9. [06.02.1979] Preimages of the quadrant $Re(z) > 0$, $Im(z) > 0$ by a map of the form $z \to \lambda W_1(z)$.

1.5 Preimage of the quadrant {Re z, Im z} under iterations of $\lambda W_1(z)$

Figure 9 is one of the most attractive in a large portfolio I devoted in 1979 to various values of λ. Here, all the limit points belonging to each of the four quadrants of the plane were pooled together and the iteration went on to 15 stages. Fewer iterations introduced all kinds of artefacts, always misleading but often visually compelling. A second quadrant is represented in gray.

05/31/79

FIGURE C14-10. [05.31.1979] Domains of attraction (in white, gray, and black) of three limit points of a map of the form $z \rightarrow \lambda W_2(z)$.

2. THE VARIANT FUNCTION $W_2(z)$ AND THE MAP $z \to \lambda W_2(z)$

Chapter C12 defined the variant Weierstrass doubling map $W_2(z)$ as

$$z \to W_2(z) = \frac{1}{4} \frac{z(z^3+8)}{z^3-1}.$$

The attractiveness of $z \to \lambda W_2(z)$ resides in the fact that its symmetries are due to the factor z^3, and therefore, they differ from those of $z \to \lambda W_1(z)$.

2.1 A Julia set for a parameter λ not close to the chaos value $\lambda = 1$

In Figure 10, three "colors," white, gray, and black, mark the values of z_0 that are attracted to each of 3 limit cycles. To be practical, the part in gray was computed and the part in black deduced by a 120° rotation.

2.2 "Advance shadows" of the Mandelbrot-set-to-be

Figure 11 includes outlines of the M set, some clear ones in the central region and other lost in the penumbra.

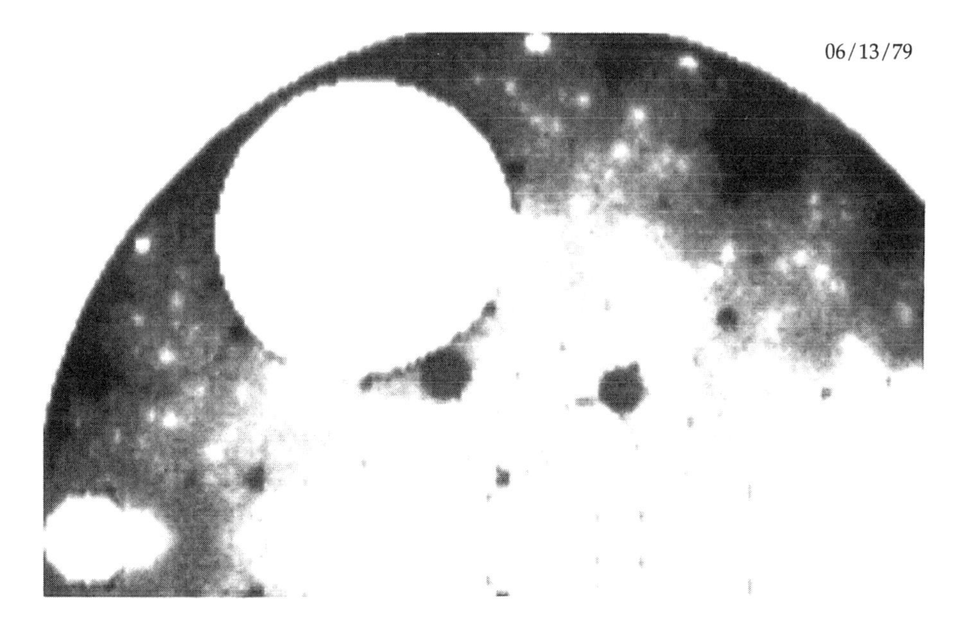

FIGURE C14-11. [06.13.1979] The counterpart of Figure 8 for $z \to \lambda W_2(z)$.

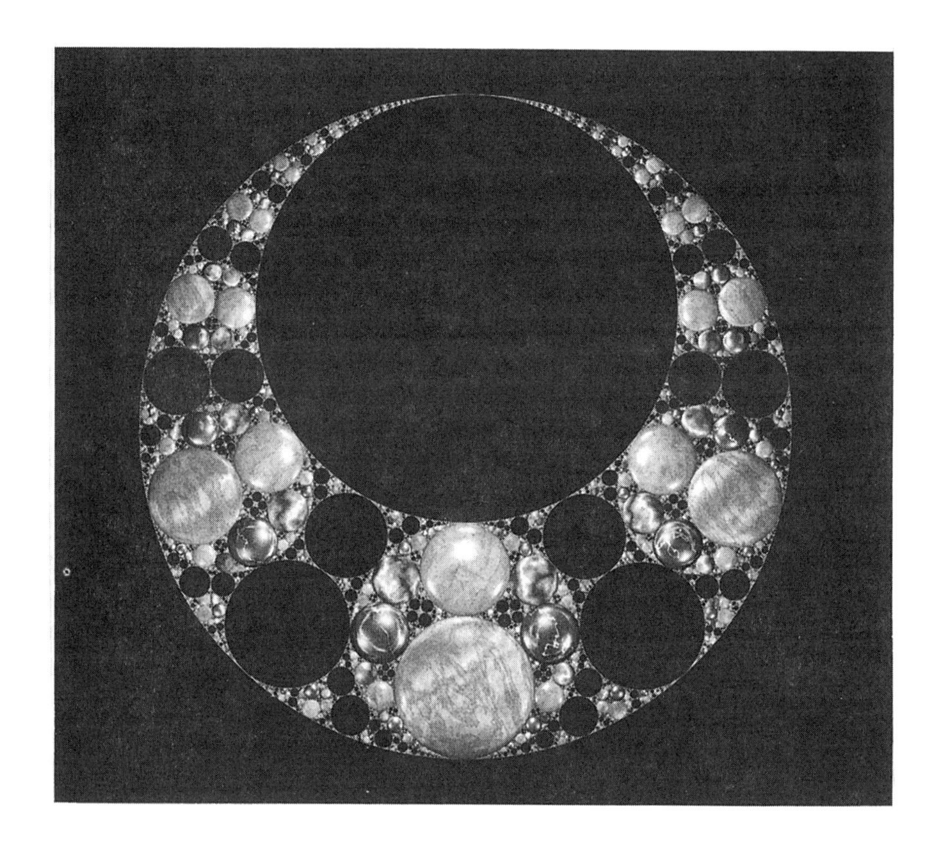

[Cover of M1999N, reduced from color to grey half-tones.] This decorative design is a further illustration of Part III of this book. It is called *Pharaoh's Breastplate* and fills the self-inverse fractal described in Figure C16.2. The algorithm described in Part III relies on eight possible kinds of "semi-precious stones," but this rendering leaves out four of those stones, for esthetic effect. The generating program was written by K.G. Monks.

PART III: ITERATED NONLINEAR FUNCTION SYSTEMS AND THE FRACTAL LIMIT SETS OF KLEINIAN GROUPS

&&&&&&&&&&&&&&&&&&&&&&&&&&&&&&

First publication

C15

Introduction to papers on Kleinian groups, their fractal limit sets, and IFS: history, recollections, and acknowledgments

IN HISTORICAL SEQUENCE, Kleinian groups came before the Fatou–Julia theory, and I explored those topics in the same sequence. But in this book, I inverted history and have put forward my best foot, or at least the best known one. This introduction will also explain the mysterious initials IFS. As in Chapter C2, some names are first printed in ***bold italics*** when it seems useful.

My later work on "Kleinian measures" is the topic of Part IV.

1. The early history of Poincaré's great innovation, one he chose to call "Kleinian" groups

The term "Kleinian" challenges the relation between deed and name. It is grossly misleading because the conception and initial study of Kleinian groups have been an early, near single-handed, and most celebrated achievement of Henri Poincaré (1854–1912).

Since ***Felix Klein*** (1849–1925) did not contribute to the early results, what led to the accepted term? He had been pestering Poincaré, insisting that the more specialized objects that Poincaré was calling "Fuchsian" should be renamed "Kleinian." Poincaré balked but finally compromised, and promised to use "Kleinian" for his (Poincaré's!) *next* good idea. And he kept his promise!!! The story lives in letters reproduced in Volume 11 of the Collected Papers, Poincaré 1916.

By no means did Poincaré disclaim credit he deserved. In the nineteenth century, credit and name were separate issues, as in the example of the Gaussian distribution. Similarly, the Dirichlet principle is not due to Peter-Gustav Lejeune-Dirichlet (1805–1859), but to his student Bernhard Riemann (1826–1866). But sloppy historians do not allow for changing customs, and today, credit is all too often given to Klein. Neither the term, nor the dispute it generated, is mentioned in the major early treatise on Fricke and Klein 1987. (In the preface, Fricke implies that he wrote the book alone and thanks his revered teacher Klein for agreeing to sign it.)

2. The study of Kleinian groups and their limit sets is filled with fractals-to-be; they did not arise only as counterexamples but also as "inevitable developments of the theories of analytic functions and of differential equations"

Yes, this fact was completely unknown to me when I let myself be carried away in M1975O to praise and scold mathematicians for creating the Cantor sets and then hiding them from all concrete use. All objective criteria suggested that my proposed link between Cantor sets and physics in real space was viewed as unprecedented and preposterous.

Yes, the fact is largely forgotten, but within a year of Cantor's 1883 paper on that set (Cantor 1932), and well before the revolutionary theories of sets and of functions of a real variable, sets close to the triadic dust and the Weierstrass function arose in perfect orthodox mathematics.

No, those applications did not go unnoticed in their time. The first was in the theory of automorphic functions, which made Poincaré and Felix Klein famous. Those applications were pursued by Paul Painlevé (1863–1933), a scholar influential well beyond the realm of pure mathematics. He was fascinated with engineering (he was Wilbur Wright's first passenger after Orville Wright's accident) and eventually entered politics, rising during World War I to the post of prime minister of France.

Yes, my ignorance was complete, founded on prejudice, and encouraged by the fact that Cantor ended up as a victim of Poincaré's comment that "Cantorism [promises] the joy of a doctor called to follow a fine pathological case." When the need arose, Poincaré *did* use the classic monsters in abstract mathematical models of physics. The next step, to use them to describe visible nature, is one he did *not* envision.

French students of my generation were familiar with the basic ideas of Kleinian groups. But I did not pay much attention until an unforgettable event: a study — first casual then passionate — of the extraordinary eulogy of Poincaré by Jacques Hadamard (1865–1963), a famous mathematician and mathematical physicist often mentioned in this book. I chanced to read Hadamard 1912 when M1997F could still be expanded. Therefore, my reacquaintance with Poincaré is mentioned there and also in M 1982F but they deserve being repeated in free translation.

From Hadamard 1912. "Poincaré was a precursor of set theory, in the sense that he applied it even before it was born, in one of his most striking and most justly celebrated investigations. Indeed, he showed that the singularities of the automorphic functions form either a whole circle or a Cantor dust. This last category was of a kind that his predecessors' imagination could not even conceive. The set in question is one of the most important achievements of set theory, but Bendixson and Cantor himself did not discover it until later.

"Examples of curves without tangent are indeed classical since Riemann and Weierstrass. Anyone can grasp, however, that deep differences exist between, on the one hand, a fact established under circumstances arranged for the enjoyment of the mind, with no aim or interest other than to show its possibility, an exhibit in a gallery of monsters, and on the other hand, the same fact as encountered in a theory that is rooted in the most usual and the most essential problems of analysis."

From Painlevé 1895. "I must insist on the relations that exist between function theory and Cantor dusts. The latter kind of research was so new in spirit that a mathematical periodical had to be bold to publish it. Many readers viewed it as philosophical rather than scientific. However, the progress of mathematics soon invalidated this judgment. In the year 1883 (which will remain doubly memorable in the history of mathematics in this century), *Acta Mathematica* alternated between Poincaré's papers on Fuchsian and Kleinian functions and Cantor's papers."

Poincaré had already sketched his results in *Comptes rendus* before Cantor's work appeared in German, and Poincaré adopted one innovation so promptly that in his first *Acta* paper he denoted *sets* by the German *Mengen*, without seeking a French equivalent.

3. Algorithms for constructing the fractal limit sets of Kleinian groups, more precisely, of groups based on inversions

Alerted by Hadamard's words on Poincaré and before facing those "Kleinian" limit sets, my understanding was brushed up by studying a book by ***Wilhelm Magnus*** (1907–1990), then a professor of mathematics at New York University; he was noted for his work on group theory, special functions, and mathematical physics. Magnus 1974 is beautifully old-fashioned and was written to draw attention to Fricke and Klein 1897, by then mostly forgotten. Hence it reproduced the original illustrations to which I had paid little attention during the Poincaré centennial in 1954.

Could it be that I was the only person old and old-fashioned enough to be thoroughly trained in the old geometry but also young and adventurous enough to have become a fluent user of the computer? Be that as it may, Magnus set me to work. Playing on the computer and examining thousands of pictures made my thinking converge within a few weeks to a very fast algorithm that is described belatedly in M 1982F {see Chapter C16} and M 1982m{C18}. My skillful programmer during those weeks was ***Peter Oppenheimer***, a Princeton student on vacation, who moved on to a career in media.

To appreciate the dumfounding simplicity of my algorithm, take a look at Figure 1 of Chapter C16, which compares side-by-side two constructions of the limit set $\mathcal{L}$ of the group based on inversions in the four large white circles top left. This limit Kleinian set is a continuous loop without double points, a Jordan curve. The Poincaré method (top panel) converges extremely slowly. My method (middle panel) converges very fast and obtains $\mathcal{L}$ as the common limit of an inner and an outer approximation. The lower panel shows that my method is strikingly close in spirit to of one E. Cesàro (1859–1906) had provided for the classical Koch curve. In its original context as Plate 43 of M1982F, this lower panel is accompanied by further explanations.

4. An unexpected and probably meaningful hundred year delay

For a hundred years, a fast algorithm had been a goal of many great mathematicians, and probably also of countless amateurs. Astonishing but true (and given the simplicity of the "trophy," perhaps almost embarrassing), all those seekers had failed. One of them was ***David Mumford,*** now at Brown University, but in 1979–1980 a professor of mathematics at Harvard, a colleague, and a warm host I wish to thank here.

In December 1980, after my first lecture on the iteration of rational functions, Mumford wondered whether I could also look for an algorithm for Kleinian limit sets. In response, I showed him a draft of M 1982m {C18}. He marveled, then observed that the tools I had used were ancient, utterly elementary, and certainly intimately familiar to Poincaré, Fricke, and Klein. He wondered aloud what made me succeed when those seekers and so many others had failed.

In a nutshell, my answer distilled — once again — the already-told story of my scientific life: When I seek, I look, look, look, and play with pictures. One picture is like one reading on a scientific instrument. One reading is never enough.

At that point in history, Kleinian groups were in a strange situation. Towering figures like Lars Ahlfors (1907–1996) at Harvard, and Lipman Bers (1914–1993) at Columbia had made great strides forward. But the impression prevailed that their act was hard to follow and hardly anyone was interested in hearing of my algorithm. But one day the overly thin wall at my Harvard office allowed me to overhear the words "Kleinian group." The speaker, who turned out to be S.J. Patterson, confirmed that there was little interest in the topic. I convinced him that this lack of interest deserved to be tested, and a seminar was organized. Perhaps thirty persons came to the first meeting!

Mumford was naturally one of them. He became very supportive of my work and (in record time!) my assistants taught him computer programming. I also introduced him to David Wright — who was the person with whom Patterson had been talking. He admitted that he was a skillful programmer but thought that Harvard graduate students should not advertise this heresy. Many persons — not all! — became enthusiastic about the power of the computer and Mumford soon moved away from algebraic geometry, a field in which he was a major figure. He experimented on the computer with Kleinian groups richer in structure than those I had looked at. One of his earliest illustrations, implemented on one of his visits to IBM, appeared in M1982F as Plate 178. He now works on a computer-based theory of vision.

A beautiful book, Mumford, Series, & Wright 2002, alludes to the events in 1979 during which I made Mumford a devotee of the power of the eye. An inevitable question arises, will this book have the effect Peitgen & Richter 1986 had in its time? The 20 years since the 1980s witnessed successive "showers" of books on iteration, but none on "Kleinian" books. This drought ought to cease.

5. Was the progress from pictures for their own sake, to pictures that open new mathematical vistas pre-ordained?

Today, long after the fact, this progress may seem pre-ordained. The many obstacles it encountered in its time are illustrated by the following episode. After discovering my algorithm, I described it during a visit to Wilhelm Magnus. He was surprised, impressed, and very supportive. Then, before I left him, he gave me a thick file of drawings of Kleinian limit sets sent by several admirers of Magnus 1974 who had access to a computer and a plotter. Those pictures were more accurate than those of the 1890s but triggered no fresh insight or inspiration of any kind. To draw pictures is today an easy mechanical process, but to "read" them remains an arcane art. This visit to Magnus was neither the first nor the last example of this fact, as the discovery of the Mandelbrot set has shown.

I also showed to L. Bers the corrected form of Figure 156 of Fricke and Klein 1897, which is "deconstructed" in M 1982m{C18}. He was curt and admitted being sorry that the old illustration of this limit set was very misleading. It had nourished his intuition, and he would continue to prefer it. Was he serious or kidding? Most probably, it did not matter, because his intuition was analytic rather than visual.

6. The notion of IFS (iterated function system or schemes) or decomposable dynamical systems

When Mumford welcomed me at Harvard in 1979, it surprised him that I always began exploring new Kleinian groups with the following method. Given a group based on the transformations G_m, attribute to each G_m a probability p_m of being chosen next. In order to study subgroups, set the values of some of the p_m to 0. An infinite sequence of indexes,

$m(1), m(2), ...m(h), ...,$ defines an "orbit" worth studying. Figure 4 of M 1980n{C3} shows that I was following the same procedure with the semigroup of inverses of $z^2 + c$, namely, the subgroup based on $\pm\sqrt{z-c}$.

I perceived that this procedure as being implicit in the late nineteenth century but unexplored and nameless. Later, it came to be universally denoted by the letters IFS and was given a name. Michael Barnsley called it "iterated function systems," while Kenneth Falconer called it "iterated function schemes." For linear IFS, many properties are known today, together with full mathematical proofs.

In 1979, however, a mathematical justification had not even been attempted for the Kleinian limit sets and I realized that those sets and the corresponding IFS orbit might conceivably differ in the limit. When Mumford objected to my cavalier attitude, I countered that it was the best way of "fishing" the easy way. Besides, in cases which the two sets actually differed, both deserved to be investigated. To approximate the IFS limit involves an easy mechanical process, and that should be undertaken first. Next, one should inspect pictures of a sufficient number of variants of the IFS limit and check for structures that may be suggested by the eye. A deep thinking mode should come later. Of course, this is precisely the strategy whose best-known success was the discovery, then about to be announced, of the Mandelbrot set.

What I really had in mind is a "generalized IFS" that was proposed in Chapter 20 of M 1982F, but never actually developed. An IFS can be viewed as a dynamical system decomposable into a collection of operations, and a "master process" ruling the operation's sequence. In an IFS, the master process is an independent random sequence $m(1), m(2), ...m(h)$, but it could be any "sufficiently mixing" ergodic dynamical systems.

7. Multifractal invariant measures on Kleinian groups

A more searching look at Kleinian groups leads to the measures investigated in this book's Part IV, particularly so in its introductory Chapter C19.

Self-inverse fractals, Apollonian nets, and soap

• *Chapter foreword (2003).* To Chapter 18 of M 1982F, this chapter adds a related figure from Chapter 20. Changing from the unusual format of M 1982F to this book's conventional format entailed a few nonlinear rearrangements and called for wording to be inserted for continuity or consistence. The last page of the original text summarized M 1983m{C18}; it was deleted. The art was downsized and the captions were integrated into the text. •

THE BULK [OF M 1982F] IS DEVOTED TO FRACTALS that are either fully invariant under similitudes or, at least, "nearly" self-similar.. As a result, the reader may have formed the impression that the notion of fractal is wedded to self-similarity.. Such is emphatically *not* the case, but fractal geometry must begin by dealing with the fractal counterparts of straight lines... call them "linear fractals."

Chapters 18 and 19 [of M 1982F] take the next step. They sketch the properties of fractals that are, respectively, the smallest sets to be invariant under geometric inversion, and the boundaries of the largest bounded sets to be invariant under a form of squaring.

Both families differ fundamentally from the self-similar fractals. Appropriate linear transformations leave scaling fractals invariant, but in order to generate them, one must specify a generator and diverse other rules. On the other hand, the fact that a fractal is "generated" by a nonlinear transformation, often suffices to determine, hence generate, its shape. Furthermore, many nonlinear fractals are bounded, i.e., have a built-in finite outer cutoff $\Omega < \infty$. Those who find $\Omega = \infty$ objectionable ought to be enchanted by its demise.

The first self-inverse fractals were introduced in the 1880's by Henri Poincaré and Felix Klein, not long after the discovery by Weierstrass of a continuous but not differentiable function, roughly at the same time as the Cantor sets, and well before the Peano and Koch curves and their scaling kin. The irony is that scaling fractals found a durable niche as material for well-known counterexamples and mathematical games, while &slfinverse. fractals became a special topic of the theory of automorphic functions. This theory was neglected for a while, then revived in a very abstract form. One reason why the self-inverse fractals were half-forgotten is that their actual shape has remained unexplored until the present chapter, wherein an effective new construction is exhibited.

The chapter's last section tackles a problem of physics in which the key structure happens to be the simplest self-inverse fractal.

Biological form and "simplicity"

The fact that many nonlinear fractals "look organic" motivates the present aside concerned with biology. Biological form being often very complicated, it may seem that the programs that encode this form must be very lengthy.

However, the complications in question are often highly repetitive in their structure. We may recall from the end of Chapter 6 [of M 1982F] that a Koch curve must *not* be viewed as either irregular or complicated, because its generating rule is systematic and simple. The key is that the rule is applied again and again, in successive loops. Chapter 17 [of M 1982F] extends this thought to the pre-coding of the lung's structure.

Chapters 18 and 19 [of M 1982F—C4 and C16 in this book] we go much further and find that some fractals generated using nonlinear rules recall either insects or cephalopods, while others recall plants. The paradox vanishes, leaving an incredibly hard task of actual implementation.

{P.S. 2003. This conspicuous reference to biological form has inspired many authors, including some who develop it with more extreme zeal than the presently available evidence justifies.}

Standard geometric inversion

After the line, the next simplest shape in Euclid is the circle. And the property of being a circle is not only preserved under similitude, but also under inversion. Many scholars have never heard of inversion since their early teens, hence the basic facts bear being restated. Given a circle C of origin O and radius R, inversion with respect to C transforms the point P into P' such that P and P' lie on the same half line from O, and the lengths $|OP|$ and OP' satisfy $|OP|\,|OP'| = R^2$. Circles containing O invert into straight lines not containing O, and conversely (see below). Circles not containing O invert into circles (third figure below). Circles orthogonal to C, and straight lines passing through O, are invariant under inversion in C (fourth figure).

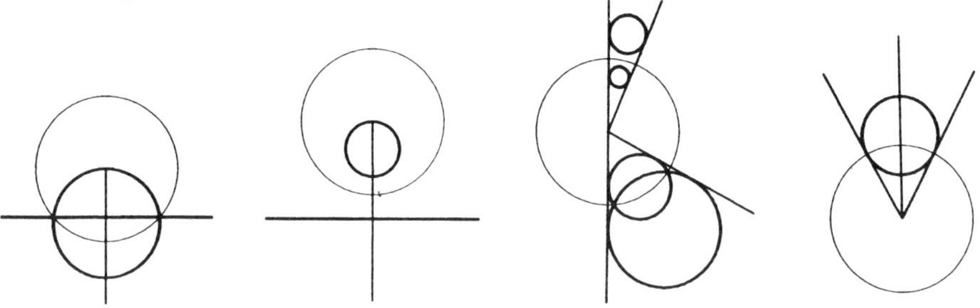

Now consider jointly the three circles C_1, C_2, and C_3. Ordinarily, for example when the open bounded discs surrounded by the C_m are non-overlapping, there exists a circle Γ orthogonal to every C_m, see above. When Γ exists, it is jointly self-inverse with respect to the C_m.

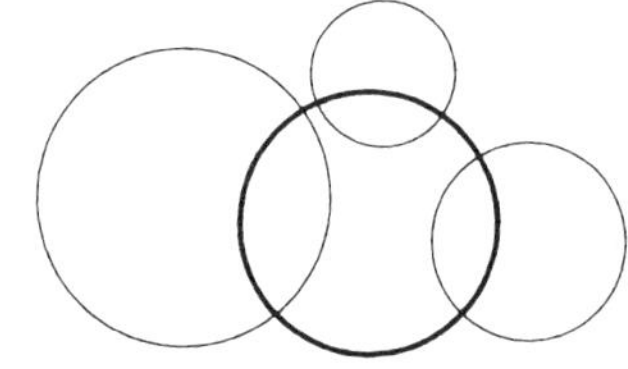

The preceding bland results nearly exhaust what standard geometry has to say about self-inverse sets. Other self-inverse sets are fractal, and most are anything but bland.

Generator. Self-inverse sets. As usual, we begin with a *generator*, which is in the present case made up of any number M of circles C_m. The transformations made of a succession of inversions with respect to these circles form what algebraists call the group generated by these inversions; call it $\mathcal{G}$. The formal term for "self-inverse set" is "a set invariant under the operations of the group $\mathcal{G}$.

Seeds and clans. Take any set $\mathcal{S}$ (call it *a seed*), and add to it the transforms of $\mathcal{S}$ by all the operations of $\mathcal{G}$. The result, to be called here the *clan* of $\mathcal{S}$, is self-inverse. But it need not deserve attention. For example, if $\mathcal{S}$ is the extended plane $\mathbb{R}^{\times}$ (the plane $\mathbb{R}^{\times}$ plus the point at infinity), the clan of $\mathcal{S}$ is identical to $\mathbb{R}^{\times} = \mathcal{S}$.

Chaotic inversion groups. Furthermore, given a group $\mathcal{G}$ based upon inversions, it may happen that the clan of every domain $\mathcal{S}$ covers the whole plane. If so, the self-inverse set must be the whole plane. For reasons that transpire in Chapter 20 [of M 1982F] I propose that such groups be called *chaotic*. The nonchaotic groups are due to Poincaré, but are called Kleinian: Poincaré had credited some other work of Klein's to L. Fuchs, Klein protested, Poincaré promised to label his next great discovery after Klein — and he did!

Keeping to nonchaotic groups, we discuss three self-inverse sets singled out by Poincaré, then a fourth set of uncertain history, and a fifth set whose importance I discovered.

Hyperbolic tessellation or tiling

Few of Maurits Escher's admirers know that this celebrated draftsman's inspiration often came straight from "unknown" mathematicians and physicists (Coxeter 1979). In many cases, Escher added decorations to self-inverse tessellations known to Poincaré and illustrated extensively in Fricke & Klein 1897.

These sets, to be denoted by $\mathcal{T}$, are obtained by merging the clans of the circles C_m themselves.

$\mathcal{G}$ being assumed nonchaotic, the complement of the merged clans of the C_m is a collection of circular polygons called "open tiles." Any open tile (or its closure) can be transformed into any other open (closed) tile by a sequence of inversions belonging to $\mathcal{G}$. In other words, the clan of any closed tile is $\mathbb{R}^{\times}$. More important, the clan of any open tile is the complement of $\mathcal{T}$. And $\mathcal{T}$ is, so to speak, the "grout line" of these tiles. $\mathbb{R}^{\times}$ is self-inverse $\mathcal{T}$ and the complement of $\mathcal{T}$ are self-inverse and involve a "hyperbolic tiling" or "tessellation" of $\mathbb{R}^{\times}$. (The Latin *tessera* = a square came from the Greek word for "four," but tiles can have any number of corners greater than 2.) In Escher's drawings, each tile bears a fanciful picture.

An inversion group's limit set

The most interesting self-inverse set is the smallest one. It is called the limit set, and denoted by $\mathcal{L}$, because it is also the set of limit points of the transforms of any initial point under operations of the group $\mathcal{G}$. It belongs to the clan of any seed $\mathcal{S}$. To make a technical point clearer: it is the set of those limit points that cannot also be attained by a finite number of inversions. Intuitively, it is the region where infinitesimal children concentrate.

$\mathcal{L}$ may reduce to a point or a circle, but in general it is a fragmented and/or irregular fractal set.

$\mathcal{L}$ stands out in a tessellation, as the "set of infinitesimally small tiles." It plays, with respect to the finite parts of the tessellation, the role the branch tips Chapter 16 play with respect to the branches. But the situation is simpler here: like $\mathcal{L}$, the tesselation $\mathcal{T}$ is self-inverse *without* residue.

Apollonian nets and gaskets

A set $\mathcal{L}$ is to be called *Apollonian* if it is made of an infinity of circles plus their limit points. In this case, its being fractal is solely the result of fragmentation. This case was understood (though in diffuse fashion) at an early point of the history of the subject.

First we construct a basic example, then show it is self-inverse. Apollonius of Perga was a Greek mathematician of the Alexandrine school circa 200 B.C. and close follower of Euclid, who discovered an algorithm to draw the five circles tangent to three given circles. When the given circles are mutually tangent, the number of Apollonian circles is two. As will be seen momentarily, there is no loss of generality in assuming that two of the given circles are exterior to each other but contained within the third, as seen to the right.

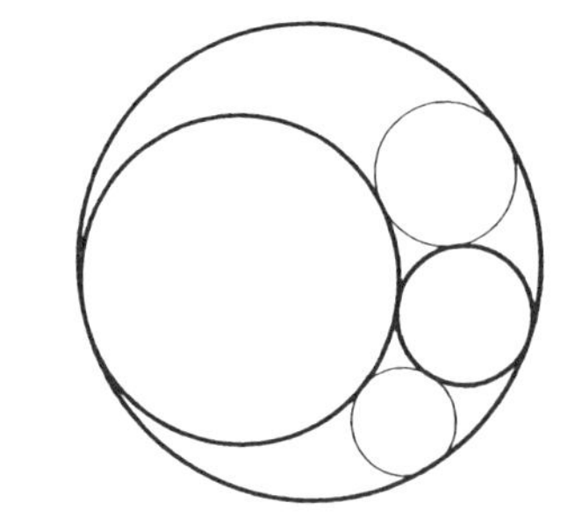

These three circles define two circular triangles with angles of 0°. And the two Apollonian circles are the largest circles inscribed in these triangles, as seen below to the right.

The Apollonian construction concludes with five circles, three given and two Apollonian, which together define six circular triangles. Repeating the same procedure, we draw the largest inscribed circle in each triangle. Infinite further repetition is called

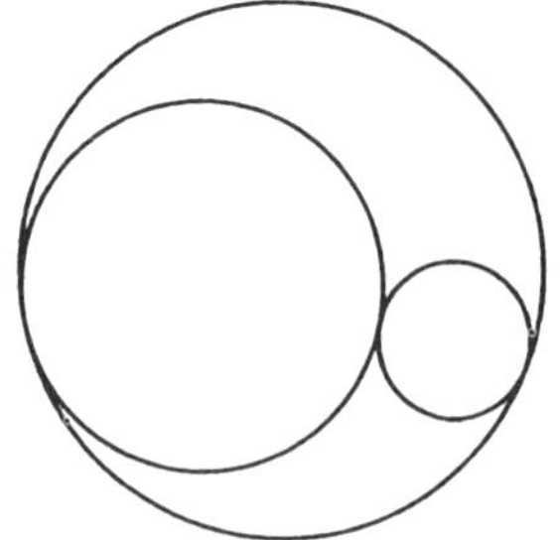

Apollonian packing. To the resulting infinite collection of circles one adds its limit points, and one obtains a set I call *Apollonian net.* A portion of net within a circular triangle, as exemplified below, is to be called *Apollonian gasket.* If a first generation Apollonian circle is exchanged for either of the inner given circles, the limit set is unchanged. If said Apollonian circle is made to replace the outer given circle, the construction starts with three given circles exterior to each other, and one of the first stage Apollonian circles is the smallest circle *circumscribed* to the three given circles. After this atypical stage, the construction proceeds as above, proving that our figures involve no loss of generality.

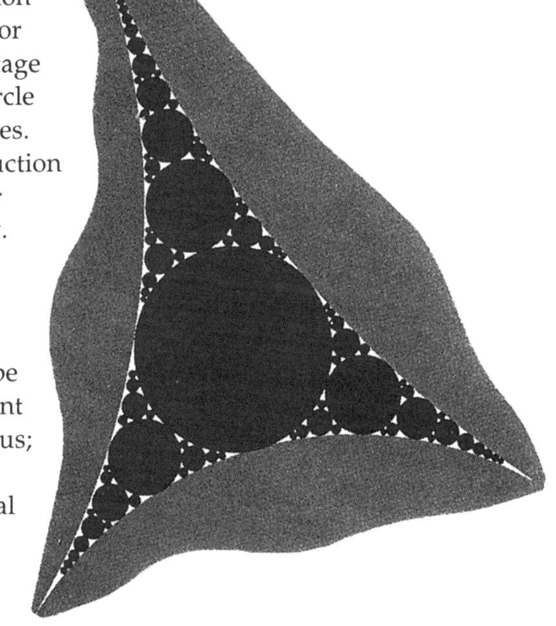

Leibniz packing. Apollonian packing recalls a construction that Leibniz described it in a letter to de Brosses: "Imagine a circle; inscribe within it three other circles congruent to each other and of maximum radius; proceed similarly within each of these circles and within each interval between them, and imagine that the process continues to infinity...."

Apollonian nets are self-inverse

Let us now return to the starting point of the construction of Apollonian net: three circles tangent to each other. Add *either one* of the corresponding Apollonian circles, and call the resulting 4 circles Γ circles. To the right, they are shown by bold curves.

There are 4 combinations of the Γ circles 3 by 3, to be called triplets, and to each corresponds a circle orthogonal to each circle in the triplet. We take these new circles as our generator, we label them as C_1, C_2, C_3 and C_4, (the diagram below shows them as thin curves), and the Γ circle orthogonal to

C_i, C_j and C_k will be labeled as Γ_{ijk}.

Having set these tedious labels, here is the payoff: Simple inspection shows that the smallest (closed) self-inverse set with respect to the 4 generating circles C_m is the Apollonian net constructed on the 4 circles Γ. Curiously, this observation is nowhere explicit in the literature I know.

A more careful inspection shows that each circle in the net transforms into one of the Γ circles through a *unique* sequence of inversions with respect to the C circles. In this way, the circles in the Apollonian net can be sorted out into 4 clans; the clan descending from Γ_{ijk} will be denoted as $\mathcal{G}\Gamma_{ijk}$.

Net knitting with a single thread

The Apollonian gasket and the Sierpinski gasket of Plate 141 [of M 1982F] share an important feature: the complement of the Sierpinski gasket is a union of triangles, a σ-triangle, and the complement of an Apollonian net or gasket is a union of discs, a σ-disc.

But we also know that the Sierpinski gasket admits of an alternative Koch construction, in which finite approximations are teragons (broken lines) without self-contact, and double points do not come in until one goes to the limit. This shows that the Sierpinski gasket can be drawn without ever lifting the pen; the line will go twice over certain points but will never go twice over any interval of line.

To change metaphors, the Sierpinski gasket can be knitted with a single loop of thread!

The same is true of the Apollonian net.

Non-self-similar cascades, and the evaluation of the dimension

The circular triangles of Apollonian packing are *not* similar to each other, hence the Apollonian cascade is not self-similar, and the Apollonian net is not a scaling set. One must resort to the Hausdorff Besicovitch definition of D (as exponent used to define measure), which applies to every set, but the derivation of D proves surprisingly difficult. Boyd 1973a,b shows that

$$1.300197 < D < 1.314534,$$

but Boyd's latest (unpublished) numerical experiments yield $D \sim 1.3058$. {P.S. 2003. The exact value remains unknown. Thomas & Dhar 1994 advances $D = 1.305686729$, and McMullen 1998 quotes $D = 1.305688$. }

In any event, the Apollonian gasket and net are fractal curves. In the present context, D is a measure of fragmentation. When, for example, the discs of radius smaller than ε are "cut off," the remaining interstices have a perimeter proportional to ε^{1-D} and a surface proportional to ε^{2-D}.

$\mathcal{L}$ in non-Fuchsian Poincaré chains

Inversions with respect to less special configuration of the generating circles C_m, lead to self-inverse fractals that are less simple than any Apollonian net. A workable construction of mine, to be presented momentarily, characterizes $\mathcal{L}$ suitably in most cases. It is a great improvement over the previous method, due to Poincaré and Klein, which is cumbersome and converges slowly.

But the older method remains important, so let us go through it in a special case. Let the C_m form a configuration one may call *Poincaré chain,* namely a collection of M circles C_m numbered cyclically, so that C_m is tangent to C_{m-1} and to C_{m+1} (modulo M, and intersects no other circle in the chain. In that case, $\mathcal{L}$ is a *curve* that separates the plane into an inside and an outside. (As homage to Camille Jordan, who first saw that it is not obvious that the plane can thus be subdivided by a single loop, such loops are called Jordan curves.)

When all the C_m are orthogonal to the same circle Γ, $\mathcal{L}$ is identical to Γ. This case, called Fuchsian, is excluded in this chapter.

Poincaré's construction of $\mathcal{L}$

The customary construction of $\mathcal{L}$ and my alternative will be fully described in the case of the following special chain with $M=4$:

To obtain $\mathcal{L}$, Poincaré and Fricke & Klein 1897 replace the original chain, in stages, by chains made of an increasing number of increasingly small links. The first stage replaces every link C_i by the inverses in C_i of the links C_m other than C_i, thus creating $M(M-1)=12$ smaller links. The top horizontal panel of Figure 1 shows them superimposed on the original links in white on gray background. Each later stage starts with a chain and inverts it in each of the original C_m. Here several stages are shown in black, each being superposed on the preceding one in white on gray background. Ultimately, the chain thins out to its thread, which is $\mathcal{L}$.

Unfortunately, some links remain of substantial size after large numbers of stages, and even fairly advanced approximate chains give a poor idea of of $\mathcal{L}$. This difficulty is exemplified in horrid fashion in M 1983m{C18}.

FIGURE C16-1. [Composite of page 173 and Plates 177 and 43 of M 1982F] Two constructions of the limit set. The top panel shows the classical one (Poincaré), the middle panel shows a proposed alternative. The latter recalls the Cesaro construction of the Koch snowflake shown in the lower panel.

The notion of fractal osculation

My alternative construction of $\mathcal{L}$ involves a new fractal notion of osculation that extends an obvious facet of the Apollonian case.

Standard osculation. This notion is linked to the concept of curvature. To the first order, a standard curve near a regular point P is approximated by the tangent straight line. To the second order, it is approximated by the circle, called *osculating*, that has the same tangent and the same curvature.

To index the circles tangent to the curve at P, a convenient parameter, u, is the inverse of the (arbitrarily oriented) distance from P to the circle's center. Write the index of the osculating circle as u_0. If $u < u_0$, a small portion of curve centered at P lies entirely on one side of the tangent circle, while if $u < u_0$ it lies entirely on the other side.

This u_0 is what physicists call a *critical value* and mathematicians call a *cut*. And $|u_0|$ defines the local "curvature."

Global fractal osculation. For the Apollonian net, the definition of osculation through the curvature is meaningless. However, at every point of the net where two packing circles are tangent to each other, they obviously "embrace" the rest of $\mathcal{L}$ between them. It is tempting to call *both* of them *osculating*.

To extend this notion to a non-Apollonian sets $\mathcal{L}$, we take a point where $\mathcal{L}$ has a tangent, and start with the definition of ordinary osculation based on criticality (= cut). The novelty is that, as u varies from $-\infty$ to $+\infty$, the single critical u_0 is replaced by two distinct values, u' and $u'' > u'$, defined as follows: For all $u < u'$, $\mathcal{L}$ lies entirely to one side of our circle, while for all $u < u''$, $\mathcal{L}$ lies entirely to the other side, and for $u' < u < u''$, parts of $\mathcal{L}$ are found on both sides of the circle. I suggest that the circles of parameters u' and u'' *both* be called *fractally osculating*.

Any circle bounds two open discs (one includes the circle's center, and the other includes the point at infinity). The open discs bounded by the osculating circles and lying outside $\mathcal{L}$ will be called osculating discs.

It may happen that one or two osculating circles degenerate to a point.

Local versus global notions. Returning to standard osculation, we observe that it is a local concept, since its definition is independent of the curve's shape away from P. In other words, the curve, its tangent, and its osculating circle may intersect at any number of points in addition to P. By contrast, the preceding definition of fractal osculation is global, but this distinction is not vital. Fractal osculation may be redefined locally, with a corresponding split of "curvature" into 2 numbers. However, in the application at hand, global and local osculations coincide.

Osculating triangles. Global fractal osculation has a counterpart in a familiar context. To define the interior of our old friend the Koch snowflake curve K as a sigma-triangle (σ-triangle), it suffices that the triangles laid at each new stage of Plate 42 [of M 1982F] be lengthened as much as is feasible without intersecting the snowflake curve.

σ -discs that osculate $\mathcal{L}$

Osculating discs and σ-discs are the key of my new construction of $\mathcal{L}$, which is free from the drawbacks [encountered by Poincaré]. This construction is illustrated here for the first time (though it was previewed in 1980, in *The 1981 Springer Mathematical Calendar*!). The key is to take the inverses, not of the C_m themselves, but of some of circles Γ_{ijk}, which (as defined on page 171) are orthogonal to triplets C_i, C_j, and C_k. Again, we assume that the Γ_{ijk} are not all identical to a single Γ.

Restriction to $M=4$. The assumption $M=4$ insures that, for every triplet i,j,k, either one or the other of the two open discs bounded by Γ_{ijk} — namely, either its inside or its outside — contains none of the [previously defined] points γ_{mn}. We shall denote this γ-free disc by Δ_{ijk}.

My construction of $\mathcal{L}$, illustrated by the middle horizontal panel of Figure 1, is rooted in the following observations: every γ-free Δ_{ijk} osculates $\mathcal{L}$; so do their inverses and repeated inverses in the circles C_m ; and the clans built using the Δ_{ijk} as seeds cover the whole plane except for the curve $\mathcal{L}$.

Poincaré chain drawn on larger scale

As is true in most cases, the first stage outlines $\mathcal{L}$ quite accurately. Later stages add detail very "efficiently," and after few stages the mind can interpolate the curve $\mathcal{L}$ without the temptation of error present in the Poincaré approach.

Left of the middle panel of Figure 1. In Poincaré chains with $M=4$, at least one of the discs Δ_{ijk}, call it Δ_{123}, is always unbounded and intersects the disc Δ_{341}. (Here, Δ_{341} is also unbounded, but in other cases it is not.) The union of Δ_{123} and Δ_{341}, shown in gray, provides a first approximation of the outside of $\mathcal{L}$. It recalls the first approximation of the outside of Koch's $\mathcal{K}$ by the regular convex hexagon in the bottom panel.

The discs Δ_{234} and Δ_{412} intersect, and their union, shown in black, provides a first approximation of the inside of $\mathcal{L}$. It recalls the approximation of the inside of $\mathcal{K}$ by the two triangles that form the regular star hexagon to the left of the bottom panel.

Middle figure of the middle panel of Figure 1. A second approximation of the outside of $\mathcal{L}$ is achieved by adding to Δ_{123} and Δ_{341} their inverses in C_4 and C_2, respectively. The result, shown in gray, recalls the second approximation of the outside of $\mathcal{K}$ in the bottom panel.

The corresponding second approximation of the inside of $\mathcal{L}$ is achieved by adding to Δ_{234} and Δ_{412} their inverses in C_1 and C_3, respectively. The result, shown in black, is analogous to the second approximation of the inside of $\mathcal{K}$ to the left of the bottom panel.

Right figure of the middle panel of Figure 1. The outside of $\mathcal{L}$, shown in gray, is the union of the clans of Δ_{123} and Δ_{341}. And the inside of $\mathcal{L}$, shown in black, is the union of the clans of Δ_{234} and Δ_{412}. Together, the black and gray open regions cover the whole plane, minus $\mathcal{L}$. The fine structure within $\mathcal{L}$ is seen for a different Poincaré chain in M 1983m{C18}.

Generalizations

Chains with five or more links. When the number of original links in a Poincaré chain is $M > 4$, my new construction of $\mathcal{L}$ involves an additional step: it begins by sorting the Γ circles into 2 bins. Some Γ circles are such that *each* of the open discs bounded by Γ contains at least one point γ_{mn} ; as a result, Δ_{ijk} is *not* defined. Such Γ circles intersect $\mathcal{L}$ instead of osculating it. But they are not needed to construct $\mathcal{L}$.

The remaining circles Γ_{ijk} define osculating discs Δ_{ijk} that fall into two classes. Adding up the clans of the Δ_{ijk} in the first class, one represents the interior of $\mathcal{L}$, and adding up the clans of the Δ_{ijk} in the second class, one represents the exterior of $\mathcal{L}$.

The same is true in many (but not all) cases when the C_m fail to form a Poincaré chain.

Overlapping and/or disassembled chains. When C_m and C_n have two intersection points γ'_{mn} and γ''_{mn}, these points jointly replace γ. When C_m and C_n are disjoint, γ is replaced by the two mutually inverse points γ'_{mn} and γ''_{mn}. The criterion for identifying Δ_{ijk} becomes cumbersome to state, but the basic idea is unchanged.

Ramified self-inverse fractals. $\mathcal{L}$ may borrow features from both a crumpled loop (Jordan curve), and an Apollonian net, yielding a fractally ramified curve akin to those examined in Chapter 14 [of M 1982F], but often much more baroque in appearance, as in Plate C7.

Self-inverse dusts. It may also happen that $\mathcal{L}$ is a fractal dust.

{P.S. 2003. An example of limit set as attractor of an IFS or "decomporable dynamical system"

The bulk of this chapter reprints Chapter 18 of M 1982F but Plate 199 of Chapter 20 concerns the closely related "Pharaoh's Breastplate." It is reproduced here as Figure 2. M1999N used a color version as its cover and added an explanatory diagram that is also shown in Figure 2. Its six circles are arranged to insure that the limit set is a collection of circles. Choosing those circles with prescribed probabilities creates what Chapter 20 of M 1982F called a "decomposable dynamical system" but is now called an IFS.}

The Apollonian model of smectics

This section outlines the part that Apollonian packing and fractal dimension play in the description of a category of "liquid crystals." In doing so, we cast a glance toward one of the most active areas of physics, the theory of *critical points.* An example is the "point" on a temperature-pressure diagram that describes the physical conditions under which solid, liquid, and gaseous phases can coexist at equilibrium in a single physical system. The analytic characteristics of a physical system in the neighborhood of a critical point are scaling, therefore governed by power laws, and specified by critical exponents (Chapter 36 [of M 1982F]). Many of them turn out to be fractal dimensions; the first example is encountered here.

We describe liquid crystals by paraphrasing Bragg 1934. These beautiful and mysterious substances are liquid in their mobility and crystalline in their optical behavior. Their molecules are relatively complicated structures, lengthy and chain-like. Some liquid crystal phases are called *smectic,* from the Greek σμηγμα signifying soap, because they constitute a model of a soap-like organic system. A smectic liquid crystal is made of molecules that are arranged side by side like corn in a field, the thickness of the layer being the molecules' length. The resulting layers or sheets are very flexible and very strong and tend to straighten out when bent and then released. At low temperatures, they pile regularly, like the leaves of a book, and form a solid crystal. When temperatures rise, however, the sheets become able to slide easily on each other. Each layer constitutes a two-dimensional liquid.

Of special interest is the focal conics structure. A block of liquid crystal separates into two sets of pyramids, half of which have their bases on one of two opposite faces and vertices on the other. Within each pyramid, liquid crystal layers fold to form very pointed cones. All the cones have the same peak and are approximately perpendicular to the plane. As a result, their bases are discs bounded by circles. Their minimum radius ε is the thickness of the liquid crystal's layers. Within a spatial domain such as a square-based pyramid, the discs that constitute the bases of the cones are distributed over the pyramid's base. To obtain an equilibrium distribution, one begins by placing in the base a disc of maximum radius. Then another disc with as large a radius as possible is placed within each of the four remaining pieces, and so on and so forth. Proceeding without end would achieve exact Apollonian packing.

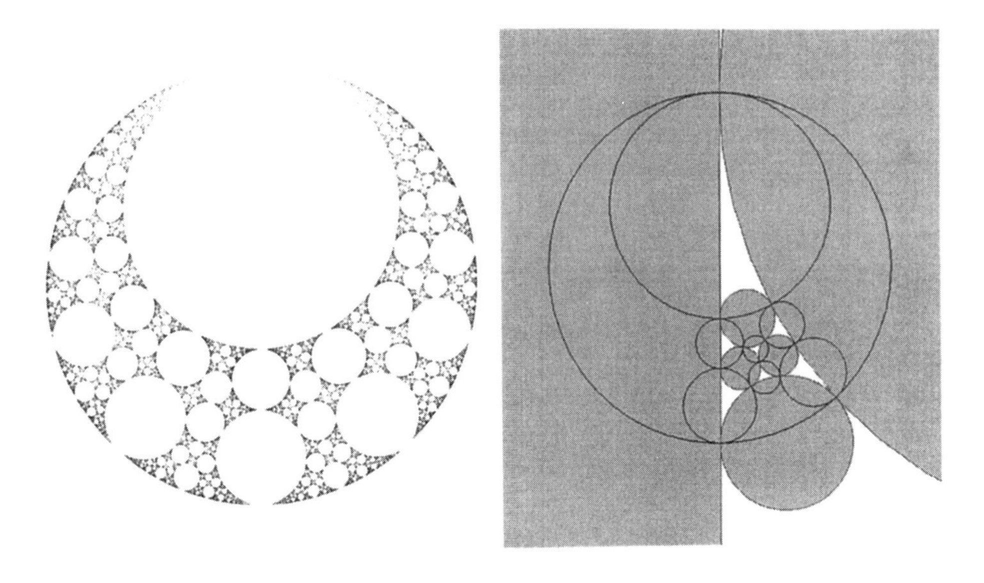

FIGURE C16-2. [Plate 199 of M 1982F and page 129 of M1999N, elaborated upon.] The "generator" part of the diagram consists in six circles filled in gray. The inversions with respect to those circles, when combined with prescribed probabilities, define a "decomposable dynamical system" also called IFS. The plot of 64000 points is an orbit of that system (the first few points were omitted). This orbit traces a self-inverse fractal but its cusps fill very slowly. (In simplified form, Part V investigates this slowness for the Minkowski measure.) The new algorithm described in this chapter relies on the diagram's remaining eight bold circles. In the decorative "Pharaoh Breastplate" on p. 170, four of those circles and their successive inverses, are represented by four kinds of "semi-precious stones."

The physical properties of this model of soap depend upon the surface and perimeter of the sum of interstices. The link is affected through the fractal dimension D of a kind of photographic "negative," namely, the gasket that the molecules of soap fail to penetrate. Details of the physics are in Bidaux, Boccara, Sarma, Sèze, de Gennes, & Parodi 1973.

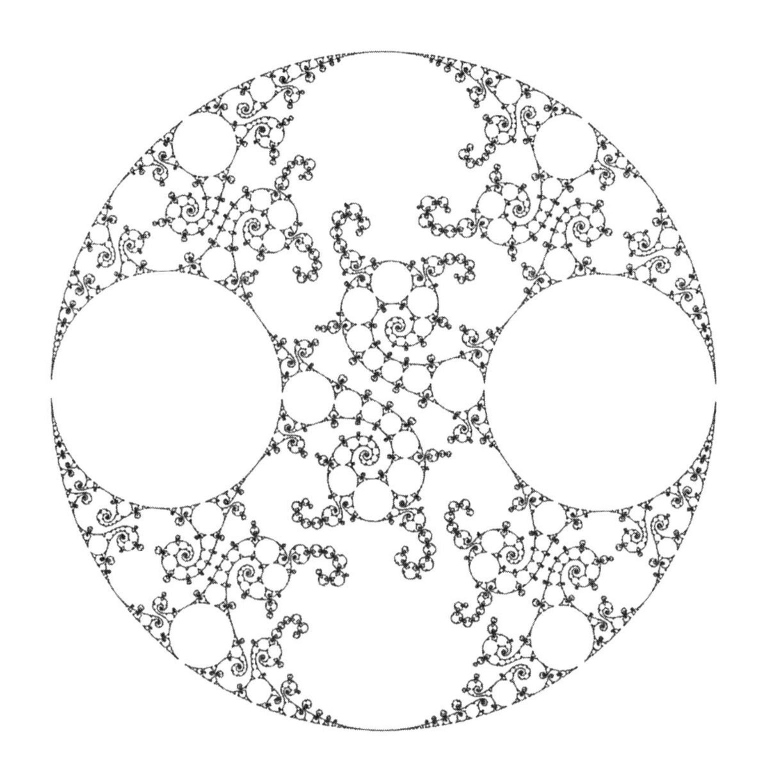

FIGURE C16-3. [Plate 178 of M1982F]. A self-homographic fractal due to David Mumford. To the mathematician, the main interest of groups based upon inversions resides in their relation with certain groups of homographies. An homography (also called Möbius, or fractional linear transformation) maps the z-plane by $z \to (az+b)/(cz+d)$, where $ad-bc=1$. The most general homography is the product of an inversion, a symmetry with respect to a line (which is a degenerate inversion), and a rotation. In the absence of rotation, the study of homographies learns much from the study of groups based on inversions, but allowing rotations brings in new riches.

This almost plane-filling figure is the limit set $\mathcal{L}$ for a group of homographies. Mumford devised it in 1982 in the course of investigations inspired by this chapter and kindly allowed its publication here.

Symmetry
Eds. I. Hargittai & T. Laurent, 2002

C17

Symmetry with respect to several circles: dilation/reduction, fractals, and roughness

• *Chapter foreword (2003).* This non-technical text serves two purposes.

The second half uses IFS to illustrate approximate self-similarity for Kleinian limit sets and for measures rather than to sets.

The first half is an easy introduction to fractals that should help at least a few readers. It could have been split off and "segregated" with C23 into this book's last part, but why bother? •

CHARACTERIZATION OF FRACTALS

Roughly speaking, *fractals* are shapes that look the same from close by and far away. Fractal geometry is famed (or notorious, depending on the source) for the number and apparent diversity of its claims. They range from mathematics, finance, and the sciences, all the way to art.

Such diversity is always a source of surprise, but this article will argue that it can be phrased so as to become perfectly natural. Indeed, the ubiquity of fractals is intrinsically related to their intimate connection with a phenomenon that is itself ubiquitous. That phenomenon is *roughness.* It has long resisted analysis, and fractals provide the first widely applicable key to some degree of mastery over some of its mysteries.

A self-explanatory term for "looking the same from close by and far away" is "self-similar". Of course, self-similarity also holds for the straight line and the plane. These familiar shapes are the indispensable starting point in the many sciences that have perfected a mastery of smoothness. The point is that the property of "looking the same from close by and far away" happens to extend beyond the line and the plane. It also holds for diverse shapes called fractals. I named them, tamed them into primary models of roughness, and helped them multiply.

The learned term to denote them is "forms invariant under *dilation or reduction.*" In the professional jargon of some thoroughly developed fields of science, such invariances are called *symmetries,* which would suffice as an excuse for this paper to be included in this book. But there is another far more direct reason: it will be shown that suitable combinations of very ordinary-looking symmetries hold a surprise for the beholder. Spontaneously, they produce self-similarity, meaning that they yield fractals.

The theme having been embedded in the order of the words in this paper's title, and every word in the title having already been mentioned (at least fleetingly) and italicized, the etymologies of two of those words are worth recalling. In ancient Greek, the scope of *summetria* went beyond a reference to self-examination in a mirror. A close synonym was "in just proportion," and it was mostly used to describe a work of art or music as being "harmonious." In time, most words' meanings multiply and diversify. In developed sciences, groups of invariances-symmetries are abstract and abstruse, but those which underlie fractals are — to the contrary — intuitive and highly visual. This paper will show typical ones to be closely related to the simplest mirror symmetry.

As to the word *fractal,* I coined it on some precisely datable evening in the winter of 1975, from a very concrete Latin adjective, *fractus,* which denoted a stone's shape after it was hit very hard. Lacking time to evolve, *fractal* rarely strayed far from the notion of roughness.

MUNDANE QUESTIONS, OLD AND NEW, AND THE UBIQUITY OF ROUGHNESS IN NATURE OR CULTURE

Poincaré remarked that there are questions that one chooses to ask and others that ask themselves. I list few examples of the latter now, in an intentionally haphazard fashion. The first few concern Nature and demonstrate that a question that had long asked itself without response tends to be abandoned to children.

- How to measure and compare the roughness of ordinary objects such as broken stone, metal, glass, or piece of rusted iron?
- How long is the coast of Britain?

• What shape is the Earth, more precisely a mountain, a coastline, a river's course, or a dividing line between two rivers' watersheds? That is, can the term "geometry" deliver what it seems to promise but does not discuss.

• How to define the speed of the wind during a storm?

• What shape is a cloud, a flame, or a welding?

• What is the density of galaxies in the Universe?

To this list, culture added many other questions recently:

• How to distinguish proper music (old or new, good or bad) from plain awful noise?

• How to measure the variation of the flow of messages on the Internet?

• How to measure the volatility of the prices quoted on financial markets?

The concrete and constructive stream of mathematics adds other questions, more removed from the proverbial "common man and woman."

• How to characterize the boundary between two basins of attraction in a chaotic dynamical system?

• How to characterize the boundary of a plane random walk?

• How to characterize self-avoiding random walks?

• How to characterize the critical clusters of percolation?

In one notable instance of the study of nature, fractal geometry became a tool of choice after it had made discovery possible.

• How to characterize the diffusion limited aggregates?

There is a strong reason for thinking of all those questions together. The objects in the last batch happens to be constructed "artificially" by algorithms that are completely described, in fact, are extraordinarily simple. On the other hand, those objects are extraordinarily complex. Any insight concerning them promises some insight on objects due to Nature or those due to uncontrollable aspects of culture.

The word, *rough,* appears *in only one* of the above questions, but the underlying concept appears *in every one.* (*Irregular* would have been a more elegant word, but rough is more telling.) All these questions were without geometric answer until fractal and multifractal geometry provided the beginnings of a workable and useful approach based on the surprising fact that, both in nature and culture, roughness is very often fractal.

In one field after another, fractal geometry became the first tool which made it possible to help shape a theory of roughness. To provide an immediate example, consider the first among the above questions. The first evidence of the fractality of metal fractures was given in M, Passoja & Paullay 1984 and previously announced in the second and later paintings of M1987f. This has since been extensively confirmed and shown to hold between a very small scale and one that is 100,000 times larger. Price records are multifractal functions. The first evidence in M 1997E has since been extensively confirmed.

An inverse question will provide contrast.

- For which natural or Man-made forms do the simplest smooth shapes of Euclid's geometry provide a sensible approximation?

To Early Man, Nature provided just a few smooth shapes: the path of a stone falling straight down, the full Moon or the Sun hidden by a light haze, small lakes unperturbed by current or wind. In sharp contrast, *homo faber* keeps adding examples beyond counting. For example, Man works hard at eliminating roughness from automobile pistons, flat walls and tabletops, Roman-Chinese-American street grids, and — last but not least — from most parts of mathematics (contributing to the widespread view that geometry is "cold and dry.")

Forgetting this last question, the others set a pattern one may extend forever. The simple reason is that roughness is ubiquitous in Nature. In the works of Man, it may not be welcome, but is not always avoided, and may sometimes be unavoidable. Examples are found in some parts of mathematics, where they were at one time described as "pathological" or "monstrous," and, once again, in the above list of questions.

Needless to say, each of these questions belongs to some specific part of science or engineering, and the practical attitude is not to waste time studying roughness but instead to get rid of it. Fractal geometry, to the contrary, has embraced roughness in all its forms and studies it for its intrinsic interest.

ROUGHNESS LAGGED FAR BEHIND OTHER SENSES, FOR EXAMPLE, SOUND, IN BEING MASTERED BY A SCIENCE

By customary count, a human's number of sense receptors is five. This may well be true but the actual number of distinct sense messages is certainly much higher.

Take sound. Even today, concert hall acoustics are mired in controversy, recording of speech or song is comfortable with vowels but not consonants, and drums are filled with mysteries. Altogether, the science of sound remains incomplete. Nevertheless, it boasts great achievements.

Let us learn some lessons from its success. Typical of every science, it went far by exerting healthy opportunism. Side-stepping the hard questions, it first identified the idealized sound of string instruments and pipes as an "icon" that is at the same time reasonably realistic and mathematically manageable — even simple. Acoustics clarifies even facts it does not characterize or explain. It builds on the "harmonic analysis" of pendular motions and the sine/cosine functions. It identifies the fundamental and a few harmonics then, in due time, builds a full Fourier series. The latter is periodic, that is, translationally invariant, which, in a broad sense, is a property of symmetry. Newton's spectral analysis of light is a related example, although the structure of incoherent white light took until the 1930s to be clarified.

More generally, the harmonies that Kepler saw in the planets' motions have largely been discredited, yet it remains very broadly the case that the starting point of every science is to identify harmonies in a raw mess of evidence.

SCALE INVARIANCE PERCEIVED AS PLAYING FOR ROUGHNESS THE ROLE THAT HARMONIOUS SOUND PLAYED IN ACOUSTICS

By contrast to acoustics, the study of roughness could not, until very recently, even begin tackling the elementary questions this paper listed earlier. My contribution to science can be viewed as centered on the notion that, like acoustics, the study of roughness could not seriously become a science without first taking the following step: it had to identify a basic invariance/symmetry, that is, a deep source of harmony common to many structures one can call rough.

Until the day before yesterday, roughness and harmony seemed antithetic. An ordering of deep human concerns, from exalted to base, would have surely placed them at opposite ends. But change has a way of scrambling up all rankings of this sort. As candidates for the role of harmonious roughness, I proposed the shapes whose roughness is invariant under dilation/reduction.

In the most glaring irony of my scientific life, this first-ever systematic approach to roughness arose from a thoroughly unexpected source: in extreme mathematical esoterica. This may account for the delay science experienced in mastering roughness.

MEEK SYMMETRIES

In the form known to everyone, the concept of symmetry tends to provoke love or loathing. My own feelings depend on the context. I dislike symmetric faces and rooms but devote all my scientific life to fractals. Let us now move indirectly, step by step, towards examples of fractals.

Among symmetries, the most widely known and best understood involves the relation between an object and its reflection in a perfectly flat mirror. To obtain an object that is invariant — unchanged — under reflection, it suffices to "symmetrize" a completely arbitrary object by combining it with its mirror reflection. As a result, there is an infinity of such objects.

A fairgrounds carousel helps explain symmetry with respect to a vertical axis. Symmetry with respect to a point is not much harder.

Now, replace the symmetry with respect to one mirror by symmetries with respect to two parallel mirrors. Any object can be symmetrized "dynamically," by being reflected in the first mirror then the second, then again the first, again the second, and so on ad infinitum. The object grows without end and its limit is doubly invariant by mirror symmetry. In addition, it becomes unchanged under translation, in either direction, whose size is a multiple of twice the distance between the planes. Like for a single mirror, the defining constraint allows an infinite variety of such objects.

Symmetry with respect to a circle is a little harder but will momentarily become crucial. One begins with an object that is symmetric with respect to a line. Then one transforms the whole plane by an operation called geometric inversion, which is a very natural generalization of the transformation of x into its algebraic inverse $1/x$. One finds that geometric inversion transforms (almost) any line into a circle. When the plane has then inverted in this fashion, an object that used to be symmetric with respect to the original line is said to have become symmetric with respect to the inverse of the line, namely, a circle.

FROM MEEK TO WILD SYMMETRIES AND ON TO SELF-SIMILARITY

So far, so good. Very elementary French school geometry used to be filled with examples of this kind. Over several grades, everything grew increasingly complex and harder, but only very gradually. Special professional periodicals made believe that they were working on an endless frontier, but it was clear that this old geometry was actually exhausted, dead.

All too many persons hastened to conclude that *all* of more or less visual geometry was dead. Actually, the germ of very different but very visual developments existed since the 1880s(!) in the work of Henri Poincaré. But it remained dormant until the advent of computer graphics and my work, as exemplified in what follows. Because of fractal geometry, Poincaré's idea now matters broadly and will provide us with a nice transition from the simplest symmetry to self-similar roughness. In the next section, the meaning of roughness is a bit stretched; in the section after the next, almost natural. Figures 1 • Figures 2 • Figure 3

THE SELF-SIMILAR "THRICE-INVARIANT DUST $\mathcal{D}$," WITH THREE PART GENERATOR SYMMETRY

The technical meaning of "dust" will become clear as we go on. Consider, in the plane, a diagram to be called a three-part "generator" that combines two parallel lines and a circle half-way between them, as shown to the left of Figure 1. Could a geometric shape be simultaneously symmetric with respect to *each* of the three parts of this generator? If this is true of more than one shape, could one identify the smallest, to be called $\mathcal{D}$? Painfully learned "intuition" tempted great thinkers to propose that one should measure the complexity of a notion by the length of the shortest defining formula or sentence. This line of thought would have implied that when the shortest defining formula doubles in length, the corresponding study becomes twice more complex.

Examined in this light, our present generator seems to involve only a small step beyond its separate parts. It seems innocuous. Fortunately for us, but unfortunately for that wrong-headed definition of complexity, this very simple combination of inversions will momentarily bring a great surprise. It will prove to involve a jump across the colossal chasm that separates consideration of very elementary geometric symmetry from the great complexity of fractals.

Both for those who do and those who do not (yet) know much about fractals, the least contrived method to understand them is to take advantage of the computer graphics technology that made all this possible.

The words "minimal thrice-invariant" in the section title might create the fear that $\mathcal{D}$ is some needle in a haystack, but the precise contrary is the case. The tactic behind the search for $\mathcal{D}$ is hardly more complicated than the "dynamics" that we used to create a mirror symmetric set: start with an arbitrary object, then add its mirror image, and so on, over an infinite number of times. The novelty is double: the dynamics that searches for a

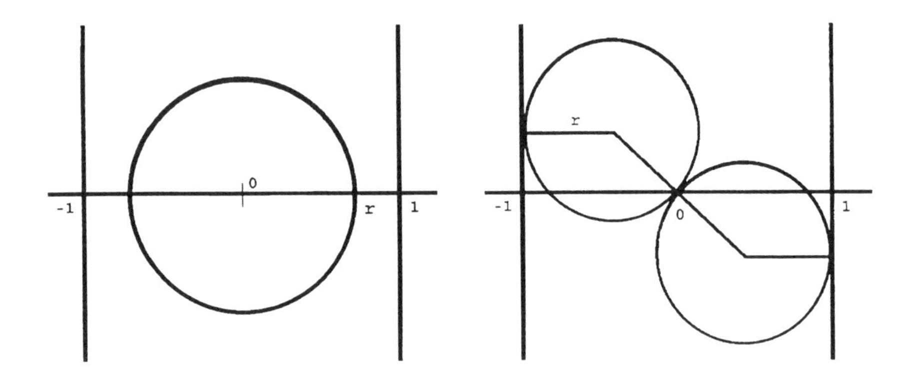

FIGURE C17-1. Figures 2 and 3 illustrate two bridges between, on the one hand, the simplest symmetries (with respect to a line or a circle, the latter being a geometric inversion) and, on the other hand, self-similarity, that is, fractality. Both figures are constructed point by point by an orbit that is an infinite random sequence of never repeating symmetries.

The generator of the dust $\mathcal{D}$ of Figure 2, shown to the left, is made of two parallel lines and, between them, a circle of radius r. The value taken for $1 - r$ is positive, which implies that there exist genuine empty gaps between the points; hence $\mathcal{D}$ is a "totally disconnected set;" I called it a dust. In Figure 2, $1 - r$ is very small, but, for the sake of clarity, the diagram to the left exaggerates it.

The generator of the curve $\mathcal{C}$ of Figure 3, shown to the right, is made of two parallel lines and two circles. Because those circles and lines are — as shown — tangent to one another the orbit merges into a gap-free curve, but does so very slowly.

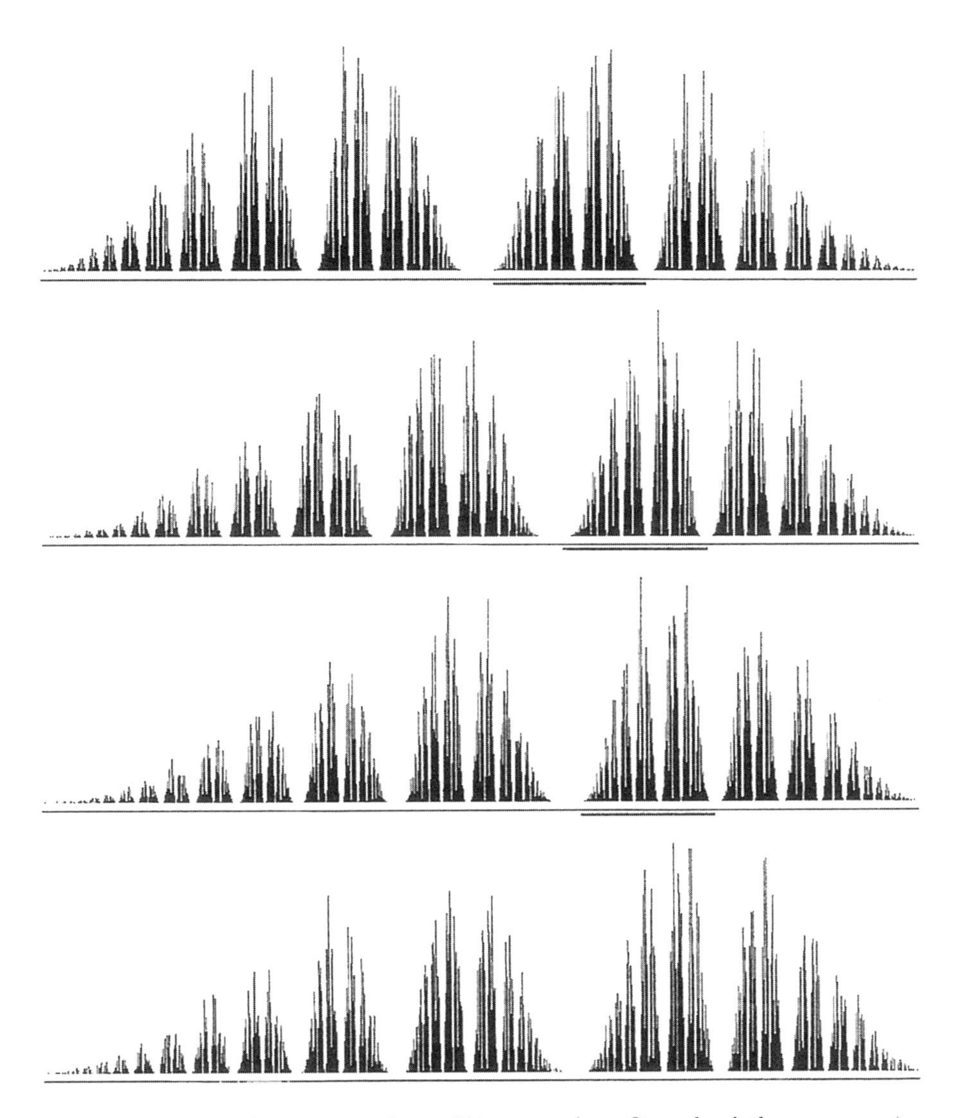

FIGURE C17-2. Histograms of a self-inverse dust $\mathcal{D}$ and of three successive "blowups." This $\mathcal{D}$ (top panel) is entirely located within the horizontal axis. The underlined portion of the top panel is blown up in the second, and so on.

To represent a dust, the best is to draw a "histogram:" divide the axis into small "bins" and represent by a vertical bar the number of points in each bin (this is the fractal counterpart of density). The point is that — except for small deformations — the blown-up histograms of this construction have very much the same form. More precisely, each panel from the top down represents an orbit whose length (therefore "density") increases. To reveal increasing detail, the overall diagram is spread horizontally and the bins are narrowed to keep their number constant.

symmetrized $\mathcal{D}$ is irresistibly "attracted" to its prey and the prey is sharply specific and extraordinarily complex.

The century-old process that yields $\mathcal{D}$ as a limit set is now called "chaos game." To appreciate what is happening with minimal notation and programming effort, it is helpful to know that $\mathcal{D}$ is entirely contained in the "horizontal" axis, which is defined as the line that crosses the generating circle's center of abscissa 0 and is perpendicular to the two generating lines of abscissas −1 and 1. Therefore, the generating circle's radius being denoted by $r < 1$, inversion simply transforms x into r^2/x.

Altogether, the points symmetric of x with respect to the generating lines and circle have the abscissas $-x-2, -x+2$, and r^2/x, respectively.

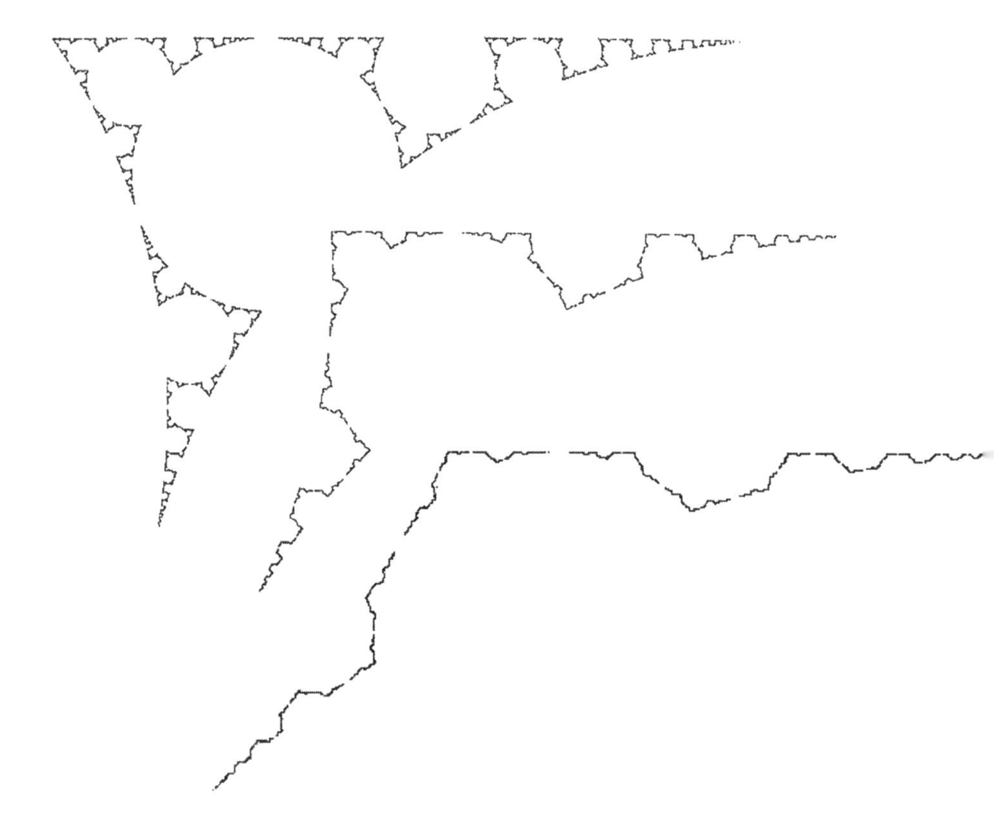

FIGURE C17-3. Three variants of a self-inverse curve $\mathcal{C}$. Contrary to the dust $\mathcal{D}$ on Figure 2, these three variants of curve $\mathcal{C}$ winds up and down and around. The "density" would have to be drawn along an axis orthogonal to the plane, therefore was omitted. While the gaps of $\mathcal{D}$ are real, those perceived in $\mathcal{C}$ are artifacts; in a longer orbit, their lengths decrease to zero.

Those formulas' simplicity is hard to beat, but the object grown from this simple seed and will soon be revealed as extremely complex.

Recall the symmetrization of an object with respect to one or two parallel mirrors. Because mirror symmetry preserves an object's size, the outcome depends on the object one started to symmetrize. That is, this form of symmetrization "reflects" — no pun is intended — the initial conditions. But reflection into the inside of a circle is different: it makes an object smaller. As a result, the search for our minimal thrice-symmetric $\mathcal{D}$ can be altogether different. The initial point of abscissa x_0 and the sequence of random moves obviously affect the orbit's first points. But those arbitrarily chosen inputs can be shown to have no perceptible effect on the orbit's limit; in practice, the limit identifies with the orbit from which the first few points have been deleted.

The algorithm simplifies if x_0 is picked outside of the circle and its first step replaces x_0 by either $-x-2$ or $-x+2$. The second and all following steps are best prevented from backtracking, therefore the throw of a coin (or the choice of 0 or 1 at random by the computer) will suffice to decide between the two possibilities other than repetition.

Seeing is believing and, to be believed, the progression of the orbit of this process has best be followed on the computer screen. Unfortunately, the fact that the orbit does not coalesce but remains a dust is hidden by the pixel structure of the screen. Hence Figure 2 must illustrate the dust through its histogram, then a blowup of a small piece and further blowups of smaller pieces. A striking observation is that those blowups come increasingly close to being identical. That is, very small parts of our limit set $\mathcal{C}$ are nearly identical — except for size — to merely small parts. That limit set is invariant by reduction; being self-similar, it is a fractal.

THE SELF-SIMILAR "FOUR TIMES INVARIANT CURVE $\mathcal{C}$," WITH FOUR-PART GENERATOR SYMMETRY

To move on from dusts to proper rough curves, it suffices to follow another narrow lane in the same conceptual "neighborhood." Preserving the two parallel lines of the previous construction, let us now add, not one but *two* circles of equal radii, each tangent to the other and to one of the straight lines, as shown in the right part of Figure 1. To preserve in this article the ideal of almost complete avoidance of formulas, let the pleasure of drawing the limit set be reserved to those who know how to program a geometric inversion in the plane, as opposed to a line. Contrary to Figure

2, Figure 3 is not contained in a straight line, and it is not a dust made of points separated by empty gaps. Now the points merge, albeit slowly, into a continuous curve that has no tangent at any point. It is loop-free, also called "singly connected," meaning that it joins two prescribed points in a single way.

Like the thrice-symmetric dust $\mathcal{D}$, this four-times-symmetric object $\mathcal{C}$ is approximately self-similar in many ways. This is especially clear-cut with respect to the "midpoint" where the curve is "osculated" therefore outlined, by two circles intersecting at an angle.

CONCLUDING REMARKS

The path this paper takes from plain symmetry to self-similarity is little-traveled but it is very attractive and worth advertising — as is now being done. Two alternatives must be mentioned.

By now, many persons know of an example that is analogous to $\mathcal{D}$ and is exactly (linearly) self-similar. It was provided by a "cascade" construction due to G. Cantor; it is simple and by now famous, but completely artificial. Many persons also know of an analog to $\mathcal{C}$, which is exactly (linearly) self-similar. It was provided by a cascade construction due to H. von Koch; it too is simple and by now famous, but completely artificial.

Figure 3 makes it obvious that one can construct $\mathcal{C}$ in analogy to the Koch curve. One proceeds by successive replacements of short arcs of a big circle by longer arcs of a smaller circle, an arc's length being measured in degrees. This construction resolves a query by Poincaré. Astonishingly, this query had been left unanswered until M 1982, 1983.

Upon discovering an object related to $\mathcal{C}$, Poincaré called it a curve, then, in an aside, commented "if you can call *this* a curve." This question was non-obvious as long as $\mathcal{C}$ was viewed as "mathematically pathological." But fractal geometry recognized shapes of this kind as models of nature; no one will any longer deprive them of the dignity of being called curves.

Acknowledgment. The figures were programmed by Aaron Benenav.

Mathematical Intelligencer 5, 9-17, 1983,

C18

Self-inverse fractals osculated by sigma-discs: the limit sets of ("Kleinian") inversion groups

• *Chapter foreword (2003).* The original paper included "tinted" figures. After much discussion, the broad-minded Editor of the *Intelligencer*, John H. Ewing, agreed to print those figures on the front and back covers of the magazine, which until then had been black-and-white. They were overlaid on a dark version of the traditional Springer-Verlag yellow. To accommodate IBM's *Script* word-processor (once a pioneer and now on its last legs), all figures had to be renumbered and captions became desirable, therefore were added. The last figure is a black-and-white reduction of the original back cover.

The Editor also added to my text an exposition of basic facts about inversion in a circle μ—mother's milk to French high school graduates in the 'thirties (as I had been), but mysterious to the bulk of his readers. •

THE DRY FACADE OF THE HARD MATHEMATICAL ANALYSIS in the style of 1900 hides a geometry of extraordinary visual beauty and suggestive power. There was a hint of it in a great old treatise, Fricke & Klein 1897, *Lectures on Automorphic Functions*, and it was fully revealed in M 1982F{*FGN*}. The figures in Fricke & Klein must rank among the most widely known of all mathematical illustrations, since they include the tessellations of the hyperbolic plane that the non-mathematical millions now credit to Maurits C. Escher. The present paper explores further a small corner of this universe, the geometry of the limit set of a special group based upon inversions illustrated in Fricke & Klein. This limit set is a fractal curve in the terminology of M 1982F{*FGN*}. It has been reproduced on faith by several famed books, thus helping to form the intuition of many generations of mathematicians.

Unfortunately, careful graphics performed on a computer revealed a rich structure that proves the Fricke & Klein illustration to be inaccurate and misleading. The conventional Poincaré algorithm used to draw the limit set (see Poincaré 1914+) is indirect and very inefficient.

Fortunately, the careful graphics also helped me discover a new algorithm that generates the limit set $\mathcal{L}$ for many groups $\mathcal{G}$ based upon inversions. This algorithm was sketched in 1981 (in Springer-Verlag's *Mathematical Calendar*) and described in M 1982F{*FGN*}, Chapter 18 (reprinted in this book as C16). It is *not* of universal validity and sorts out the groups based upon inversions as either being or not being "directly osculable." (The limit set may become osculable after a change of basis.)

The details of these new distinctions are, however, beyond the scope of this paper. Its sole purpose is to demonstrate this algorithm and its efficiency by describing in detail its application to the most striking of the Fricke & Klein examples.

Given a group of geometric transformations, it is interesting to identify the sets that are invariant under the action of the group, and especially the smallest among these sets. Such a question was first raised by Leibniz, who suspected that the only shapes invariant under all similarity transformations of the plane are straight lines and the whole plane (see M 1982F{*FGN*}, p. 419). Under the assumptions of smoothness, he was correct: these are the only connected and smooth invariants, called *self-similar sets.* Waiving the standard conditions of smoothness and restricting the similarities, however, one finds many other self-similar sets, all of which are not shapes from standard geometry, but fractal sets; for example, Cantor sets (totally disconnected sets; dusts in the terminology of M 1982F{*FGN*}), the boundaries of Koch snowflakes (nonsmooth curves), and Brownian motion (the best known self-similar random curve).

The present paper is devoted, in the same spirit, to a nonlinear group of transformations. Given $M \geq 3$ circles $C_m (1 \leq m \leq M)$ in the plane, to be called generating circles, we consider the group $\mathcal{G}$ generated by inversions with respect to these circles. Choosing the center of a circle C of radius R as origin, the inversion with respect to C is described in polar coordinates as the map $(r, \theta) \rightarrow (R^2/r, \theta)$. In what follows, the key fact about inversions is that they map circles to circles—a straight line being a circle through the point at infinity.

The sets invariant under the action of this group can be called *self-inverse sets* under $\mathcal{G}$. The closed plane, including its point at infinity, is self-inverse under every group. And for many groups it is the only standard solution. One notable exception—called Fuchsian—is when all the C_m are orthogonal to a common circle Γ. In this case (which includes most cases with $M=3$), the circle Γ is self-inverse under $\mathcal{G}$, since any circle orthogonal to C is invariant under inversion in C.

It is also instructive to side track to a case excluded by the above definition. When $M=2$ and the finite discs bounded by C_1 and C_2 do not intersect, two points are mutually inverse with respect to both C_1 and C_2 : (If Γ is disc bounded by C_1 and γ_{21} in the finite disc bounded by C_2, γ_{12} in the finite a circle orthogonal to C_1 and C_2 then the points γ_{12} and γ_{21} belong to Γ.) Therefore, the set $\{\gamma_{12}, \gamma_{21}\}$ is self-inverse. Some very special Fuchsian groups also have a self-inverse set reduced to 2 points.

On the other hand, the first student of this topic, Poincaré, observed a hundred years ago that waiving the standard assumptions of smoothness can yield strange subsets of the plane as closed self-inverse sets. Shortly before (!) Cantor introduced his set in 1884, and well before Koch introduced his nondifferentiable snowflake curve in 1904, Poincaré noted that in typical configurations of the generators C_m, the self-inverse set $\mathcal{L}$ can be either a totally disconnected set ("dust"), or a curve that is nondifferentiable (either without tangent or with a tangent but no curvature). (He commented that one must assume that "one can call *that* a curve".... Later studies of the concept of curve have confirmed that one can and that one should.)

The limit set

How do we find a self-inverse set? The answer is that one can start with any set S, and enlarge it just enough to *make* it self-inverse. For any set S the *clan* of $\mathcal{G}S$ is defined to be the union of all transforms $g(S)$ for $g \in \mathcal{G}$ —this is usually called the orbit of S, but I prefer the term "clan". Of course the closure of $\mathcal{G}S$ is a closed, self-inverse set. So is the subset $(\mathcal{G}S)'$ consisting of all limit points of $\mathcal{G}S$. Recall that a point P is a limit point of a set if every deleted neighborhood of P —i.e., the neighborhood minus P itself—intersects the set; the set of limit points is called the derived set.

In particular, one can start with any point P_0, then form the clan $\mathcal{G}P_0$ and take the derived set $(\mathcal{G}P_0)'$. The amazing fact is that, under wide conditions, the derived set is (a) independent of the point P_0, so that it can be denoted by $\mathcal{L}$; and (b) of zero area (= planar Lebesgue measure.) The derived set $\mathcal{L}$ has several very important characteristic properties. Not only is it self-inverse, but it is the *minimal* self-inverse set. It is (by construction) the *limit set* of $\mathcal{G}$. Furthermore, it is the set on which the group $\mathcal{G}$ is *continuous*, and outside of which $\mathcal{G}$ is discontinuous. We give one final useful characterization: it is clear that $\mathcal{G}$ includes an infinity of inversions in addition to its base, and that $\mathcal{L}$ is the derived set of the centers of these inversions. To my knowledge, this last statement is not put in this form in the literature, but it can be seen to be equivalent to Poincaré's construction of $\mathcal{L}$.

What is the shape of $\mathcal{L}$? Our knowledge of the Cantor and Koch sets benefits from the availability of a multiplicity of direct and transparent constructions. To the contrary, the shape of the minimal self-inverse $\mathcal{L}$ has remained elusive. To close this gap, I have devised a new construction which involves the self-inverse open sets obtained as clans of open discs; I call them "sigma-discs" or "σ-discs", where the letter σ is self-explanatory; it indicates that a σ-disc is the denumerable union of discs. The complement of a self-inverse σ-disc is also a self-inverse set, and the minimal self-inverse $\mathcal{L}$ can ordinarily be represented as the complement of a finite union of σ-discs. Each of these σ-discs can be said to *osculate* $\mathcal{L}$. This *osculating* σ-disc construction can be made recursive: as the recursion advances, it tends to outline $\mathcal{L}$ very rapidly— much more rapidly than the classical construction of Poincaré.

Fuchsian groups with $M = 3$ for which $\mathcal{L}$ is the circle Γ

It was noted that when $\mathcal{G}$ is Fuchsian, the circle Γ is self-inverse. A case when Γ is the minimal self-inverse set occurs when $M = 3$ and each of three circles C_1, C_2 and C_3 is tangent to the other two (Figure 1.1). By inverting the plane about the point of tangency γ_{12} of C_1 and C_2, one achieves the situation of Figure 1.2, where C_1 and C_2 are parallel straight lines and C_3 is tangent to both. The group is also transformed: it is now generated by reflections in C_1 and C_2 and inversion in C_3. It is easily seen that $\mathcal{L}$ is Γ. The group of Figure 1.2 is a special "modular" subgroup.

Digressive former footnote. Take the center of C_3 as origin and its radius as 1. Then each real x_1 can be written uniquely as $i_1 + p_1$, where i_1 is a signed even integer and p_1E lies in the open interval $]-1, 1[$. Similarly, $1/p_1 = i_2 + p_2$. When x_1 is rational, every p_k is also rational; as k increases, the denominators of p_k decrease and eventually, for $k = K$, reach $p_k = 0$ or 1. Call K the depth of x_1. For irrational numbers, $K = \infty$; for integers $K = 0$. It is easy to prove that every rational x_1 with $p_k = 0$ is the transform of $x_K = 0$ by a word of the group $\mathcal{G}$, while every rational x_1 with $p_K = 1$ is the transform of $x_K = 1$ when $i_k/4$ is an integer, and of

$x_K = -1$ otherwise. Furthermore, every x_1 with $p_K = 0$ is obviously the limit point of x_1s with $p_K + 1 = 1$. Conclusion: whenever the abscissa of P_0 is rational, the set $\mathcal{G}P_0$ is dense in Γ.

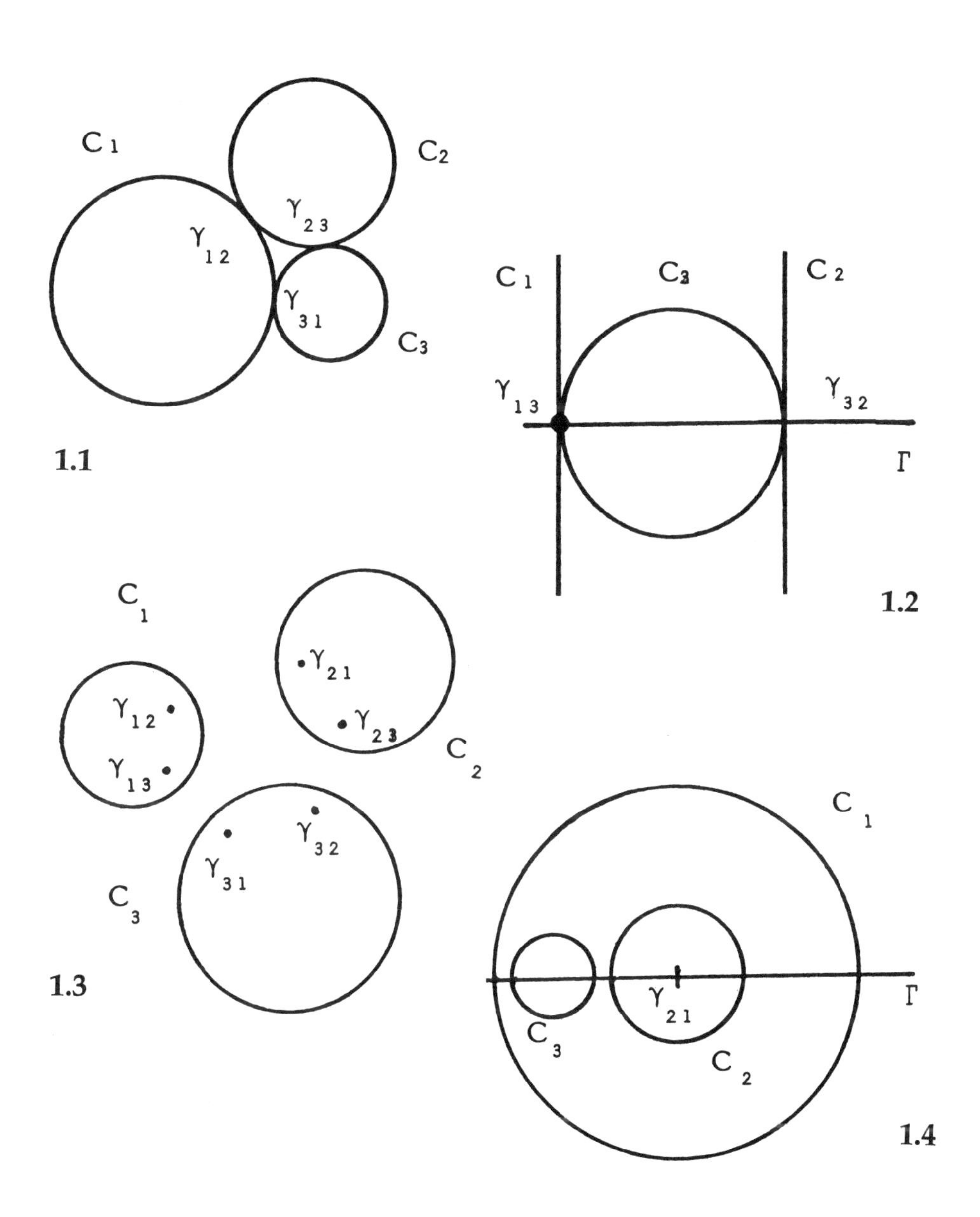

FIGURE C18-1. Illustration of some aspects of inversion in circles.

Some Fuchsian groups with $M = 3$ for which $\mathcal{L}$ is a fractal dust; osculating σ-intervals

Figure 1.3 gives an illustrative example where $M = 3$ and the group is Fuchsian but $\mathcal{L}$ is a very small subset of Γ: we let Γ be the orthogonal circle and assume that the finite discs bounded by C_1, C_2 and C_3 are nonintersecting. It was noted that there are points γ_{12} and γ_{21} on Γ that are mutually inverse with respect to both C_1 and C_2. Now invert the plane about γ_{12}, obtaining Figure 1.4. Since γ_{12} goes to infinity, the circle Γ becomes a straight line; C_1 and C_2 become two circles with the common center γ_{21}. It is easy to see that the half of Γ that does not include the points γ_{13} and γ_{31}, and the points γ_{23} and γ_{32}, fails to belong to $\mathcal{L}$. We use this half line to define the open interval $]\gamma_{12}, \gamma_{21}[$.

Thus, the self-inverse set $\mathcal{G}]\gamma_{12}, \gamma_{21}[$, which is a σ-interval (a union of nonoverlapping open intervals) lies entirely in the complement of $\mathcal{L}$. The same is true of the σ-intervals $\mathcal{G}\ ]\gamma_{13}, \gamma_{31}[$ and $\mathcal{G}]\gamma_{23}, \gamma_{32}[$. Also, but not quite so obviously, the complement of these three σ-intervals is $\mathcal{L}$. It is a fractal dust of zero length (linear Lebesgue measure), and provides a self-inverse version of the usual self-similar Cantor set.

One can delineate the shape of $\mathcal{L}$ with rapidly increasing accuracy by injecting the intervals of the complement of $\mathcal{L}$ in order of decreasing length. One can also use the order of increasing length of the shortest words in $\mathcal{G}$ that obtain these open intervals from one of the $]\gamma_{ij}, \gamma_{ji}[$.

Although I find it hard to believe that the preceding algorithm is new, I do not recall having seen it described anywhere. {P.S. 2003. No earlier reference has come up.}

Fricke & Klein: mislead and misleading

The limit set is relatively easy to find for a group generated by inversions in three circles, but what about larger configurations? That is the point of the new algorithm for $\mathcal{L}$. It too, is so completely elementary that it might have been (but was not!) recognized in the 1880s, when Henri Poincaré and Felix Klein first tackled this topic. Hence, it is best presented against the classic background of those illustrations in Fricke & Klein 1897 that purport to represent the limit sets $\mathcal{L}$ of several special inversion groups.

The main discrepancy does *not* lie in the fact that the true $\mathcal{L}$s involve detail that no one would attempt to draw by hand. Even if detail is erased (which is best done by stopping the new algorithm after a small number of stages) the "old $\mathcal{L}$" seems cruder: structureless and lawless.

Perhaps it is best to tackle a single example in detail in order to illustrate my construction. Our point of departure is Figure 2.1, which reproduces the 5 circles used in Figure 156 of Fricke & Klein. Next, Figure 2.2 reproduces the corresponding "old $\mathscr{L}$" as claimed by Fricke & Klein: it consists of a wiggly curve together with a collection of circles, shown in heavy outline. The "true $\mathscr{L}$" is shown in diverse guises in many of the later figures in this article. On Figure 3, the true $\mathscr{L}$ is shown as the boundary of a black background domain. Figure 4

It now seems obvious that the hapless draftsman preparing Figure 2.2 (according to legend he was an engineering student in Fricke's class) determined a few points of $\mathscr{L}$ exactly, and then drew "some very wiggly

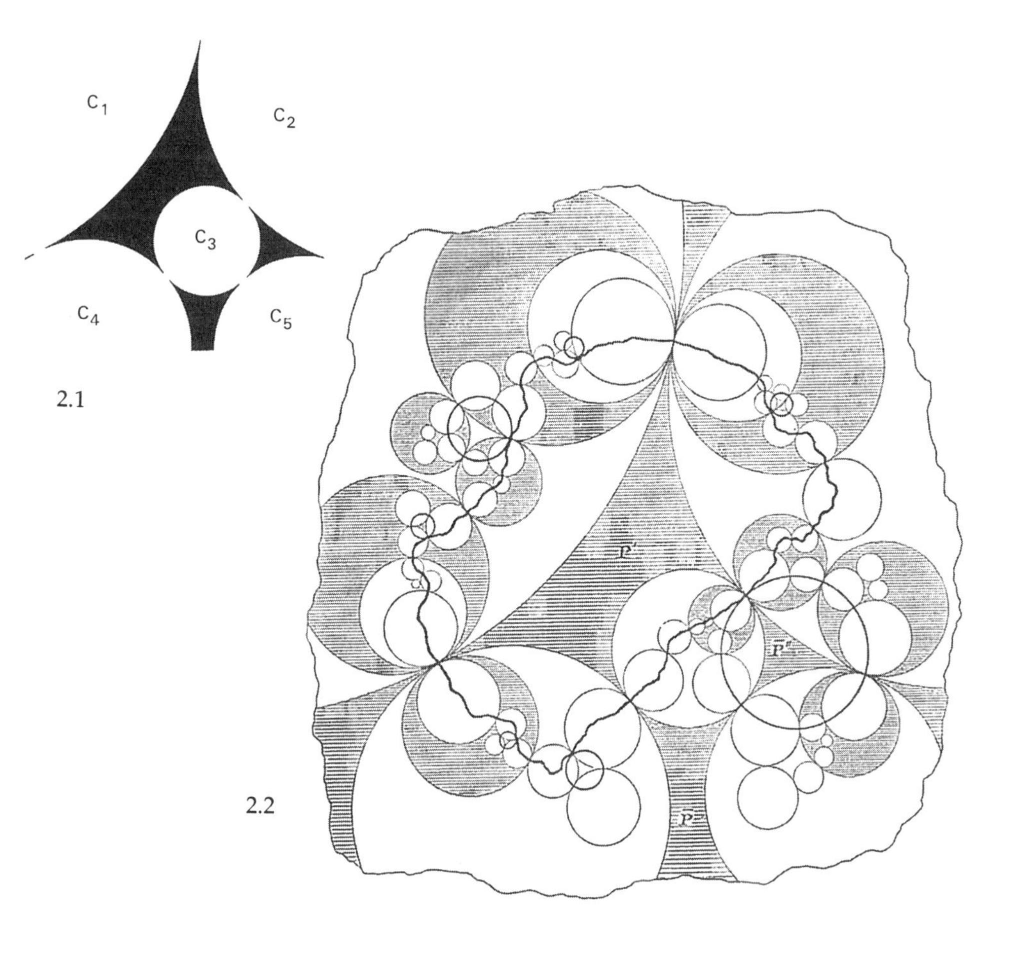

FIGURE C18-2. According to Fricke & Klein 1897, Figure 2.2 represents the limit set under inversions with respect to the five circles shown in Figure 2.1.

and complicated curve" passing through these points. As Fricke did not know what to expect, the draftsman received no explicit directions.

Observe that the black domain on Figure 2.1, viewed as open (not including the generating circles), splits into three disjoint maximal open domains, to be denoted as $\mathcal{D}_3$ (a triangle), $\mathcal{D}_4$ (a quadrilateral), and $\mathcal{D}_5$ (a pentagon containing the point at infinity). This suggests that we might begin to investigate $\mathcal{L}$ by investigating the limit sets of two subgroups of $\mathcal{G}$: first, the Fuchsian subgroup $\mathcal{G}_3$, generated by the inversions in the 3 circles bounding $\mathcal{D}_3$, and second, the subgroup $\mathcal{G}_4$, generated by the inversions in the 4 circles bounding $\mathcal{D}_4$. The limit set of a subgroups is, after all, contained in the limit set of the group.

FIGURE C18-3. The boundary between black and white is the correct form of the invariant curve (mis)represented by Figure 2.2.

The Fuchsian subgroup $\mathcal{G}_3$

One point at which the old $\mathcal{L}$ and the true $\mathcal{L}$ agree is that both include a large circle, to be denoted as $\mathcal{L}_3$, and smaller circles. The circle $\mathcal{L}_3$ is orthogonal to the 3 generating circles C_2, C_3, and C_5 that bound the domain $\mathcal{D}_3$, and from our previous discussion it is the limit set of the subgroup $\mathcal{G}_3$.

Now suppose we take $\mathcal{L} = (\mathcal{G}P_0)'$, where P_0 is an arbitrary point of $\mathcal{L}_3$. *A fortiori*, we have the representation $\mathcal{L} = (\mathcal{G}\mathcal{L}_3)'$, which is extravagantly redundant, yet very useful in describing the limit set. An approximate idea of $\mathcal{L}$ based upon this representation is given by Figure 4. It limits itself to the elements of $\mathcal{G}$ that are obtained as products of fewer than a certain large number K of inversions.

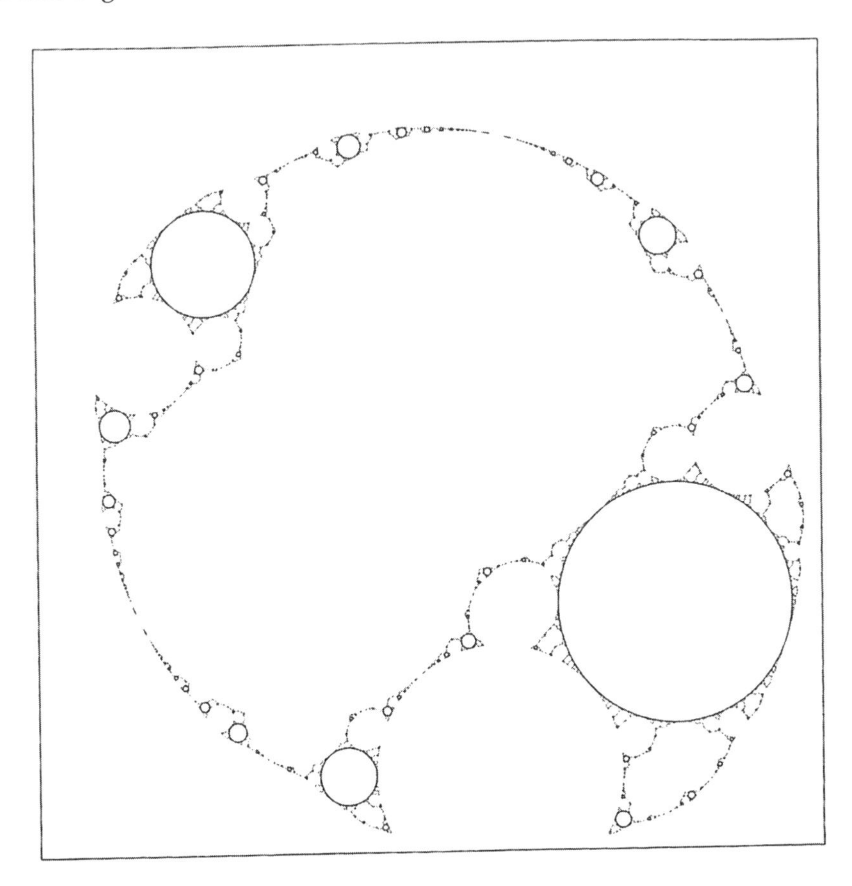

FIGURE C18-4. Fast converging approximation of the invariant curve in Figure 3 by images of a circle under a sequence of inversions.

Large circles are represented by thin curves, and circles whose radius is smaller than the large circles' thickness, are represented by points. Were the algorithm pushed further, all these points would bleed together to form an infinitely ramified curve. Had we abstained altogether from plotting the large circles, the result would not change perceptibly, because every point on every circle can be obtained in alternative fashion as the limit of points belonging to smaller circles, that is, of points which stand for small circles.

The formula $\mathcal{L} = (\mathcal{G}\mathcal{L}_3)'$ suffices to represent $\mathcal{L}$, but it is cumbersome. The sequence $\mathcal{G}_K\mathcal{L}_3$, where $\mathcal{G}_K$ is the collection of products of, at most, K inversions, converges to $\mathcal{L}$ more rapidly than the sequence $\mathcal{G}_K P_0$ relative to any single point P_0. Nevertheless, it converges very slowly: without the exceptional computer facilities marshalled for Figure 3, this algorithm would only yield a very loose idea of $\mathcal{L}$.

The chain-based subgroup $\mathcal{G}_4$

The most striking discrepancy between the old $\mathcal{L}$ and the true $\mathcal{L}$ concerns the limit set $\mathcal{L}_4$ corresponding to the subgroup $\mathcal{G}_4$ based on the 4 circles, C_1, C_2, C_3, and C_4, that touch the domain $\mathcal{D}_4$. In the terminology of M 1982F{*FGN*}, these circles form a connected "Poincaré chain", in which each link is tangent to exactly two neighbors. (As a result, the limit set $\mathcal{L}_4$ is a Jordan curve.) The old $\mathcal{L}_4$, separated from the rest of Figure 2.2, is shown on Figure 5.1, and the true $\mathcal{L}_4$, as constructed by my new algorithm, is shown on Figure 6, Figure 7.

To determine $\mathcal{L}_4$, the idea is (once again) to determine what is *not* in $\mathcal{L}_4$.

We begin with the observation that for any three circles C_i, C_j, C_K there is a common orthogonal circle Γ_{ijk} that passes through the points of tangency of C_i, C_j, and C_K. In this case, the Γ_{ijk} are distinct; that is $\mathcal{G}_4$ is not Fuchsian.

Each circle Γ_{ijk} divides the plane into two discs—one bounded and one unbounded. One of these two open discs contains no points of tangency of the four circles; we denote it by Δ_{ijk}. To see this, we can invert the plane about a circle centered on the point of tangency γ_{ij}. The configuration of circles become that in Figure 5.2, and Γ_{ijk} becomes a horizontal line: the transform of Δ_{ijk} is the half-plane above or below the transform of Γ_{ijk}.

Now it is not hard to see that an inversion in one of the four circles cannot carry a point of tangency inside Δ_{ijk}. It follows that no transform of Δ_{ijk} contains a transform of a tangent point. If we use the tangent point of the four discs, Δ_{ijk} must be in the complement On the other hand, the boundary of Δ_{ijk} (the circle Γ_{ijk}) contains the limit set of the Fuchsian subgroup generated by inversions in C_i, C_j, and C_k. Therefore, any open disc containing Δ_{ijk} must intersect $\mathcal{L}$.

Now consider the 4 open discs Δ_{ijk} and their clans separately. One of the initiators, namely the disc Δ_{124}, is unbounded, and it intersects the disc Δ_{234}. (In the present configuration, Δ_{234} is also unbounded—in fact, Γ_{234} is nearly a straight line—but in other configurations of 4-link Poincaré chains Δ_{234} may be bounded.) Together, the discs Δ_{124} and Δ_{234} easily identified on Figure 8, provide a first approximation of the outside of $\mathcal{L}_4$.

This and later approximations are analogous to the approximations of the Koch snowflake curve $\mathcal{K}$ in Plate 43 of M 1982F{*FGN*}.

The other initiator discs Δ_{ijk}, namely, the discs Δ_{123} and Δ_{143}, are bounded and intersect each other. They are easily identified on Figure 7. Together, they provide a first approximation of the inside of $\mathcal{L}_4$.

A second approximation of the outside of $\mathcal{L}_4$, also clearly seen in Figure 8, is achieved by adding to Δ_{124} and Δ_{234} their inverses in C_3 and C_1, respectively. The corresponding second approximation of the inside of the $\mathcal{L}_4$ is achieved by adding to Δ_{123} and Δ_{143} their inverses in C_4 and C_2, respectively.

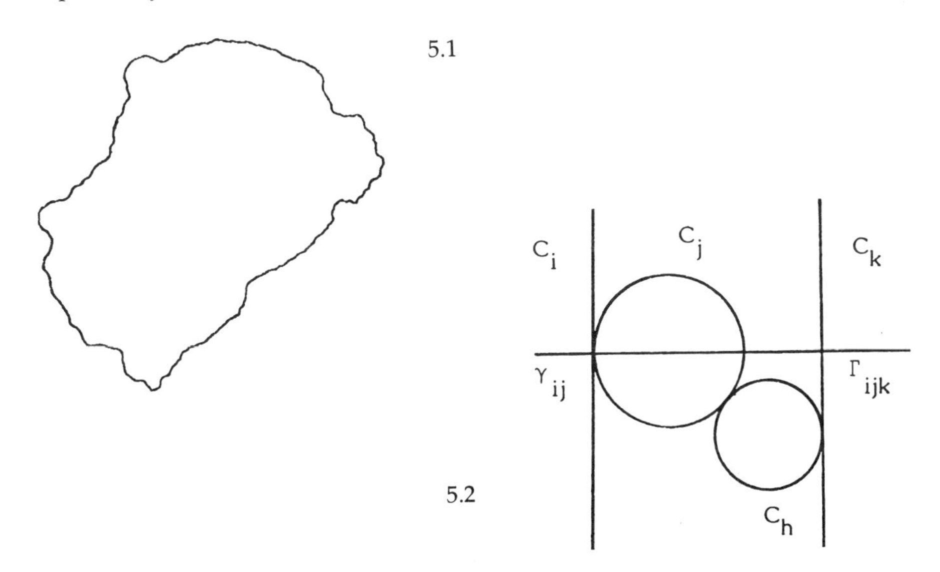

FIGURE C18-5. Figure 5.1 is extracted from Figure 2.2. Figure 5.2 is obtained by inverting four circles in a Poincaré chain.

The complement of the σ-disc (denumerable union of discs) made up of the four clans $\mathcal{G}\Delta_{ijk}$ squeezes down to the curve $\mathcal{L}_4$. The union of the four "initiator discs" alone provides a useful approximation of the complement of $\mathcal{L}_4$. The approximations using the product of K or fewer inversions converge rapidly to $\mathcal{L}_4$.

Figure 8, which reproduces the back cover of the original, generalizes the construction of Figure 7 to the whole group $\mathcal{G}$.

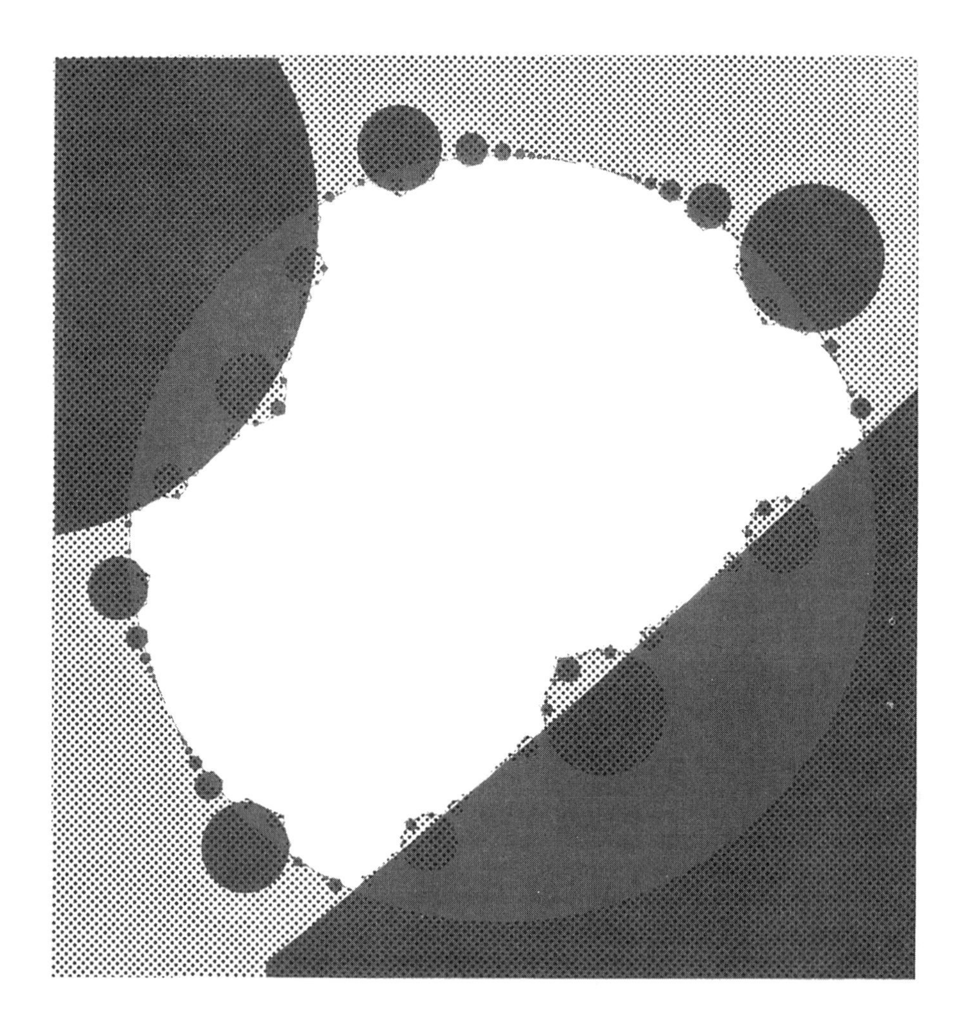

FIGURE C18-6. New fast algorithm for the construction of the exterior of the corrected Figure 5.1.

Fractal osculation; osculating discs

The fact that $\mathcal{L}_4$ is not intersected by any of the four open discs Δ_{ijk}, with indices associated with $\mathcal{L}_4$, but is intersected in more than one point by the circle bounding every Δ_{ijk}, suggests that $\mathcal{L}_4$ and Δ_{ijk} be called *osculating*.

In its standard context in differential geometry, the notion of osculation is linked to the concept of curvature. To the first order, a standard curve near a regular point P is approximated by the tangent line. To the second order, it is approximated by the circle, called "osculating", which has the same tangent and the same curvature.

The circles tangent to the curve at a point P can be indexed by u, the inverse of the distance from P to the circle's center. The index of the osculating circle will be written as u_0. If $u < u_0$, a small portion of curve centered at P lies entirely on one side of the tangent circle, except for P

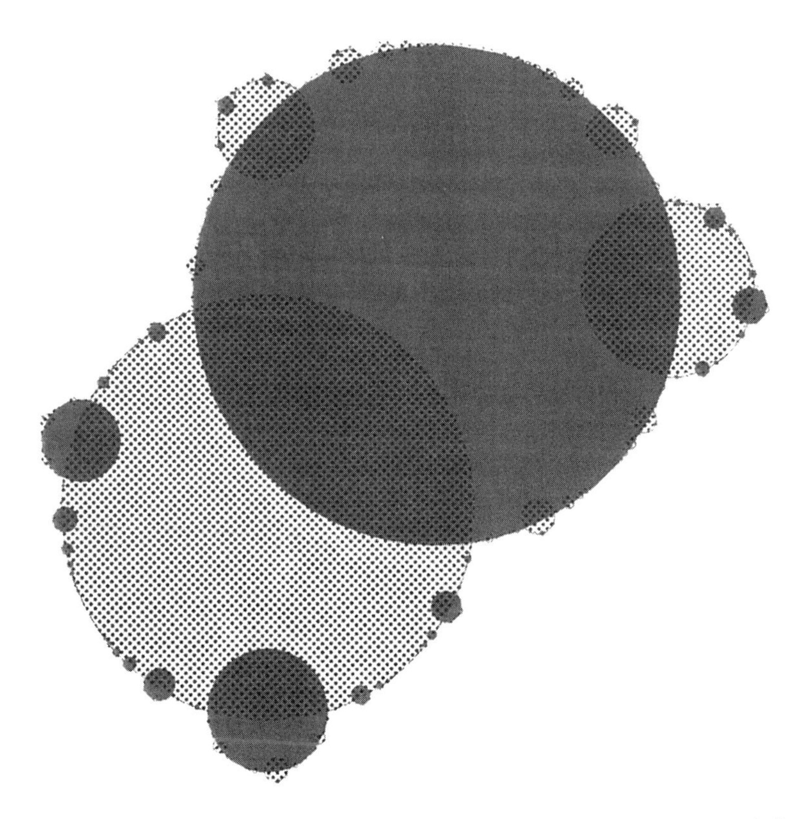

FIGURE C18-7. New fast algorithm for the construction of the interior of the corrected Figure 5.1.

itself, while if $u > u_0$, it lies entirely on the other side, except for P. We say that u_0 is a *critical value* or a *cut*.

For fractals, the definition of osculation by curvature is meaningless. However, there is an infinity of points where the limit set $\mathcal{L}$ of any Poincaré chain squeezes between two discs tangent to each other. For example, the point of tangency of the generating circles C_i and C_j belongs to $\mathcal{L}$, and $\mathcal{L}$ squeezes between two discs Δ_{ijk} and Δ_{ijk}. It is tempting to call both of these discs *osculating*.

FIGURE C18-8. [Back cover of M1983m changed to grays]. New fast algorithm for the interior of the correct invariant curve (mis)represented by Figure 2.2.

To pinpoint this notion, we take a point P where $\mathcal{L}$ has a tangent, and start with the definition of ordinary osculation based on critically (= cut). The novelty is that, as u varies, the single critical u_0 is replaced by two distinct values, u' and $u'' > u'$, defined as follows: for all $u < u'$, $\mathcal{L}$ lies entirely on one side of our circle, except for P, while for all $u > u''$, $\mathcal{L}$ lies entirely to the other side, except for P; and for $u' < u < u''$, parts of $\mathcal{L}$ are found on both sides of the circle. I suggest that the circles of parameters u' and u'' be *both* called *fractally osculating* to $\mathcal{L}$. The open discs bounded by the osculating circles and not intersection $\mathcal{L}$, will be called osculating discs. It may happen that one or two osculating circles degenerate to a point.

As is well known, standard osculation is a local concept, since its definition is independent of the curve's shape away from P. By contrast, I have defined fractal osculation globally, because this is all that was needed here; a local version is defined without difficulty.

The group $\mathcal{G}$; classification of the circles Γ_{ijk} as either osculating or intersecting the set $\mathcal{L}$

We found that, for the subgroup $\mathcal{G}_4$, every one of the $4!/3!1! = 4$ circles Γ_{ijk}– defined as orthogonal to 3 of the generating circles C_m– is osculating to $\mathcal{L}_4$. For other configurations of generating circles C_m, including the configuration in Figure 5 and Poincaré chains with $M > 4$, the situation is more complex.

For example, consider the complement of the limit set $\mathcal{L}$. We know already that one needs the clan of Δ_{235} under $\mathcal{G}_3$ to generate the inside of $\mathcal{L}_3$, and that one needs the clans of Δ_{123} and Δ_{234} under $\mathcal{G}_4$ to generate the inside of $\mathcal{L}_4$, and we shall see that one needs the clans of Δ_{235}, Δ_{123} and Δ_{234} under $\mathcal{G}$ to generate the inside of $\mathcal{L}$ (Figure 6). Similarly, we shall see that one needs the clans of Δ_{345}, Δ_{145}, and Δ_{125} under $\mathcal{G}$ in order to generate the outside of $\mathcal{L}$ (Figure 7).

However, the circles based on the triplets 234 and 135 behave in a *totally different* way: in either case, some of the remaining circles C_m are positioned, respectively, in the interior disc $\mathcal{I}\Gamma_{ijk}$ and the exterior disc $\mathcal{E}\Gamma_{ijk}$. Two open discs, namely $\mathcal{G}\Gamma_{ijk}$ and its inverse in the circle C_m located within $\mathcal{I}\Gamma_{ijk}$, cover the whole plane; the same is true of $\mathcal{E}\Gamma_{ijk}$. *A fortiori*, the clans $\mathcal{G}\mathcal{I}\Gamma_{ijk}$ and $\mathcal{G}\mathcal{E}\Gamma_{ijk}$ are both identical to the whole plane. Hence, when $ijk = 234$ or 135, neither the interior nor the exterior of Γ_{ijk} can serve as the initiator disc of an osculating σ-disc. It continues to be true, however, that the complement of $\mathcal{L}$ is a σ-disc, and that it is obtained as the union of clans of the form $\mathcal{G}\Delta_{ijk}$.

Moreover, whenever Δ_{ijk} *can be* defined, *it is needed* for the construction, and whenever Δ_{ijk} *cannot be* defined, *one can do without it.* (Nevertheless, when C_i, C_j and Ck form a Fuchsian subgroup having a full circle Γ_{ijk} as its limit set, the construction of $\mathcal{L}$ is made much faster by including $\mathcal{G}\Gamma_{ijk}$.)

A *general rule.* In order for the triplet i, j, k to be needed in the case of the group $\mathcal{G}$, the key facts are as follows: (A) There exists a unique circle orthogonal to C_i, C_j, and C_k; (B) Either the inside of Γ_{ijk}, or the outside of Γ_{ijk} fails to contain any of the points of tangency between any two circles C_m and C_n.

More general non-Fuchsian groups

This last rule applies to more general groups $\mathcal{G}$ based upon inversions. The first step is to replace the points of tangency when C_m and C_n are not tangent. As noted early in this paper, when C_m and C_n fail to overlap, there are two points, γ_{mn} in C_m and γ_{nm} in C_n, which are mutually inverse with respect to both C_m and C_n. When C_m and C_n overlap, let γ^*_{mn} and γ^*_{nm} denote their intersection points. Now requirement (B) at the end of the preceding section generalizes as follows: the circle Γ_{ijk} is osculating if, for all m and n, either its inside or its outside fails to include γ_{mn}, and includes either the point γ^*_{mn} or the point γ^*_{nm}, but not both.

Fuchsian groups and osculating intervals

When there is a circle Γ orthogonal to all the C_m, my algorithm always yields Γ itself, but $\mathcal{L}$ may be either Γ, or a fractal dust subset of Γ.

In addition, the group may be everywhere continuous, a case which I call chaotic and which is not investigated in this paper; if so, $\mathcal{L}$ is the whole plane. In order to construct $\mathcal{L}$ when it is not the plane, we use osculating intervals.

When both intervals bounded by γ_{ij} and γ_{ji} contain no other γ_{mn}, then $\mathcal{L}$ reduces to γ_{ij} and γ_{ji}. This case plays for the Fuchsian groups the same degenerate role as the Fuchsian groups themselves play for the other groups based upon inversions.

Acknowledgements. The illustrations in this paper were drawn using computer programs written for this purpose by Dr. V. A. Norton.

PART IV: MULTIFRACTAL INVARIANT MEASURES

&&&&&&&&&&&&&&&&&&&&&&&&&&&&&&

First publication

C19

Introduction to measures that vanish exponentially almost everywhere: DLA and Minkowski

WHEN A NEW MATHEMATICAL DEVELOPMENT IS EVALUATED, some purists allow the computer but are offended by strong physical motivation. This part is a new opportunity to discuss an example.

I. A CASE OF CONTINUING INFLUENCES, BACK AND FORTH, BETWEEN MATHEMATICAL ESOTERICA AND PHYSICS

In previous parts of this book, the role of physics was very indirect, each topic being triggered by allowing the computer to assist the internal logic of existing mathematics. While that mathematics had originated in physics, this had happened in a dim and safely forgotten past. Also, while it happened to be reviving in the form of chaos theory, I stated repeatedly that my motivation came directly from mathematics circa 1900.

This part is different, as this introductory chapter proposes to show, and the fit of Chapter C22 within this book may seem questionable but will improve upon consideration. Gutzwiller & M 1988 {C20} arose naturally when the authors, investigating two areas of physics far removed from each other and from Kleinian groups, were both forced into a path far from traditional "applied mathematics." Using prefractal technical terms, we investigated an "example of a nonrandom singular measure on the unit interval that is anomalous because it vanishes exponentially almost everywhere." Quite a mouthful.

Is this not an object fit for esoterica? Indeed, as Yuval Peres informed us, the same mathematical structure had been considered long ago, by ***Hermann Minkowski*** (1864-1909), who defined it and then stopped. Much later, ***Arnaud Denjoy*** (1883-1974) heard it mentioned by J. Hadamard, paid attention, and observed that the Minkowski measure is invariant under some very special Kleinian groups. Those mathematicians, and scattered followers, are cited in M1993s{C21}.

Minkowski being one of my heroes makes me comfortable with the term "Minkowki measure," and "Minkowski–Denjoy" would have been acceptable. But esoterica are esoterica, and this measure did not become known in the wide world. Had Gutzwiller and I been parochial physicists, we would have proceeded in blissful ignorance of the past.

While Gutzwiller went straight from Hamiltonian systems to the Minkowski measure, I reached it indirectly. My interest was triggered by the study of DLA, the mysterious diffusion-limited aggregates, which were discovered by Witten & Sander 1981 and belong to computational condensed-matter physics. Section 4 is meant to help orient the reader not acquainted with this topic: Figure 1 is a large sample of plane DLA, and Figure 2 is a piece of cylindrical DLA from M & Evertsz 1991. It is small enough to show clearly the "fjords" that form trees with branches that become extraordinarily thin.

Figure 2 also represents the isolines of the Laplacian potential, showing that it becomes extraordinarily small in the bottom of the thinnest fjords. So does the harmonic measure defined as the gradient of the potential along the boundary. In my terminology, my ultimate goal was to represent the Laplacian harmonic measure on DLA, an example of a random multifractal measure supported by a random fractal curve. To achieve some degree of "traction" on this forbiddingly difficult topic, I approximated brutally by giving up randomness and adopting a linear support. The Minkowski measure emerged as a very rough "cartoon," sharply focused on very low values of the harmonic measure. The tenuous link between Kleinian limits (Section 3) and DLA (Section 4) goes through multifractals (Section 2).

2. MULTIFRACTAL/HÖLDER SPECTRUM: FOR SMALL AND LARGE α; ZEROS AND EXPONENTIAL ZEROS

2.1 The barest sketch of multifractals (M1969b, M1972j and M1974f)

Given a nondecreasing function $F(t)$, its differential dF in an interval dt is a measure. In multifractals, an exponent now denoted by α enters through a relation of the form

measure in an interval of length $dt \rightarrow (dt)^{\alpha}$.

I was aware that the Hölder–Lipschitz exponent of a function $f(t)$ had entered mathematical esoterica around 1870 as the minimum value $\alpha_{min} > 0$ in an interval. For the needs of physics, a singular measure had to be described far more precisely by considering the value of α for every value of t. Specifically, in order to handle a problem of fluid mechanics called intermittent turbulence, I put forward and developed a notion later renamed "multifractality" and a technique that grew into a "multifractal formalism" that I prefer to call "Hölder analysis." It is centered on a function $f(\alpha)$ consisting, in effect, of a suitable probability distribution for α. Those early papers, with comments, are collected in M1999N.

Multifractals became widely known through Halsey et al. 1986, a paper that one coauthor, L.P. Kadanoff, correctly describes (see page 73 of M1999N) as an exposition of my ideas. Unfortunately, the pedagogical excellence of the heuristic approach of Halsey et al. misled many users by creating the widespread impression that the function $f(\alpha)$ always has a graph shaped like the mathematical symbol $\cap$ stretched between $\alpha_{min} > 0$ and $\alpha_{max} < \infty$.

2.2 The $\alpha_{max} = \infty$ "anomaly"

In terms of multifractal formalism being applied to the Minkowski measure, mathematical esoterica from before 1990 can, in hindsight, be viewed as having examined only α_{min} and $\alpha = 1$. Similarly, early studies of DLA concerned other small values of α near $\alpha_{min} > 0$. But when $\alpha_{max} < \infty$, it too is an important descriptive characteristic of a measure. Hence, in all cases, extensive attempts were also made to estimate α_{max}.

Those efforts led nowhere for DLA. For the Minkowski measure, Gutzwiller & M 1988{C20} found the value of α_{max} to be elusive. In contrast, the inverse Minkowski measure was found to yield $\alpha_{min} = 0$. Not until M 1993s{C21} did I fully realize that $\alpha_{min} = 0$ for the inverse Minkowski measure meant $\alpha_{max} = \infty$ for the Minkowski measure itself, and it became clear why α_{max} was elusive. This explanation made Chapter C20 obsolete, but it remains a useful warning against the dangers that lurk where as it was in the early study of DLA, the multifractal formalism is used thoughtlessly. For the latter, M & Evertsz 1991{C22} concluded that the elusiveness of α_{max} was deeply rooted in "anomalies" corresponding to the bottoms of the deep "fjords" clearly seen in Figures 1 and 2. This observation set the authors to test the possibility that the harmonic measure may be smaller than any "power law" expression of the form $(dt)^{\alpha}$ with finite α.

2.3 Measures with exponential zeros almost everywhere: definition

Thinking of DLA and its Minkowski measure "cartoon," led me to conjecture that $\alpha_{max} = \infty$, imply a measure that vanishes exponentially on a set of ts of measure 1. This concept will now be defined.

When $[t, t + dt]$ falls in an open interval such that $\mu([t, t + dt]) = 0$, the Hölder exponent is usually not evaluated. Yet it is good to observe that its definition applied mechanically yields $\alpha(t) = \infty$. More generally, when there is a positive measure in every interval arbitrarily close to t, as is the case for the Minkowski measure μ, we shall find that there exist points such that $\alpha(t) = \infty$.

These possibilities suggest that the notions of "zero" and "exponential zero" should be generalized from functions to singular measures. More precisely, we shall say that t_R is a right zero of the measure μ when $\alpha_R(t) = \infty$. We shall say that t_R is a *right exponential zero* when if c' and c'' are two constants greater than 0, one has $\mu([t, t + dt]) < c' \exp(-c''/dt)$ for small enough $dt > 0$.

The definitions relative to α_L are similar. When t is both a left and a right zero or exponential zero, it will be called a *zero* or *exponential zero*.

Slight extensions of the arguments in Chapters 20 and 21 show that for the Minkowski measure μ almost every t is an exponential zero.

3. PLACING THE MINKOWSKI MEASURE AMONG OTHER MULTIFRACTAL INVARIANT MEASURES GENERATED BY SOME SPECIAL KLEINIAN GROUPS

The possibility that $\alpha_{max} = \infty$ is understood as being "critical" within a broader context where $\alpha_{max} < \infty$. The notion of IFS was described in Chapter C15. Plotting the successive positions of an IFS amounts, in effect, to weighing each previously massless point with a theoretically infinitesimal mass or measure. We now proceed from Kleinian limit sets to invariant measures on those sets. For this discussion, Poincaré's original construction (nicknamed "Indra's pearls") is the appropriate one. The slowness of its convergence was a deep handicap in Part III. But here it serves to clarify (and be clarified by) the meaning of very large values of the Hölderian α.

3.1 A basic construction: three equal and equally spaced circles

Given three circles C_1, C_2, and C_3, nonoverlapping and of common radius ρ, draw the circle C that is orthogonal to all three; the radius of C will be used as the unit of length. Assume further that this pattern is invariant by a rotation of $2\pi/3$, and distribute a mass of 1 along C with constant density. The portions of C within C_1, C_2, and C_3 will each carry the mass 1/3.

Part of the first stage of the Indra's pearls construction inverts the circle C_1 with respect to C_2 and C_3. Let the inversion transfer the mass 1/6 into each of 2 pearls. Altogether, the first stage generates 6 pearls, each with a mass of 1/6. The kth iteration creates a kth-order necklace made of 2^k pearls, each containing a uniformly distributed mass equal to $(1/3)2^{-k}$.

Within each pearl, a coarse density along C is defined as the total mass divided by the length of the intercept of the pearl by C. While the numerator is $(1/3)2^{-k}$ for every pearl, the denominator is anything but uniform. Hence the coarse density is large (small α) in small pearls but small (large α) in large ones. As k increases, both the pearl sizes and the coarse densities become increasingly unequal. One can think of increasingly small pearls as being made of increasingly valuable "stuff" used increasingly sparingly to ensure that all pearls are of identical value.

3.2 Noncritical cases; multifractality of the invariant measure and Hölder analysis

As $\rho \to 0$, inversion with respect to a circle becomes practically linear. It is easily verified that the originally uniform mass is restricted to a fractal dust of very small dimension very close to a Cantor dust with a uniform measure on it. The inequality pits against one another a group of Indra's pearls roughly equal to one another and empty portions of C.

In the supercritical case, $\rho > \rho_{crit} = \sqrt{3}$, the circles C_1, C_2, and C_3 overlap, which raises special problems.

In the subcritical case when ρ is between 0 and $\sqrt{3}$, the degree of inequality is reminiscent of what is observed for the Cantor dust.

Figure 2 of Chapter C17 illustrates a case in which ρ is slightly below $\sqrt{3}$ to achieve an esthetic effect, and the inversion prevents ρ from being read directly from the figure.

3.3 The "critical case" when the common radius ρ of C_1, C_2, and C_3 takes the value $\rho_{crit} = \sqrt{3}$

Chapters C20 and C21 investigate the critical case $\rho = \sqrt{3}$. For this value, each circle touches its two neighbors. After inversion with respect to the point where C_2 and C_3 touch, the two circles touching each other have become parallel straight lines. This case yields the Minkowski measure, and hence is of greatest interest in this part.

4. SKETCH OF THE DIFFUSION-LIMITED AGGREGATES

4.1 Construction of DLA

A DLA cluster is generated by allowing an "atom" to perform Brownian motion until it hits an initial "seed." In Figure 1, the seed is also an atom; in Figure 2, it is the (opened up) bottom of a half-cylinder. When the atom and the seed hit, they are "fused," and a fresh Brownian atom is launched against the enlarged target. By a classic result of Kakutani 1944, the distribution of the hitting points is governed by the "harmonic measure" relative to the Laplacian potential. In the alternative dielectric breakdown model (DBM), the growth rules are based explicitly on this measure (Niemeyer et al. 1984).

Overwhelming evidence from computer simulations shows that the arrival of many atoms transforms the seed into a cluster that shows about the same degree of complexity at all scales of observation. Hence any mathematical definition of the concept of fractal must be constrained to include DLA.

The simplicity of the growth rules of DLA and its basic role in understanding many physical phenomena have motivated extensive quantitative studies (Pietronero & Tosatti 1986, Feder 1988, Meakin et al. 1988, Vicsek 1989, Aharony & Feder 1989). However, a full theory is not available. Even a more informal understanding of the resulting complex structure is still lacking. One reason, in my long-held opinion, resides in definite divergences from strict self-similarity.

FIGURE C19-1. A very large sample of plane DLA, called "circular" because it is grown from a seed reduced to one atom.

At an early stage, those deviations were thought to be no worse than those relative to critical phenomena. The latter has a well-developed theory, and it was hoped that a theory of DLA could be achieved in the absence of a careful and complete description. This optimistic view is no longer widely held, and a careful description cannot be neglected.

FIGURE C19-2. A smallish sample of plane DLA, called "cylindrical" because it is grown from the bottom of a half-cylinder (opened up). This DLA is small enough to compute the Laplacian potential and draw its isolines. The latter are a graphic device but also much more: an essential tool of study. A curious visual resemblance is thereby created between DLA and the Mandelbrot set. Of the two, DLA has proven the more resistant to analysis.

4.2 An ironic story: accepting certain "anomalies" in the harmonic measure around DLA as "real" spurred the search for analogous anomalies in mathematical structures and led to Minkowski's casual plaything

Also, Halsey et al. 1986 proposed that the harmonic measure is multifractal and that the proper tool of study is provided by the multifractals that originated in M1974f. The discussion of DLA has tended to split between considerations of small and large values of α.

The bulk of the attention went to the tips where the cluster grows. There the harmonic measure is highest, and by a general theorem of Makarov 1997, the tips form a dust of dimension 1. At the other extreme, the value of α_{max} would have served to quantify the "degree of inaccessibility" of the deepest points within the DLA "fjords." Therefore, an active search for α_{max} ensued. Unfortunately, different investigators—often the same one in successive attempts—kept reaching contradictory conclusions.

Carleson & Jones 1992 had made me aware that $\alpha_{max} < \infty$ follows from self-similarity. Self-similarity is a geometric property that cannot be tested directly, only through quantitative symptoms. This led me to the wild thought that α_{max} might be actually infinite, meaning that the harmonic measure is exponentially negligible with probability one. This and the failure of strict self-similarity to hold for DLA would have been a very significant conclusion had time and collaboration been available to help me investigate DLA further.

4.2 The physical validity of the motivation from DLA has been questioned

The mathematical reader who will not even glance at Chapter C—may welcome an "advance epilogue" here.

The tests described in M & Evertsz 1991 suggest $\alpha_{max} = \infty$. This implies that the harmonic measure seems to vanish exponentially almost everywhere along the cluster boundary. This is particularly so in the far ends of the very thin "fjord branches."

One must not rush to ask, "why should it be so?" This possibility is so extreme that it deserves further tests to be either confirmed or shown misleading. In fact, there has been recent talk of an even more extreme possibility. Deep in the fjords, the harmonic measure may actually be even less than exponentially negligible. It may actually vanish. This would mean that the $f(x)$ graph has an upper bound less than 1 at infinity.

This sequence of influences and opinions exemplifies the special difficulties and risks associated with the practice of physics along a murky frontier. Once a hint has been transmitted from science to mathematics, repeated experience suggests that it continues on its own: an idea's success in one field neither guarantees nor hinders its success elsewhere.

Be that as it may, my original interest in DLA extended our knowledge of the Minkowski measure, but only by adding a few wrinkles.

4.3 DLA is a very difficult but fascinating topic

Careful evaluations of the harmonic measure are extraordinarily computer-intensive and demand a great deal of experience. More generally, my work on the structure of DLA contributed to showing that this is an extraordinarily subtle topic. But — to my great regret — the same work has all too often led to this topic being viewed as risky and unpromising. The interested reader may want to go beyond Chapter 22 and consult my other papers in the bibliography as well as the works they cite. In any event, I still view DLA and its departure from self-similarity as fascinating and well worth continuing study. However, limited time and resources made my latest papers favor an approach not based on α_{max}. Instead, I studied the distribution of gaps bounded by the intersections of the cluster branches with a large circle centered at the cluster's seed.

Our present understanding of DLA is murky. A popular scenario takes for granted that self-similarity will prevail, not only in the limit but for sizes beyond some practically accessible threshold. This implies that α_{max} will increase until it stops increasing — a supposition that, at present, cannot be confirmed or refuted by experiment.

A confirmation that DLA requires $\alpha_{max} = \infty$ would consolidate the opinion that DLA is a very difficult topic indeed.

Physical Review Letters: 60, 22 February, 1988, 673-676.

C20

Invariant multifractal measures in chaotic Hamiltonian systems and related structures (Gutzwiller & M 1988)

✦ **Abstract.** The coding of chaotic trajectories in Hamiltonian systems, and the stochastic reflection of points on circles examplify a new kind of multifractal measure, whose Hölder α ranges from 0 to ∞. ✦

1. Introduction

This letter introduces and investigates some unusual multifractal measures on the real line which we have found independently, and for quite different reasons. Their salient feature is that $\alpha_{min} = 0$, and/or $\alpha_{max} = \infty$, where α (in the notation of M 1982{*FGN*}, p 373, which is now widely accepted) is a Hölder exponent. These multifractal measures have the same general geometric origin, and we believe that they are the simplest representatives of a large class. Gutzwiller encountered his examples in classical Hamiltonian systems that display "hard chaos." Mandelbrot encountered his in stochastic reflections of points on three or more circles, which may be cyclically tangent. Our first example, which is even simpler, involves elementary arithmetic. It arose originally in the study of the third example, yet has turned out to be equivalent to the second. We will present our multifractals and their properties and explain the underlying physics and the geometric motivation, mixing mechanics and geometry with arithmetic and analysis. Beyond the new examples of multifractal measure, the analysis provides a better understanding of the structure of phase space in Hamiltonian systems with hard chaos.

2. First example: a simple problem from arithmetic

Let us represent a real number η by its continued fraction

$$n_0 + \cfrac{1}{n_1 + \cfrac{1}{n_2 + \dots}}.$$

Suppose that $0 < \eta < 1$, i.e., $n_0 = 0$ and $n_k \geq 1$ for $k > 0$. Then the sequence n_k attaches to η the real binary number β (with $0 < \beta < 1$) made of $n_1 - 1$ times 0, followed by n_2 times 1, followed by n_3 times 0, and so on.

The functions $\beta(\eta)$ and $\eta(\beta)$ are continuous and monotone increasing. Since $\beta(1-\eta) = 1 - \beta(\eta)$, it suffices to represent $\beta(\eta)$ for $0 \leq \beta \leq 0.5$ (Figure 1) (The x and μ scales are explained in the second example).

However, the graph of $\beta(\eta)$ is not a devil's staircase, defined as the integral of a measure carried by a Cantor dust. In fact, $\beta(\eta)$ is singular and its derivative does not vanish in whole intervals but on an everywhere dense set. The bottom of the curve shows this feature quite clearly: small values of η yield $n_1 \simeq 1/\eta$, and therefore

$$\beta \simeq \exp(-n_1 \log 2) \simeq \exp[-(\log 2)/\eta].$$

The same thing happens every times η gets close to a rational number, since there is then a large integer n_i in its continued fraction.

Neither is the curve $\beta(\eta)$ the integral of the classic binomial multifractal measure (M 1982{*FGN*}, p 277), which refers to the set of points for which the digits 0 and 1 in binary representation have some specified limit frequencies q and $1-q$. Nevertheless, the increments of $\beta(\eta)$ define a multifractal measure.

For each interval $(\eta, \eta + \Delta\eta)$, one defines the Hölder α

$$\alpha = \log[\beta(\eta + \Delta\eta) - \beta(\eta)] / \log \Delta\eta = \log \Delta\beta / \log \Delta\eta.$$

When the interval is chosen at random, the first descriptive characteristic of a multifractal is the distribution of the probability density of α. An important observation was made by Frisch & Parisi 1985; see M1999N, pages 94 and 96. It implies that the logarithm of this density is a fractal dimension. Recent papers describing applications of multifractals, beginning with Halsey *et al.* 1986, denote this logarithm of density by $f(\alpha)$.

We have evaluated $f(\alpha)$ directly, from the numbers of occurrences of various $\Delta\beta = (\Delta\eta)^\alpha$ corresponding to a preselected $\Delta\eta$. The largest $\Delta\beta$ for a given $\Delta\eta$ was found to arise when η is the golden mean γ, and $\beta = 2/3$. Thus $\alpha_{min} = \log 2 / \log(1/\gamma^2) = 0.7202$, as confirmed numerically. The plot of $f(\alpha)$ has an extremely long tail as shown in Figure 2. In fact, about 10 of data are not plotted, because the $\Delta\beta$ were approximated by 0 in quadruple accuracy (precision of 112 bits $< 10^{-33}$). For rational η, one finds

$$\alpha = \lim_{\Delta\eta \to 0} \frac{\log \Delta\beta}{\log \Delta\eta} = \frac{(\log 2)/\Delta\eta}{\log \Delta\eta} = \infty.$$

{P.S. 2003. By developing this argument a bit further, one finds that the zeros of $d\beta$ are exponential.}

The inverse curve $\eta(\beta)$ has very steep sections; its $f(\alpha)$ has the complementary shape: a long "head" in which $f(\alpha)$ is very close to α (observe that α is always an upper bound for $f(\alpha)$), and an ordinary short tail with $\alpha_{max} = \log(1/\gamma^2)/\log 2 = 1.3885$, as confirmed numerically. Much larger samples were calculated; but double accuracy (8 bytes) is quite sufficient in this case; no $\Delta\eta$ are missing from Figure 3.

3. Second example, motivated by Fuchsian groups

In its simplest form, this example involves three maps of the real line on itself: $x \to f_0(x) = 1/x$, $x \to f_R(x) = 2 - x$ and $x \to f_L(x) = -2 - x$. The square of each of these maps is the unit map. The group based on these three maps associates to every word, w, i.e., to every ternary sequence of letters 0, R

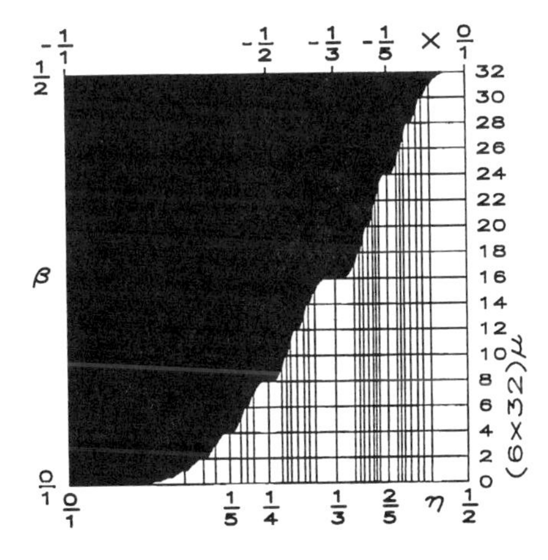

FIGURE C20-1. A "slippery devil's staircase". This Figure has two interpretations. The η and β scales concern the function $\beta(\eta)$ in the first example, computed for $0 < \eta < 1/2$. The abscissas of the vertical lines form the first levels of a Fairey series. The x and μ scales concern the function $\mu([-1, x])$ in the second example. This graph demonstrates the occurrence of very small values of the Hölder α.

and L, the product of a sequence of the maps f_0, f_R and f_L. This group is the reduction to the real line of a bare-bones "Fuchsian group" whose fundamental domain is the singular triangle bounded by the circle of radius 1 centered at 0, and the vertical lines $x = 1$ and $x = -1$. The required knowledge of Fuchsian groups reduces to two facts.

A) When the present group is extended to the complex plane by replacing x by z, it has the whole real line as its limit set. That is, given the orbits z_0, under the actions of all the words w, the set of limit points of these orbits is the whole real line for every z_0.

B) The group extended to the complex plane is continuous on the real line, and is discrete elsewhere. The orbit x_k of a real x_0 under the word w can return arbitrarily close to x_0 without returning exactly to x_0.

Now we move away from the Fuchsian background. We restrict the words so that two successive letters are different and the first letter is not 0 if $x \in [-1, 1]$, is not L if $x < -1$, and is not R if $x > 1$. Then each application of f_0 on an interval is contracting, and the application of f_R and f_L remain length-preserving. Therefore, there exists an attractor that is independent of x_0, hence only depends on the word w. It can be any of the points $x = 1$, $x = -1$, or $x \pm \infty$; it can be a cycle of period ≥ 3 (if the word is periodic with the same period, after a finite number of arbitrary letters); it can be chaotic. Independently of x_0, the randomly chosen word w generates a random orbit, and the $x_k(w)$ form a sample from an underlying invariant measure, which is multifractal. This measure depends on the

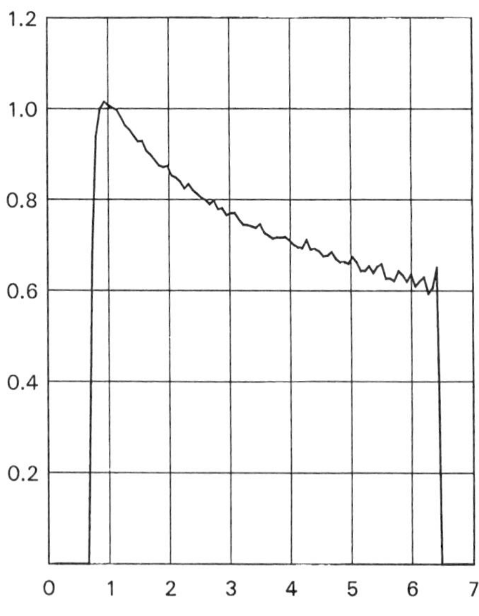

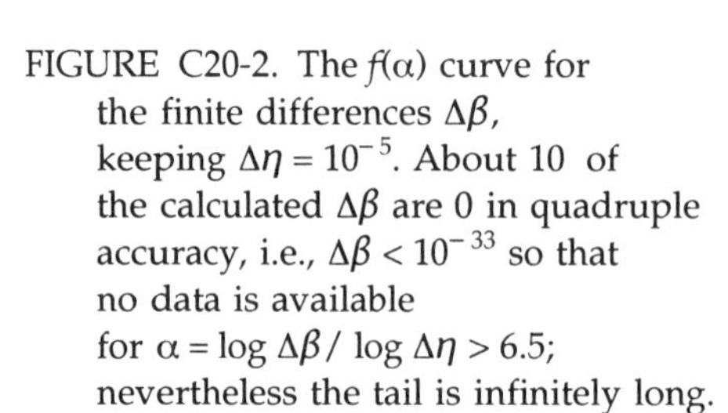
FIGURE C20-2. The $f(\alpha)$ curve for the finite differences $\Delta\beta$, keeping $\Delta\eta = 10^{-5}$. About 10 of the calculated $\Delta\beta$ are 0 in quadruple accuracy, i.e., $\Delta\beta < 10^{-33}$ so that no data is available for $\alpha = \log \Delta\beta / \log \Delta\eta > 6.5$; nevertheless the tail is infinitely long.

random process ruling w. Figure 4 shows how this measure is distributed for a very long random word.

Examine for $0 < x < 1$ the function $\mu([-1, x]) = (1/3)\beta(1/2 + x/2)$, where $\beta(\eta)$ is the function in the first example. It will seen to be the multifractal μ corresponding to symmetric binomial words, in which, given the previous letters, the two non excluded possibilities for the next letter are of probability 1/2. This explains the scales to the bottom and to the right of Figure 1.

By obvious symmetries $\mu([-\infty, -1]) = \mu([-1, 1]) = \mu([1, \infty]) = 1/3$, and the distribution of μ is expressed most compactly in terms of $\beta(\eta)$. The rule is that if $\eta(p2^{-k}) = n'/d'$ and $\eta((p+1)2^{-k}) = n''/d''$ with p, k, n', n'', d' and d'' being integers >0, one has $\eta((p+1/2)2^{-k}) = (n'+n'')/(d'+d'')$. This is the Fairey subdivision.

It is easy to expand to asymmetric binomial words, and to letter sequences ruled by Markov or other process without long memory.

4. Third example, motivated by a Hamiltonian system

This example starts with the singular quadrangle D_2 in the upper (x, y) plane whose boundaries are the circles of radius 1/2 centered on $x = -1/2$ and $x = +1/2$, and the vertical lines $x = -1$ and $x = +1$. If one identifies the opposite sides, this quadrangle becomes a torus. This torus has one exceptional point, however, because the vertex of the quadrangle is infinitely far

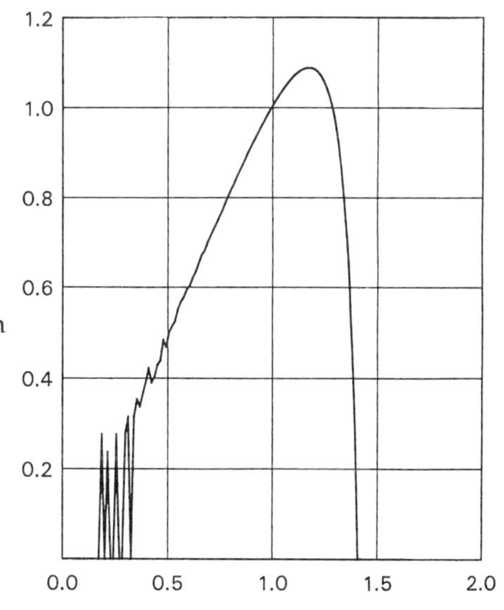

FIGURE C20-3. The $f(\alpha)$ curve for the finite differences $\Delta\eta$ for constant values of $\Delta\beta = 2^{-26}$. No data are missing; the curve starts linearly with slope 1, indicating very small values of α. This linear dependence for $\alpha < 1$ is obscured because very little data is available for small α. The largest jumps in η occur at $\beta = 0$ and 1, where $\Delta\eta = 1/26$, so that $\alpha = \log \Delta\eta / \log \Delta\beta \geq \log 26/26 \log 2 = 0.1786$, as is indeed observed.

away in the hyperbolic metric $ds^2 = (dx^2 + dy^2)/y^2$. (One can think of the torus as a closed box, and of the exceptional point as a narrow opening for the particle to enter and to exit, as if it were scattered from some molecule with an internal degree of freedom, Gutzwiller 1983).

The next step is inspired by Series 1985, 1986. She considers a singular triangle D_1, the right-hand half of D_2, which is bounded by the circle of radius 1/2 centered in $x = 1/2$, and by the vertical lines $x = 0$ and $x = 1$. A geodesic is represented by an Euclidean circle, centered on the x-axis, which goes from $x = \xi$ (past) to $x = \eta$ (future). Consider the special case $\xi < 0 < \eta$; the geodesic enters D_1 through the vertical side $x = 0$. If $0 < \eta < 1$, the geodesic exits D_1 through the circle, and is brought back into D_1 by the transformation $z' = z/(1-z)$. If $1 < \eta < \infty$, the geodesic exits D_1 through the vertical $x = 1$, and is brought back into D_1 by $z' = z - 1$. In the first case, the geodesic is said to turn right, while in the second case, it makes a left turn.

Continuing in this manner, the future of the geodesic is associated with a sequence of letters R and L, or equivalently with a sequence of integers $(n_0, n_1, n_2, \ldots)$ which indicates n_0 times L, followed by n_1 times R, followed by n_2 times L, and so on. The letter sequence is best represented by a real β with the same binary expansion, while, most remarkably,

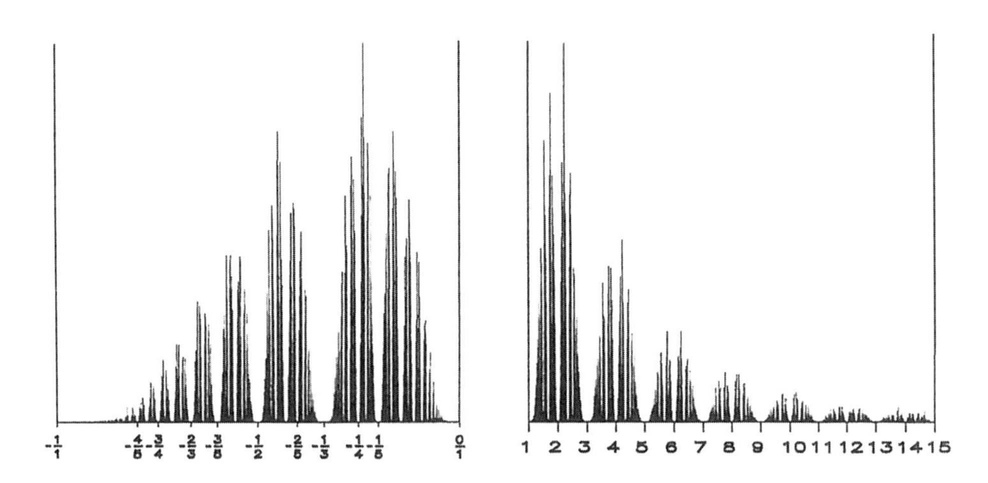

FIGURE C20-4. Sample measures of small intervals of x for the symmetric binomial multifractal measure defined by inversions in 3 circles. Figure 4.1: plotted for $-1 < x < 0$. Figure 4.2: plotted for $1 < x < 15$.

$$\eta = n_0 + \cfrac{1}{n_1 + \cfrac{1}{n_2 + \cfrac{1}{\cdots}}}.$$

Therefore, the functions $\beta(\eta)$ and $\eta(\beta)$ at the beginning of this Letter connect the intuitive description of the geodesic β and its coordinate with respect to the singular triangle η.

Functions with the same unusual characteristics appear in the singular quadrangle D_2, and can be interpreted as physical properties of chaotic Hamiltonian systems. The starting point ξ and the endpoint η of a geodesic can be used as coordinates in a Poincaré surface of section with the invariant element of area $d\xi \; d\eta/(\xi - \eta)^2$. The four sides of the quadrangle correspond to the intervals $\{1\} = (-\infty, 1)$, $\{2\} = (-1, 0)$, $\{3\} = (0, 1)$, and $\{4\} = (1, \infty)$ on the x-axis. Entry and exit of the geodesic are determined by the intervals to which ξ and η belong.

Opposite sides of D_2 are mapped into one another by $z \to z' = (az + b)/(cz + d)$, written as $(a, b; c, d)$ in flattened notation, where a, b, c, d are real and $ad - bc = 1$. Side $\{1\}$, the vertical $x = -1$, is mapped into $\{3\}$, the circle around $x = 1/2$, by $A = (\delta, \delta; \delta, (1 + \delta^2)/\delta)$; side $\{4\}$, the vertical $x = 1$, is mapped into $\{2\}$, the circle around $x = -1/2$, by $B = (\delta, -\delta; -\delta, (1 + \delta^2)/\delta)$. The parameter $\delta > 0$ describes a family of non-equivalent geometries on D_2. A further parameter γ could be introduced, but would require a less symmetric singular quadrangle than D_2.

The "story" of a trajectory in D_2 is told by the sequence of sides it crosses. This story is encoded in a word with an alphabet of three letters,

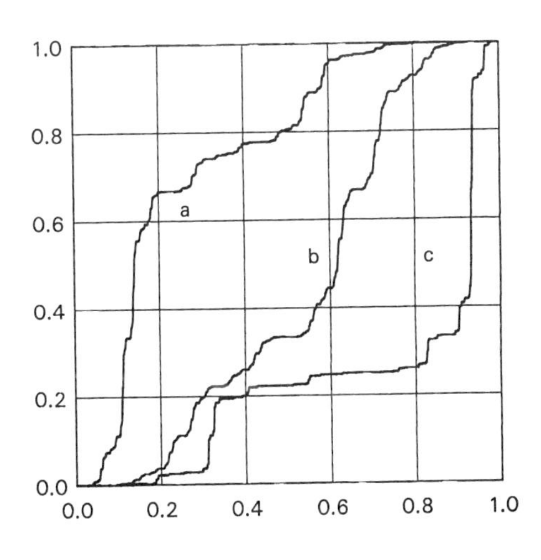

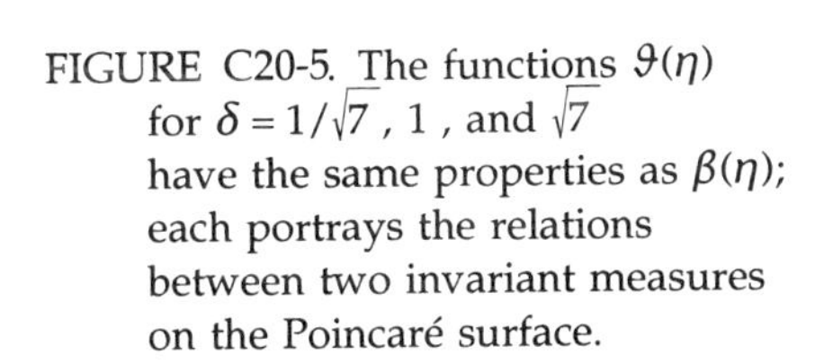

FIGURE C20-5. The functions $\vartheta(\eta)$ for $\delta = 1/\sqrt{7}$, 1, and $\sqrt{7}$ have the same properties as $\beta(\eta)$; each portrays the relations between two invariant measures on the Poincaré surface.

$R(=\text{right}=0)$, $S(=\text{straight}=1)$, and $L(=\text{left}=2)$ in the obvious manner: following {1}, side {2} is right, {3} is straight, and {4} is left; etc. The computation for the future sequence depends only on η, while the past sequence depends only on ξ. Thus, a function from $0<\eta<1$ to $0<\vartheta<1$ is defined where the ternary expansion of ϑ gives the story of the trajectory. Figure 5 shows $\vartheta(\eta)$ for the three values $\delta=1/\sqrt{7}$, $\delta=1$, and $\delta=\sqrt{7}$.

The multifractal analysis of $\vartheta(\eta)$ gives results like $\beta(\eta)$. The infinitely long tail in the $f(\alpha)$ curve is now due to the existence of parabolic elements in the group which is generated by A and B; e.g. the sequence 0000 leads to $BAB^{-1}A^{-1}=(-1,0;\Gamma,-1)$ with $\Gamma=2(2+\delta^{-2})$. A simple argument then shows that $\vartheta(\eta)$ starts with $\vartheta \cong \exp(-\Gamma \log 3/4\eta)$.

The presence of almost flat pieces in the graph of $\vartheta(\eta)$ indicates that the trajectory gets trapped in the exit-entry, as opposed to wandering around the torus, i.e. interacting with the internal degree of freedom. Thus, ϑ gives the physically more informative description than η; but both define invariant measures on the Poincaré surface of section; they can be reduced to the one-dimensional $\vartheta(\eta)$ curve because past and future are decoupled. The $f(\alpha)$ curve characterizes the chaos in this system, because it relates the coding of the trajectories to their physical parameters (initial conditions). This connection is important in the sum over all trajectories (classical approximation to Feynman's path integral), or in the sum over all periodic orbits as in Selberg's trace formula. As shown in Gutzwiller 1980 for the Anisotropic Kepler Problem where the coding is binary, such sums can be computed explicitly as sums over code words.

Several of the Figures were prepared by Réjean Gagné.

(Received 27 October 1987)

Chaos in Australia
Eds. G. Brown & A. Opie, 1993.

C21

The Minkowski measure and multifractal anomalies in invariant measures of parabolic dynamic systems

• *Chapter foreword (2003).* This reprint combines M1993s with an earlier slightly revised reprint, namely, M1995s. •

✦ **Abstract.** The author has recently shown that an important singular non-random measure defined in 1900 by Hermann Minkowski is multifractal and has the characteristic that α is almost surely infinite. Hence $\alpha_{max} = \infty$ and its $f(\alpha)$ distribution has no descending right-side corresponding to a decreasing $f(\alpha)$. Its being left-sided creates many very interesting complications. Denjoy observed in 1932 that this Minkowski measure is the restriction to [0, 1] of the attractor measure for the dynamical system on the line based on the maps $x \rightarrow 1/2 + 1/4(x - 1/2)$, $x \rightarrow -x$ and $x \rightarrow 2 - x$. This paper points out that it follows from Denjoy's observation that the new "multifractal anomalies" due to the left-sidedness of $f(\alpha)$ extend to the invariant measures of certain dynamical systems.

The author's original approach to multifractals, based on the distribution of the coarse Hölder α, also injects approximate measures $\mu_\varepsilon(dt)$ that have been "coarsened" by replacing the continuous t by multiples of $\varepsilon > 0$. This theory therefore involves a sequence of observable approximant functions $f_\varepsilon(\alpha)$; their graphs are not left-sided. ✦

THIS SHORT PAPER primarily means to describe an experience I lived through, and to address a warning to the specialist in multifractals. There is a widespread expectation that the $f(\alpha)$ distribution of every multifractal measure μ satisfies $f > 0$ and has a graph shaped like the mathematical symbol $\bigcap$. From classical results from the 1930s and the 1940s (by Besicovitch, Eggleston et al.), this expectation is fulfilled when μ is the binomial measure, or near binomial.

But for many measures — some theoretical, other obtained from nature — this expectation grossly *fails* to be fulfilled, in one way or another. This creates a variety of so-called "anomalies." Some anomalies are relative to measures in real space; examples include the distributions of turbulent dissipation (the notion of multifractal first arose in that context) and of the harmonic measure around a DLA cluster. These anomalies are beginning to be well recognized; in particular, the references show that I have contributed several papers to their investigation.

I expect that many of the same anomalies will also be encountered for multifractal measures encountered in dynamical sytems. The goal of this paper is to make the students of measures on attractors aware of the preceding references. To do so, I shall discuss the Minkowski measure μ, which is most attractive but extremely anomalous.

When applied to this measure, my approach to multifractals (to be called the *method of distributions*) will be seen to involve necessarily two distinct aspects of $f(\alpha)$. First, there is a theoretical "population" function $f(\alpha)$. For the Minkowski μ, this $f(\alpha)$ is *not* $\bigcap$ -shaped; instead, it is left-sided, i.e., monotone increasing toward a maximum $f(\infty) = 1$ and without a right (decreasing) side. But let the measure μ be coarse-grained to intervals of length ε, in order to become observable. If so, the method of distributions defines an empirical "sample" $f_\varepsilon(\alpha)$ for each ε. This $f_\varepsilon(\alpha)$ is observable, and for the Minkowski μ its shape happens to be altogether different from that of $f(\alpha)$.

As $\varepsilon \to 0$, one has $f_\varepsilon(\alpha) \to f(\alpha)$, which expresses that the theory behind the method of distribution is logically consistent. But the convergence is excruciatingly slow and extremely singular. To extrapolate the shape of $f(\alpha)$ from that of the $f_\varepsilon(\alpha)$ demands very great care.

Those results illustrate a limitation of the "thermodynamical" theory behind $f(\alpha)$. In some cases $f(\alpha)$ is very poorly approximated by the pre-thermodynamical results one obtains when $\varepsilon > 0$, however small ε may be. Contrary to the left-sided multifractals investigated in the Minkowski measure is not obtained by a multiplicative cascade. Nevertheless, several properties of μ were first conjectured on the basis of an approximation of μ by a multiplicative multifractal and later proved to be correct.

When thinking of attractor measures, there is good reason to think first of those supported by strange attractors. But an extraordinary confusion continues to characterize much of the literature on multifractals. This shows that the topic is more delicate than many realize, and that it is best to tackle each issue after every extraneous difficulty has been eliminated.

1. The Minkowski measure on the interval [0, 1]

The *Minkowski measure* μ and the *inverse Minkowski measure* $\tilde{\mu}$ are, simple to define and work with, but exhibit very interesting and totally unexpected peculiarities. They are respectively, the differentials of two increasing singular functions: $M(x)$ and its inverse $X(m)$.

It is easier to start by defining the inverse Minkowski function $X(m)$, which is constructed step by step, as follows. The first step sets $X(0) = 0$ and $X(1/2) = 1/2$. The second step interpolates: $X(1/4)$ is taken to be the Fairey mean of' $X(0)$ and $X(1/2)$, where the Fairey mean of two irreducible ratios (a/c) and (b/d) is defined as $(a+b)/(c+d)$. More generally, the k-th step begins with $X(m)$ defined for $m = p2^{-k}$, where p is an even integer, and uses Fairey means to interpolate to $m = p2^{-k}$, where p is an odd integer. Finally, $X(m)$ is extended to the interval $[1/2, 1]$ by writing $X(1-m) = 1 - X(m)$. The resulting function $X(m)$ is continuous, increases in every interval, and is singular; that is, it has no finite derivative at any point. It has an inverse function $M(x)$ with the same properties, illustrated by the "slippery staircase" in Figure 1.

The differentials $\tilde{\mu}$ and μ of the functions $X(m)$ and $M(x)$ are singular measures. The two parts of Figure 2 illustrate the measure μ, as evaluated for intervals of length 10^{-5}.

Minkowski 1911 (Vol 2, p. 50-51), had called $M(x)$ the "?(x) function," function," a term with few redeeming features. Little (if anything) was written about the measure μ until Denjoy 1932 observed that it has the following property. It is the restriction to [0, 1] of the attractor measure for the dynamical system on the line based on the maps

$$x \to \frac{1}{2} + \frac{1}{4(x-1/2)}, \quad x \to -x \ \text{ and } \ x \to 2 - x.$$

To transform such a collection of functions into a dynamical system, the standard method is, of course, to choose the next operation at random.

This method was used in Plates 198 and 199 of M1982F, and the current (and recent) term for it is IFS: *iterated function system.*

Thanks to this interpretation, $M(x)$ proves to have deep roots in number theory (modular functions) and in Fuchsian or Kleinian groups. From this paper's viewpoint, however, the main virtue of the above dynamical system lies in its extraordinary simplicity. I surmise that any complication or difficulty encountered in the study of its invariant measures will *a fortiori* appear in more complex systems grounded in physics. Moreover, one must keep in mind that Gutzwiller & M1988{C20} had two motivations. I was concerned with the above maps, but my co-author was concerned with an important Hamiltonian system in which x is the Liouville measure and m a second invariant measure yielding equally interesting information about individual trajectories.

2. The functions $f(\alpha)$ and $f_\varepsilon(\alpha)$ of the Minkowski measure

2.1. The theoretical function $f(\alpha)$. For a derivation of the $f(\alpha)$ functions of the Minkowski measure μ and of the inverse Minkowski measure $\tilde{\mu}$, we must refer the reader elsewhere (unpublished). A first basic fact is that

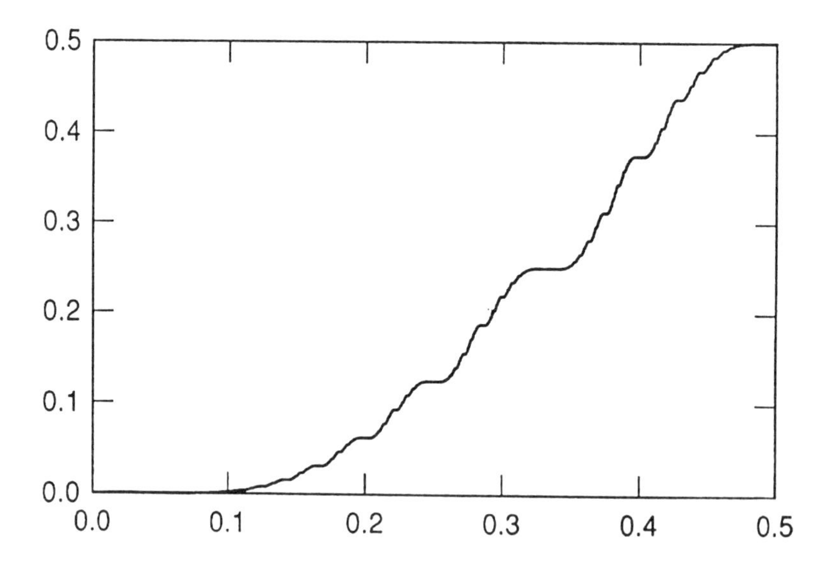

FIGURE C21-1. Graph of the Minkowski function $M(x)$ for $0 < x < 1/2$. Contrary to the well-known Cantor devil staircase (M1982F, Plate 83), the graph of $M(x)$ has no actual steps, only *near steps* that led Gutzwiller & M1988{C20} to call it a *slippery staircase.*

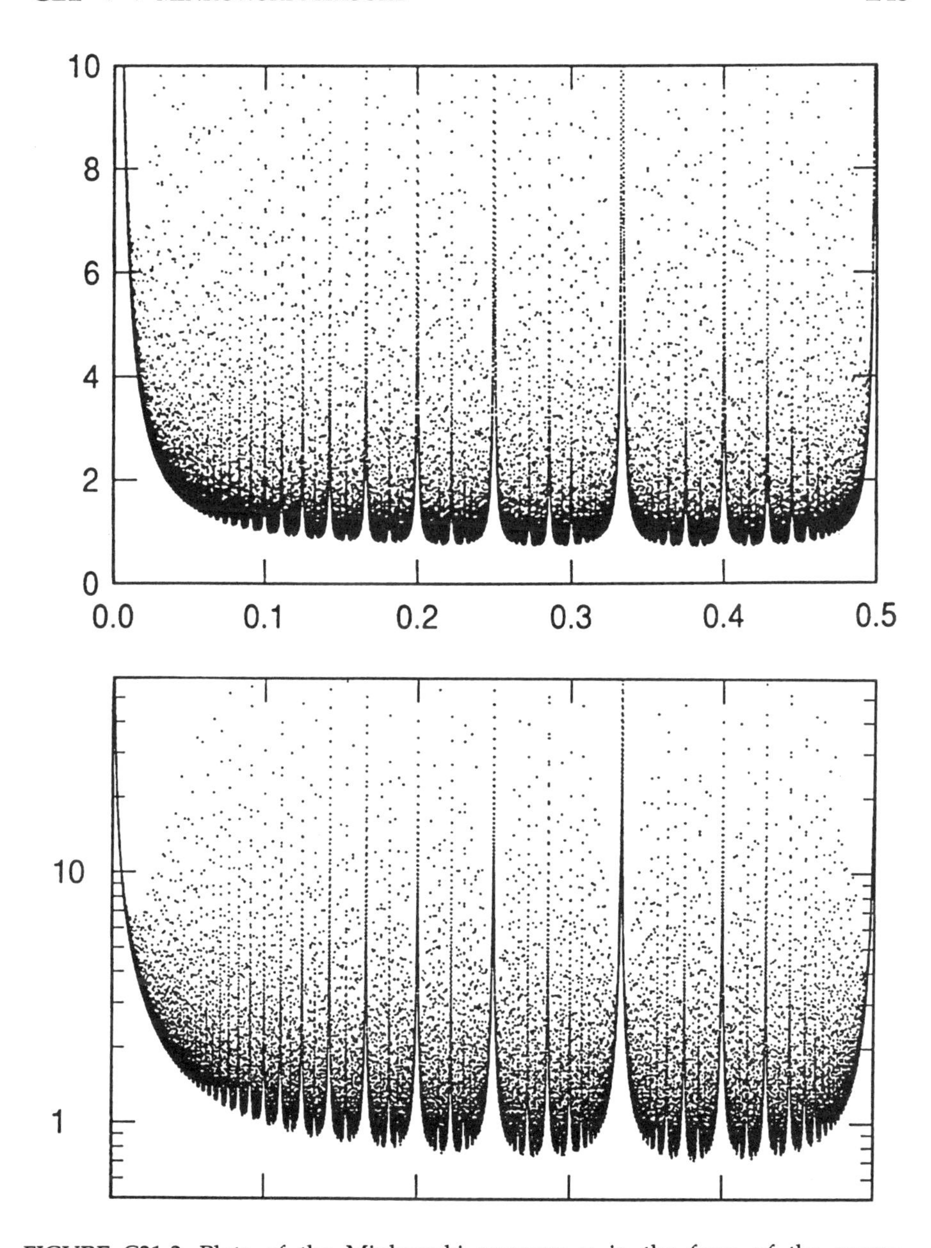

FIGURE C21-2. Plots of the Minkowski measure μ in the form of the coarse-grained Hölder exponent and of its logarithm. Taking $\varepsilon = 10^{-5}$, we evaluated the increments $\Delta M = M[(k+1)\varepsilon] - M[k\varepsilon]$. Figure 2a (top) plots the coarse-grained Hölder $\alpha = \log \Delta M / \log \varepsilon$, and Figure 2b (bottom) plots $\log \alpha$. The theory described later in the paper shows that the fine-grained (local) Hölder almost everywhere exceeds any prescribed α. In this figure, this property altogether fails to be reflected.

there is a theoretical $f(\alpha)$ for these measures: $f(\alpha)$ is the Hausdorff dimension of the set of points x such that the Hölder exponent $H(x)$ takes the value α. In the case of μ, the graph of $f(\alpha)$ has the following properties:

- $f(\alpha)$ is defined for $\alpha \geq \alpha_{\min} = -1/\log_2 \gamma^2 \sim .7202...$ where γ is the golden mean $\sim .6180....$ (obtained in Salem 1943 and — independently — in Gutzwiller & M 1988)
- $\alpha_1 = [2\int_0^1 \log_2(1+x)dM(x)]^{-1} = 0.874...$ (obtained in Kinney 1960)
- $f(\alpha) \to 1$ as $\alpha \to \infty$. This property of $f(\alpha)$ has a strong bearing on the nature of the set of points x where $H(x) > \tilde{\alpha}$: this set is of measure 1.

2.2. The problem of inferring the shape of $f(\alpha)$ from data. After $f(\alpha)$ has been specified analytically, one cannot rest. One must continue by asking whether or not this function can also be inferred when the mechanism of our dynamical system is unknown and only some "empirical data" are available. In this context, "data" may mean one of two things. It may denote the "coarse" (or coarse-grained or quantized) form of the function $M(x)$, as computed effectively for values of x restricted to be multiples of some quantum $\Delta x = \varepsilon$. "Data" may also denote a long orbit of the above dynamical system, that is, a long series of successive values of x; they too must be recorded in coarse-grained format. For many physical quantities in real space, coarse-graining is physically intrinsic; i.e., they are not defined on a continuous scale, but only for intervals whose length is a multiple of some $\Delta x = \varepsilon$ due to the existence of atoms or quanta; in other physical quantities, there are intrinsic limits to useful interpolation; for example, a turbulent fluid is locally smooth. In the present case, coarse-graining is the result of the necessary finiteness of actual computations and of observed orbits.

Given coarse data, there are at least two ways of seeking to extract or estimate $f(\alpha)$.

2.4. The method of moments as applied to the Minkowski distribution. The better-known way to estimate $f(\alpha)$ (Frisch & Parisi 1985 and Halsey et al 1986) deserves to be called the *method of moments.* It starts with the coarse-grained measures $\mu_\varepsilon(x)$ contained in successive intervals of length ε and proceeds as follows: a) evaluate the collection of moments embodied in the "partition function" defined by $\chi(\varepsilon, q) = \Sigma \mu_\varepsilon^q(x)$; b) estimate $\tau(q)$ by fitting a straight line to the data of $\log \chi(\varepsilon, q)$ versus $\log \varepsilon$, and c) obtain $f(\alpha)$ as the Legendre transform of $\tau(q)$.

When applied mechanically to the Minkowski μ, the method of moments yields either nothing or nonsense. More precisely, the more prudent mechanical implementations of the method do not fit a slope $\tau(q)$ without also testing that the data are straight (this can be done by eye). But the Minkowski data for $q < 0$ are not straight at all. Therefore, the prudent conclusion is that "there is no $\tau(q)$. " Since no such difficulty arises for $q > 0$, conclusions of this sort are is often accompanied by the assertion that the data are "not quite" multifractal. The less prudent mechanical implementations of the method of moments simply forge ahead to fit $\tau(q)$. Depending on a combination of the rule used to fit and of the details of how quantification is performed, those methods may yield an estimated $\tau(q)$ that is not convex. The resulting "Legendre transform" is not a single-valued function, and $f(\alpha)$ is a mystery. In other mechanical methods, the difficulty in estimating $\tau(q)$ is faced by first "stabilizing" the estimate in one way or another; such stabilization may yield some sort of $f(\alpha)$, but one can hardly say what it means and what purpose it serves.

A central feature of the method of moments should be mentioned at this point. The limit process $\varepsilon \to 0$ is invoked in estimating $\tau(q)$ from the data. But the preasymptotic data corresponding to $\varepsilon > 0$ do *not* define an approximate $f_\varepsilon(\alpha)$.

2.5. *The method of distributions*. A second way to estimate $f(\alpha)$ is the *method of distributions,* which is used in all my papers listed as references. I have been using it since 1974 and every new development motivates me to recommend it more strongly. The key is simple. While the method of moments rushes to compute the moments of $\mu_\varepsilon(x)$ embodied in the partition function $\chi(\varepsilon, q)$, the method of distributions considers, for every ε, the full frequency distribution of the $\mu_\varepsilon(x)$. These distributions are embodied in graphs statisticians call *histograms.*

First, the range of observed α's is subdivided into equal "bins." and one records the number of data in each bin. If the number of bins is too small, information is lost, but if there are too many bins, many are empty. In the case of the Minkowski μ, there are many α's a little above α_{min} and few α's strung along up to very high values.

Denote by N_b the number of data in bin b. When N_b is large, $N_b/\Delta\alpha$ serves to estimate a probability density for α. When $N_b = 1$ and the neighboring bins are empty, one estimates probabilities by averaging over a suitably large number of neighboring bins; these probabilities are very small.

Having estimated the probability density $p_\varepsilon(x)$, one forms

$$f_\varepsilon(\alpha) = \frac{\log p_\varepsilon(\alpha)}{\log \varepsilon}.$$

Thus, the method of distribution creates a sequence of functions $f_\varepsilon(\alpha)$. Because $f_\varepsilon(\alpha)$ is the normalized logarithm of a measure, each $f_\varepsilon(\alpha)$ is nothing but a histogram that was replotted in doubly logarithmic coordinates and was suitably weighted. These histograms should be evaluated for a series of values of ε. When the measure is multifractal, $f_\varepsilon(\alpha)$ converges to a limit $f(\alpha)$. That is, the function $f(\alpha)$ enters the theory as

$$f(\alpha) = \lim_{\varepsilon \to 0} f_\varepsilon(\alpha).$$

A typical reaction to histograms is, "Why bother? We all know that the information they contain is also found—and in a better organized form—in the moments. Besides, moments are familiar and far easier to handle than histograms." Unfortunately, this typical reaction does not address the complexity of fractals and multifractals.

In the study of fractals, the typical probability distributions follow a power law, and population moments of high order are infinite. The corresponding sample moments — sometimes even the sample average — behave in totally erratic fashion; they bring out no useful information and can be thoroughly misleading.

Now proceed to multifractals. When $f(\alpha)$ is truly $\cap$ -shaped, with $f > 0$, moments raise no major issue, the method of moments works well, and the method of distributions is a less efficient way to obtain $f(\alpha)$. But in all delicate cases, the sample moments embodied in the partition function are treacherous. The method of distributions is the only way to go.

2.6. The method of distributions as applied to the Minkowski measure. Gutzwiller & M1988 used histograms, and Figure 2 (reproduced here as Figure 3) reproduces the empirical $f_\varepsilon(\alpha)$ we obtained. To obtain this graph, we coarse-grained x, then (in effect) we coarse-grained M. The "quantum" of M was tiny, because it was simply the smallest $M(x+\varepsilon) - M(x)$ our computer allowed in quadruple precision. Thus, the values that the computer could not distinguish from 0 (10% of the whole) were not used.

The resulting data-based curve is utterly different from the theoretical left-sided $f(\alpha)$. It begins with an unquestionably cap-convex left side — as usual. The middle part satisfies $f_\varepsilon(\alpha) > 1$, which cannot be true of $f(\alpha)$, but was expected; this is one of the inevitable biases of the method of distributions, and can be handled. Finally, there is cup-convex right side. This was totally unexpected, because a theoretical $f(\alpha)$ is necessarily cap-convex throughout,

We gave up seeking a better test of this cup-convexity. We did not come close to testing my further hunch, that the estimated $f_\varepsilon(\alpha)$ — if extended far enough — would become < 0 for large enough α. We showed that for $q < 0$ the moment $\chi(q, \varepsilon)$ was not a power law function of ε. But, to our disappointment, we did not succeed in evaluating $f(\alpha)$ analytically. We did conjecture the correct actual form of $f(\alpha)$, (but did not write it

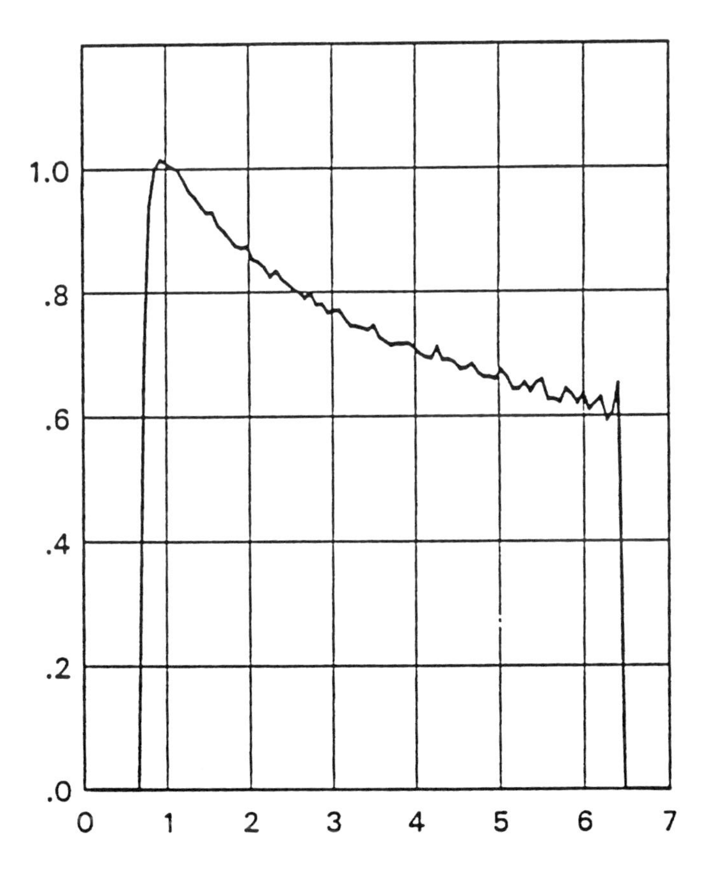

FIGURE C21-3. Early plot of the estimated function $f_\varepsilon(\alpha)$ for the Minkowski measure. Reproduced from Gutzwiller & M 1988{C20}.

down) and were concerned by functions $f_\varepsilon(\alpha)$ and $f(\alpha)$ that differ to such extreme degree.

Recently, I have returned to this problem. Figure 4 was prepared using a method that computes $M(x+dx) - M(x)$ directly, not through $M(x)$. This can be done with arbitrary relative precision, therefore we can reach huge values of α. Figure 4 gives resounding confirmations of the earlier conjectures concerning the existence of a cup-convex right side in the empirical $f(\alpha)$ and of a negative tail.

This sharp mismatch between the theory and even the best experiments spurred me to a rigorous derivation of the theoretical $f(\alpha)$ and of the predicted $f_\varepsilon(\alpha)$. The shape of $f(\alpha)$ has already been mentioned. For $f_\varepsilon(\alpha)$, it suffices to say that, for large ε,

$$f_\varepsilon(\alpha) \sim 1 - \text{ (a constant) } \log \alpha/\log\varepsilon.$$

Figure 4 verifies this dependence on the data.

2.7. Is $f(\alpha)$ a useful notion in the case of the Minkowski μ ? Once again, our recent evaluations of $f_\varepsilon(\alpha)$ did not come close to reproducing the true shape of the graph of $f(\alpha)$, despite the fact that they involved precision that is totally beyond any conceivable physical measurement. Even the early evaluations in Gutzwiller & M 1988{C20} were well beyond the reach of physics.

Given the difficulties that have been described, should one conclude, in the case of the Minkowski measure, that $f(\alpha)$ is a worthless notion? Certainly this measure confirms that I have been arguing strenuously for a long time: that $f(\alpha)$ is a delicate tool. Its proper context is distributions, that is, probability theory. Moreover, it does not concern the best known and more "robust" parts of that theory, namely, those related to the law of large numbers and the central limit theorem. Instead, it concerns the probability theory of large deviations, which is a delicate topic.

3. Remarks

3.1. On continuous models as approximations and on "thermodynamics." Why do physicists study limits that cannot be attained? Simply because it is often easier to describe a limit than to describe a finite structure that can be viewed as an approximation to this limit. In particular, this is why it is often taken for granted in the study of multifractals that a collection of "coarse-grained" approximations can be replaced by a continuous "fine" or "fine-grained" description. The latter involves Hausdorff dimensions and introduces the functions $\tau(q)$ and $f(\alpha)$ directly, not as limits.

For the Minkowski μ, however, the actual transition from coarse to fine-graining (as $\Delta x \to 0$ and $\Delta m \to 0$) is extraordinarily slow and many aspects of the limit differ qualitatively from the corresponding aspects of even close approximations. Therefore, the role of limits demands further thoughts, which I propose to describe elsewhere.

The continuous limit approximation has been described as ruled by a "thermodynamical" description. Thus, the fact that the convergence is slow and singular in the case of the Minkowski μ reveals a fundamental practical limitation of the thermodynamic description.

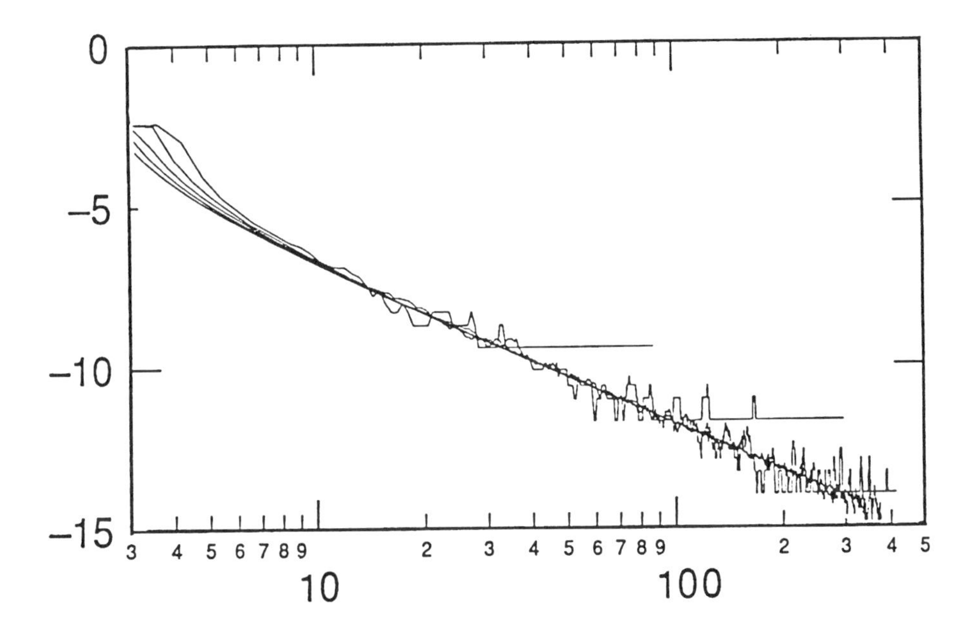

FIGURE C21-4. Plot of the estimated functions $f_\varepsilon(\alpha)$ for the Minkowski measure restricted to the interval [1/10, 1/9]. We started with $f_\varepsilon(\alpha) - 1$ as the vertical coordinate and — in order to straighten $f_\varepsilon(\alpha)$ — we chose $\log \alpha$ as the horizontal coordinate. Then we collapsed the five graphs that correspond to $\varepsilon = 10^{-5}, 10^{-6}, 10^{-7}, 10^{-8}$ and 10^{-9}. This figure strongly confirms the cup-convexity suspected in Figure 3. It shows the occurrence of $f < 0$. Finally, the weighting rule shows that $1 - f_\varepsilon(\alpha)$ decreases as $\varepsilon \to 0$.

3.2. Parabolic versus hyperbolic dynamical systems. To pinpoint the essential ingredient of our special dynamical system, it is important to see how its properties change if the system itself is modified. If one wants $f(\alpha)$ to become $\bigcap$ -shaped, it suffices to replace the first of our three maps by

$$x \rightarrow \frac{1}{2} + \frac{\rho}{x - 1/2},$$

with $\rho < 1/4$. As $\rho \rightarrow 1/4$, the right side of $f(\alpha)$ lengthens and is pushed away to infinity, and the anomalies disappear asymptotically. Formally, the system changes from being hyperbolic to parabolic. Hence, the anomalies we have investigated are due to the system's parabolic. In terms of the limit $f(\alpha)$, the differences between parabolic and hyperbolic cases increase as $\rho \rightarrow 1/4$. But actual observations lie in a preasymptotic range; for a wide range of ε, $f_\varepsilon(\alpha)$ will be effectively the same for ρ close to $1/4$ as it is for $\rho = 1/4$.

3.3. In lieu of conclusion. The apparent "strangeness" of the facts described in this paper *must not* discourage the practically minded reader. Repeating once again a pattern that is typical of fractal geometry, it turns out that what had seemed strange should be welcomed and not viewed as strange at all.

4. Acknowledgements. The work that led to this paper began in 1987. Over the years, I had invaluable discussions with M. C. Gutzwiller, C. J. G. Evertsz, T. Bedford and Y. Peres. Figures 1, 2, and 4 were prepared by J. Klenk. Peres, whose Ph.D. concerned this topic, informed us that our definition was anticipated by Minkowski and Denjoy. Thus, Gutzwiller and I have independently rediscovered μ before we joined forces, and (however imperfectly) the $f(\alpha)$ function of μ was not discussed until our joint paper.

Physica: A 177, 1991, 386-393

C22

Harmonic measure on DLA and extended self-similarity (M & Evertsz 1991)

• *Chapter foreword (2003).* This chapter's fit within the book is non-obvious, hence explained in Chapter C19. •

✦ **Abstract.** DLA is nearly self-similar, but departures from simple self-similarity are unquestionable and quantifying their statistical nature has proven to be a daunting task. We show that DLA follows a surprising new scaling rule. It expresses that the screened region, in which the harmonic measure is tiny, increases more than proportionately as the cluster grows. This scaling rule also gives indirect evidence that the harmonic measure of lattice DLA follows a hyperbolic probability distribution of exponent equal to 1. This distribution predicts that sample moments behave erratically, hence explains why the common restricted multifractal formalism fails to apply to DLA. ✦

TWO NEW SCALING PROPERTIES OF THE HARMONIC MEASURE μ of plane DLA are described in this text. The need for two distinct methods of scaling implies that the notion of self-similarity actually subdivides into several distinct sub-notions. Our more significant new scaling property of μ is unusual, and indicates that DLA satisfies an "extended form" of self-similarity, but not the form that is ordinarily postulated. We compare small DLA clusters with sized-down and coarse-grained clusters, and find them to differ from each other in a systematic and unexpected way. In rough terms, the "screened" region, in which the harmonic measure is tiny, increases far more than proportionately as the cluster grows. It was hoped for a time that the harmonic measure could be represented within a restricted form of the notion of multifractal (Frisch & Parisi 1985, Halsey et al. 1986). This would have implied that μ is self-similar in a strong sense. With local irregularities of μ characterized by the classical Hölder exponent α, the structure of strong self-similarity is characterized either by a function $\tau(q)$ defined for all values $-\infty < q < +\infty$, or by a function $f(\alpha)$, whose graph is shaped like an asymmetric form of the symbol $\cap$. Unfortunately, except perhaps for the sites with the highest growth probability, the results of this multifractal analysis of DLA have been mutually contradictory or otherwise unacceptable. Together with many other authors, we interpret these difficulties as strong evidence that the power-law scaling relations that characterize the restricted multifractals fail to apply to DLA. These observed "anomalies" are intimately related to the behavior of small harmonic measures.

However we have recently shown by explicit examples that the failure of $\tau(q)$ to exist for all q does *not necessarily* imply that a measure is not self-similar. The measures described in M et al. 1990 fit in the context of the more general theory of multiplicative multifractals (M 1974f{N15}, M 1988c, M 1989g), yet the right-hand side of the $f(\alpha)$ is altogether absent, so that such a μ can be called a "left-sided multifractal." But the principal point of this paper goes beyond the evaluation of $f(\alpha)$. We argue that a one-sided $f(\alpha)$ fails to adequately describe the distribution of a left-sided measure μ. Beyond the scaling needed to define $f(\alpha)$, the characterization of other significant aspects of μ requires additional scaling relations.

This paper attacks the problem of the distribution of the harmonic measure μ using a direct probabilistic method (M 1974f{N15}, M 1988c, M 1989g). The main reason for selecting this method is that the attempts to fit the harmonic measure on DLA within the restricted multifractal formalism seem to fail because small μ's are characterized by high statistical scatter. One consequence is that it is hard to draw inferences from small measured μ's. Thus, the exponential decay of the minimum probability (as postulated in Blumenfeld & Aharony 1989) would be difficult either to confirm by direct methods, or to confront with other alternatives. Another consequence is that statistical techniques such as the partition function become unreliable at best.

Our numerical work was done for both circular and cylindrical DLA. Our figures concern circular geometry, but cylindrical geometry gives similar results. We grew 10 clusters of $N = 50000$ particles using a random walker algorithm by Y. Hayakawa, which grows the more common site version of DLA. The potential was computed by solving the discrete Laplace equation iteratively on the square lattice underlying the growth, with boundary conditions 0 on the cluster and 1 on a circle with radius equal to 3/2 the overall size of the clusters. The harmonic measure μ at a site on the boundary of the cluster is theoretically proportional to the gradient of the potential, but we approximate it by the potential at the nearest neighbors of the cluster, and then normalize (Niemeyer et al. 1984). For each cluster, μ was determined at the successive stages of growth $N = 781, 3125, 12500, 50000$; i.e., for $N_k = 4^k N_0$, with $N_0 = 781$ and $k = 0,1, \ldots 3$. At each stage, k, we estimated the probability density $p_k(\alpha)$ of the Hölder $\alpha = \log\mu_i / \log N_k$ by determining the sample frequency of α for each cluster and then averaging over the 10 clusters.

If the harmonic measure on DLA had been a restricted multifractal, one would expect the quantities $C_N(\alpha) = (1/\log N) \log p_N(\alpha)$ to converge to the better known quantities $f(\alpha)/D_0 - 1$ as $N \to \infty$ (M et al. 1990, M 1974f). The plots of $C_N(\alpha)$ would therefore provide an approximation of $f(\alpha)$. (These rescaling rules are usually expressed in terms of the size L of the cluster, but in this case $L \sim N^{1/D}$). Also note that the actual convergence to $f(\alpha)$ may be extremely slow and the approximation of $f(\alpha)$ by $C_N(\alpha)$ may be poor. Nevertheless, Figure 1 shows a high level of bunching for low α. But the right sides show no sign of converging as $N \to \infty$. The difficulties encountered by Legendre estimation of $f(\alpha)$, which starts with $\tau(q)$, appear to originate solely with these tails.

Figure 2 shows the same data replotted in nonstandard fashion in physics, which we call *positive Cauchy rescaling*. The abscissa is taken to be $\log \mu / \log N - \log \log N$, and the ordinate is taken to be $\log p_N(\alpha)$ itself, without any renormalization. Now, it is the right sides of the resulting graphs that collapse into a single one.

The seemingly peculiar rescaling used to obtain Figure 2 was picked neither by chance, nor by trial and error, but was inspired by a limit theorem that is little known in physics, in which the limit is the asymmetric or positive Cauchy random variable. This theorem enters via the theory of multiplicative multifractals (M 1974f{N15}). Before we restate the main ideas of that theory, let us acknowledge that in the case of DLA, the successive stages we shall describe are conjectural.

Consider a regular grid of base 2 on the unit interval [0, 1] and denote by $I(\beta_1, \beta_2, \ldots, \beta_k)$ the interval $[t, t+\varepsilon]$, where $\varepsilon = 2^{-k}$ and $t = 0.\beta_1\beta_2 \ldots \beta_k$ (in binary representation), with β_i equal to 0 or 1. The structure of a multiplicative multifractal is determined by a random variable M, whose expectation satisfies $EM = 1/2$ (to insure conservation) and other conditions (M 1989g, Stanley & Ostrowsky 1988). The first stage of the multiplicative cascade begins with mass equal to 1 on [0, 1] and redistributes it by giving to the subinterval $I(\beta_1)$ the mass $M(\beta_1)$. The second stage gives to the interval $I(\beta_1,\beta_2)$ the mass $M(\beta_1)M(\beta_1,\beta_2)$. After k stages, the interval $I(\beta_1,\beta_2,\dots,\beta_k)$ contains the mass

$$\mu_t(\varepsilon) = M(\beta_1)M(\beta_1,\beta_2)\dots M(\beta_1,\beta_1,\dots,\beta_k).$$

Each realization of the cascade (i.e., each seed in the "random" generator) yields a different "sample." The rules of the multiplication are statistically the same at each step of fine-graining, in the sense that, given t, the multipliers $M(\beta_1)$, $M(\beta_2)$ etc., are independent and identically distributed (i.i.d.). Therefore, the measures resulting from these infinite cascades are *statistically* self-similar and multifractal in the more general sense (M et al. 1990, M 1989g, Stanley & Ostrowsky 1988, M 1974f).

For convenience, we now introduce the random variable $V = -\log_2 M$, whose probability density, $p(v)$, is obtained from that of M, and we write

$$\begin{aligned} H_k &= \log \mu(\varepsilon)/\log \varepsilon \\ &= \frac{1}{k}(V(\beta_1) + V(\beta_1,\beta_2) + \dots + V(\beta_1,\dots,\beta_k)) = (1/k)\sum_{n=1}^{k} V_n. \end{aligned}$$

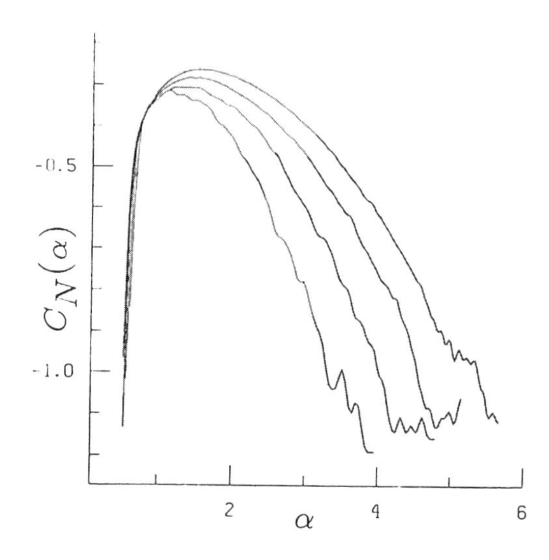

FIGURE C22-1. Cramèr rescaling of the densities of the distributions of the αs estimated from 10 clusters of masses $N = 781$, $N = 3125$, $N = 12500$, and $N = 50000$. This rescaling plots the the ordinate $C_N(\alpha) = (1/\log N)\ln p_N(\alpha)$ and the abscissa $\alpha = -\ln \mu/\ln N$. For a restricted multifractal, the limit of $C_N(\alpha)$ for $N \to \infty$ would be $f(\alpha) - 1$. Here, on the contrary, the right tails of the distributions fail to converge.

This H is a random variable whose sample value, denoted α, is simply the Hölder exponent ("singularity strength"). Here, this exponent is simply the sample average of k i.i.d. random variables $V_1, \ldots, V_k$, with probability density $p(v)$. Sums of such variables are extensively studied in probability theory (Gnedenko & Kolmogorov 1954-1968).

By construction, the above multiplicative cascades yield statistically self-similar measures. Hence one may hope that the probability densities $p_k(\alpha)$ of H_k, corresponding to successive prefractal levels k, can be renormalized (or collapsed) in such a way, that a suitably renormalized version of the density p_k converges to a limit other than 0 or ∞. If a limit exists, it characterizes the fractal properties of the multifractal measure.

First, suppose that the first and second moments of the distribution $p(v)$ exist. This seems almost obvious, but turns out to be a special case. Then two familiar rules of probability theory are valid (Gnedenko & Kolmogorov 1968, Gnedenko 1967). From the law of large numbers and the central limit theorem the sample average $\Sigma V_n / k$ converges to the population (ensemble) expectation $\mathbf{E}V$, and the distribution of $\Sigma\{V_n - \mathbf{E}V\}/\sqrt{k}$ converges to the Gaussian of variance $\mathbf{E}(V^2) - (\mathbf{E}V)^2$.

However, a description of multiplicative multifractals requires far more detailed knowledge. The appropriate limiting distribution is given by a little known rule, namely, *Harald Cramèr's theorem on large deviations* (M 1974c{N16}, M 1988c, M 1989g). It asserts that the probability density $p_k(\alpha)$ of H_k is such that $C_k(\alpha) = (1/k) \log_2 p_k(\alpha)$ converges to a limit $C(\alpha)$. In

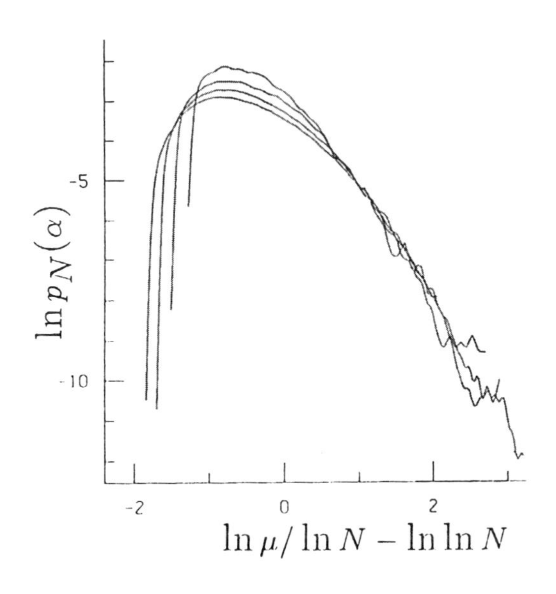

FIGURE C22-2. The same estimated densities as in Figure 1 but with positive Cauchy rescaling. Here the left sides fail to converge, but the right sides do converge.

the case that $C(\alpha)$ is neither 0 nor ∞, this function provides a characterization of the fractal properties of the measure. In the case of restricted multifractals, the function $f(\alpha)$ is equal to $f(\alpha) = D_0 + C(\alpha)$, with $D_0 = 1$ on the interval (M et al. 1990, Frisch & Parisi 1985, Halsey et al. 1986).

This terminates our exposition of basics. However, as M et al. 1990 points out, all the above rules may either fail, or yield trivial results. As a significant preliminary illustration, suppose that the random variables V_n are Cauchy distributed, that is, have the "Lorentzian" probability density $p_1(v) = 1/[\pi(1 + v^2)]$. Probabilists are familiar with the following easily verified fact: in the Cauchy case, $\Sigma V_n / k$ has precisely the same distribution as each of its addends V^n, i.e., $p_k(\alpha) = p_1(\alpha)$. On the other hand $\mathbf{E}V = \infty$ and $\mathbf{E}(V^2) = \infty$, so that the expressions that enter into the law of large numbers and the Gaussian central limit are both meaningless. As to $C_k(\alpha) = (1/k) \log p_k(\alpha)$, it takes the form $C_k(\alpha) = -(1/k) \log [\pi(1 + \alpha^2)]$. Therefore, $C(\alpha) = \lim_{k \to \infty} C_k(\alpha) \equiv 0$, and hence $f(\alpha) \equiv D_0$. In other words, not only do the usual (Gaussian) scaling properties fail altogether, but the Cramèr rescaling yields a degenerate result.

Whenever such is the case, one hopes for a new renormalization scheme and an alternative to the functions $C(\alpha)$ or $f(\alpha)$. It may even happen that, a more complete characterization of the fractal properties of μ, will involve more than one normalization. This is the case in a left-sided fractal measure (M et al. 1990) and — according to the results in this paper — for DLA. For DLA, Cramèrian $f(\alpha)$ normalization produces collapse for low α's (see Figure 3), while positive Cauchy rescaling produces collapse for the high α's shown in Figure 2.

For Cauchy distributed Vs, the fact that the distribution of H_k is independent of k is in itself an unexpected alternative scaling property. It implies that, for different levels of coarse graining, the plots of the densities $p_k(\alpha)$ automatically collapse to a Cauchy distribution. This is a very strong property, because it is known to identify the Cauchy distribution uniquely, among all possible limits of sums of i.i.d. random variables.

In our context, however, the quantities V_n cannot be Cauchy distributed, because $M < 1$ implies $V = -\log M > 0$. But suppose that $\Pr \{V > v\} \sim v^{-1}$, like in the Cauchy case. If so, the variable V is in the domain of attraction of the positive Cauchy law (Gnedenko & Kolmogorov 1968), which can be shown to be intricately related to the case $\lambda = 1$ of the multifractals examined in M et al. 1990. The main implication is that the densities of H_k can be collapsed by subtracting a quantity proportional to $\log k$. In our clusters, $k = \log N$, we therefore recover the rescaling $\alpha - \log \log N$, which is used in Figure 2. Conversely, if a distribution is known to be a limit under the Rescaling procedure leading to Figure 1, that distribution is perfectly determined. Furthermore, one can show that the tail of $f(\alpha)$ for $\alpha \to \infty$ behaves like in the $\lambda = 1$ case in M et al. 1990. That is, $f(\alpha) \simeq D_0 - c \exp(-c'\alpha)$, c and c' being positive constants.

Individual errors in a sample of a Cauchy random variable are of the same order of magnitude as the average error over many samples. This is also nearly the case for the positive Cauchy. This is why sample moments for the harmonic measure on DLA behave erratically, and, thus, why the method of moments fails in estimating $f(\alpha)$.

It is known that $\exp(-L^2)$ is an absolute lower bound for the behavior of the smallest growth probability in lattice DLA (Evertsz et al. 1991, Lee et al. 1989). Assume as the above findings suggest, that H_k is in the domain of attraction of the positive Cauchy. Then $\Pr\{H_k > \alpha\} \sim \alpha^{-1}$. The largest value $\alpha_{max}(N)$ in a sample of N such random variables is expected to satisfy $\Pr\{H_k > \alpha_{max}(N)\} \approx 1/N$, i.e., $\alpha_{max}(N) \approx N$. In a cluster of N sites, one expects the smallest probability to behave like $\exp(-L^D)$, D being the fractal dimension of the cluster. This behavior was assumed for DLA in Blumenfeld & Aharony 1989.

When moments are of no help to describe a distribution, statisticians turn to "quantiles." The tail quantile Ω_r of order r of H_k (with $0 < r < 1$) is defined by $\Pr\{H_k > \Omega_r\} = r$. In the positive Cauchy case, all Ω_r behave like $\log k$. Therefore, imagine that the observed μ's are censored systematically, by erasing the lower values up to a proportion r. Then, the censored minimum $\mu_{min}(r)$ would satisfy $\log(\mu_{min}(r)) \sim -\log L \log\log L$. If censorship is unsystematic, one expects "μ_{min}" to fall somewhere between the quantile's decay and that of the absolute minimum.

The above limiting distribution was obtained by varying cluster size. It is also possible to study the limiting behavior of the density of the probabilities obtained by coarse-graining the harmonic measure on a single large cluster. We have coarse-grained 10 clusters of mass $N = 12500$ and size $L \sim N^{1/D}$, with square boxes of sizes 2^k with $k = 0, 1, 2, \ldots$, and have determined the densities $p_k(\alpha)$. Figure 3 shows the results of Cramèr rescaling of $p_k(\alpha)$ for $k = 1, 2, 3$ and 5. The collapse is very good for $k = 1, 2, 3$, while a positive Cauchy rescaling (which need not be shown here) would yield no collapse at all. Furthermore, comparing the density p_k with the density $\tilde{p}_0$ of clusters of size $2^{-k}L$, one finds that p_k has a much longer right-tail than $\tilde{p}_0$. Thus, the screened regions in which the α's of the harmonic measure are huge seem to increase more than proportionately as the cluster grows.

We conclude that the harmonic measure on DLA can be described as self-similar, both from the point of view of growth and from the point of view of coarse-graining. However, two notions of self-similarity must be involved. Under coarse graining, one may be content with the standard notion that underlies the restricted multifractals. But growth demands an alternative rescaling rule, and an extended notion of self-similarity.

Of course, our experimental discovery that DLA satisfies an extension of self-similarity is not contingent on the theoretical argument that led us to test for positive Cauchy rescaling. This paper raises but does not answer a challenging question concerning the applicability to DLA of the theory of random multiplicative processes. Since, as already noted, the cascade postulated by this theory is still conjectural, why does this theory prove so effective? The same question also arises in the context of turbulence, where restricted multifractals are also insufficient.

This text reproduces IBM research report RC-16595 of August 1990, with minor editorial corrections. The same experiment has now been redone with off-off lattice DLA. The resulting yields graphs are practically indistinguishable from those included in this paper.

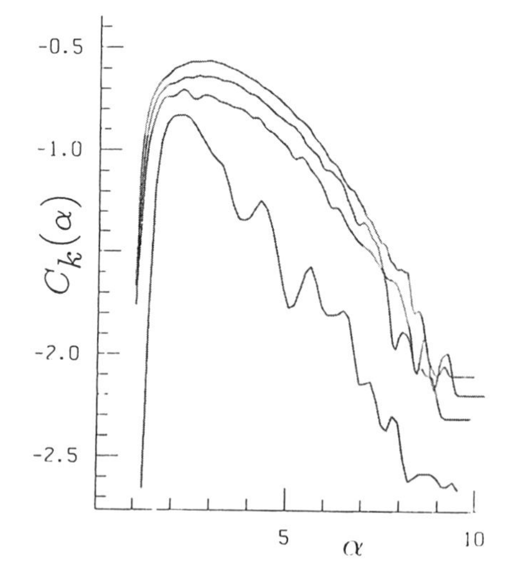

FIGURE C22-3. Cramèr rescaling for the densities of the distributions of the α's estimated after coarse-graining the harmonic measure on 10 clusters of mass $N = 12500$ and radius ≈ 180, with square boxes of sizes 2^k, where $k =$ 1, 2, 3, 5 . Here, Cramèr rescaling consists in plotting $C_k(\alpha) = (1/k) \log_2 p_k(\alpha)$ versus $\alpha = -(1/k)\log_2 \mu$.

PART V: BACKGROUND AND HISTORY

Some chapters in this part are introductions whose aim is to assist even the non-expert in gaining something from this book; the first half of Chapter C_ is also of that kind. Other chapters are a kind of miscellany.

&&&&&&&&&&&&&&&&&&&&&&&&&&&&&&

Draft written about 1982; first publication

C23

The inexhaustible function z squared plus c

• ***Chapter foreword (2003).*** The "buzz" created by the publication of M1982F was so intense that a feature article was considered by the monthly *Scientific American*. They promised space for color graphics that were beginning to gather dust in my bulging filing cabinets. To make a long story short, this feature was never completed, its thrust being pre-empted when Dewdney 1985 presented in the *Mathematical Games* column of the same magazine an advance sampling of the art in Peitgen & Richter 1986. The caption of the cover picture was "Exploring the Mandelbrot set" and the title mentioned "the most complex object in mathematics." Neither of these terms was part of my vocabulary, but I could not complain.

Be that as it may, I had prepared notes to help the editors of the magazine ghostwrite a story under my name. This text was returned after

Dewdney's piece appeared, and a shortened version is included here as an informal introduction for the sake of readers who may welcome one. •

✦ **Abstract.** Allow the very elementary transformation, from z to z^2+c to be repeated indefinitely. The process reveals geometric shapes—fractal attractors, fractal repellers and other fractal sets—that astonish by their number, their variety, and their beauty. ✦

IN OUR SOCIETY, IT SEEMS NECESSARY TO KNOW THAT $2\times 2=4$ and wise to know that the area of a square of side x is $x\times x=x^2$. This is why the expression x^2 is read as "x squared" and also why squaring is widely viewed as an elementary operation. Furthermore, the equation of a parabola is $y=ax^2+c$, where y is the height measured from the horizontal axis to a point, and x the distance measured horizontally. This equation is of practical importance, since it describes the trajectory of a stone that falls in empty space after being thrown in a horizontal direction. After this result has been derived, the squaring operation may seem effectively exhausted from the viewpoint of the nonmathematician.

But it is not exhausted at all. In fact a seemingly minor modification suffices to reawaken it beyond all expectation. To explain this modification, observe that to obtain the parabola one performs the quadratic operation $x\to x^2+c$ once, and the result is then used in ways unrelated to the operation. But let us subject the outcome of the quadratic operation to the very same operation, and do so again and again.

Such a repetition of an arbitrary operation is known as "iteration," and in the present case suffices to transform seeming dullness into boundless excitement. The motivation for iteration and many of its most important properties come straight from physics. In iteration, physicists see a simplified view of the "state" of a dynamical system and its evolution in time. By definition, a dynamical system is a transform that obtains its state today from its state yesterday, and its state tomorrow from its state today. Therefore, to obtain the state tomorrow from the state yesterday, one may apply the basic transform twice in succession. But one may also apply the iterated transform once.

More generally, physics has now adopted wholeheartedly a view that the great Henri Poincaré pioneered around 1880: A dynamical system's evolution is ruled by the iterates of the day-to-day transformation. This is the reason why the iterates' behavior deserves to be studied. In some cases, as physicists have been expecting since Newton, the iterates behave very smoothly. But in other cases, their behavior is quite unexpectedly involved, a fact that turns out to be extremely significant.

For this second possibility to be appreciated fully, it is best first to recall that the likes of Galileo, Newton, and Laplace found that simple dynamical systems behave perfectly smoothly and predictably. This finding had a profound effect on the whole of Western thought. Since time immemorial, most aspects of nature and of man were viewed as largely unpredictable and uncontrollable. However, mechanics provided a big exception, and many persons, including laymen, philosophers, and scholars, came to view mechanics as *the* ideal model to which every other intellectual activity should aspire. Of course, thinking persons necessarily acknowledged that most fields were extremely far from implementing predictability and full controllability. Yet both features continued to be viewed as remote but shining ideals.

Actually, as is often the case, the seed of the future destruction of this ideal had already appeared at a time when it seemed to be triumphant, that is, in the late nineteenth century.

First symptom: The motion of fluids is surely a prototypical problem of mechanics, yet turbulent motion is neither predictable nor controllable. This was a source of deep worry to the masters of nineteenth-century rational mechanics, like Sir Horace Lamb, and close to our time to many great minds, including Enrico Fermi.

Second symptom: Poincaré's dynamics centered on an attempt to prove that the solar system is stable. However, not only did he fail to find an actual proof, but he had the extraordinary foresight to see that something might be seriously amiss. Pure mathematics often disappoints physicists by doing little beyond putting a stamp of approval on what had already seemed "obvious." But Poincaré observed that in the case of planetary systems, many very unstable possible behaviors could *not* be excluded a priori.

However, Poincaré did not pursue this line of thought, and it found few followers in the self-satisfied scientific scene of his time or immediately thereafter. Scientists had a plethora of "nicer" problems to tackle first, and the prevailing belief in stability as the norm was also to be influential.

In the 1970s, however, this particular work of Poincaré became very influential. Mechanics turned around and was quite suddenly taken by a near-obsession with chaos. It follows, incidentally, that mechanics thereby relinquished (at least implicitly) its claim to be a model for the less-developed sciences.

But even the study of complexity had better seek inspiration from the study of simple transforms. Hence many scholars independently thought of studying the quadratic transform $x \to x^2 + c$. Two French mathematicians of the 1910s, Gaston Julia and Pierre Fatou, went further, replacing the real x by a complex z. A real number is the position of a point on a straight line, while a complex number is basically a point in the plane, namely, a combination of the coordinates of a point P. The replacement of z^2 by $z^2 + c$ might seem a minor change, but that is very far from being the case. Quite to the contrary, the iterates' behavior depends crucially on the value of c.

A later and unexpected reward came when Metropolis, Stein, & Stein, followed by Feigenbaum, found that the properties of the real quadratic transform are shared by many other much more complicated ones. The results were given the grandiose appellation of "universality theory."

Given the above results, it is good to note here that this book abstains from referring to this form of universality, except perhaps in asides addressed to the experts. The most important results, due to the author, consist in extensions of the Fatou–Julia theory. That theory used to be viewed, even by the specialists, as extremely subtle and impossible to extend. However, accurate computer-assisted illustrations took away most of the mystery and unleashed striking developments.

Let us now describe a key step. When x is an ordinary number, a "real number" in the mathematicians, peculiar terminology, the iteration of $x \to x^2 + c$ involves only geometric sets on the line. Such sets are important, but they necessarily lack variety, and one reason why they tend to look complicated is that they are hard to visualize. However, every reader of mystery novels recalls many cases in which the sequence of visits of a suspect to a house looks complex, but the sequence of the suspect's displacements about town obeys simple rules. Therefore, it is not totally surprising that when one is faced with phenomena restricted to the line, a frequently effective way to simplify consists in interpreting them as the trace left upon the line by corresponding phenomena residing in the plane. Using again the peculiar vocabulary of mathematicians, many mathematical theories can be simplified by being "complexified" by which a real number x is changed to a complex number z.

A glance through the illustrations in this book shows that the geometric implications of iteration in complex numbers are truly staggering. To many readers, mathematics and physics are of little interest, but my work on iteration had an impact well beyond mathematics and physics. It turns out that manipulation of very elementary mathematical transformations has a totally unexpected power to generate structure that is very complex and very widely felt to be surprisingly beautiful.

A new form of art, fractal art, is thereby revealed. Since it can be described fully by short mathematical sentences, this art is more "minimal" than any of the works that art galleries have been exhibiting. And it raises a new issue: The traditional view, almost a cliché, distinguishes mathematics and the sciences from art by arguing that in the former, an achievement is a discovery, and in the latter, it is a creation. However, an achievement in the new fractal art is best considered a discovery.

The reader might hesitate to believe that complexity can emerge from such simple equations. As a matter of fact, the philosophy of classical mechanics (to which we have already alluded) had long supported a contrary preconceived notion, that "simple" mathematical manipulations must have simple outcomes, while complicated outcomes can be achieved only by complicated operations. It is necessary to dispel that notion, and to encourage the reader to read the text and not to neglect it altogether for the pictures.

Therefore, let me briefly sidetrack to examine a few loose similes. First, consider the operation of hitting hot steel with a hammer; it can be called an elementary operation, in the sense that the outcome shows little variety, but everyone knows that by repeated hits a skilled smith can create very intricate swords or ploughs. Actually, this simile does not do justice to the findings that follow. Indeed, mindless repetition of the same stroke of the hammer yields a result without interest, and a smith who wants to fashion an involved object must vary his gestures in very precise fashion that demands great skill. In contrast, we shall see that even a "totally mindless" repetition of the quadratic operation, if it is chosen suitably, leads to shapes of totally unexpected complexity.

A second simile is closer to the third and final one to come. Consider a circular lathe used to reduce a big and rough log into a smooth one with a final radius equal to one unit of length. We shall not take account of the wood's cellular structure. Precise cutting knives being slow and fragile, one is well inspired to proceed by successive steps. A first knife whose operation is fast and robust but rough removes the bulk of excess radius, leaving a nearly circular but irregular shape. Then a different knife leaves a somewhat smaller and smoother shape. If high precision is needed, the best lathing will take several stages. Late stages will attempt only to remove an amount of the order of magnitude of the irregularities one is prepared to tolerate in the final product.

Now let us consider an all-purpose program of lathing that always replaces a radius $1+r$ by a smaller radius equal to $1+Vr$. If the initial radius is high, successive stages will first move fast, then slowly. The instruction "reduce the radius from r to Vr" remains unchanged, hence plays the same role as the instruction "hit with a hammer" used in the first simile. But the new instruction also has an exceptional virtue: exactly the same terms also rule the gradual change from rough to fine tools.

To improve on the above similes, one needs computers. Even the slightest knowledge of computers identifies iteration as a very special example of a logical loop. Let us therefore translate our lathe into a mental tool that will be easy to generalize. We describe each successive lathing stage as eliminating all the points P such that the representative complex number z, once squared, lies outside of the piece left by the preceding stage. This immediately leads to a generalization that consists in introducing an additional parameter c, and eliminating all the points z such that z^2+c lies outside of the piece at the beginning of this stage.

The changes brought in by the introduction of c would have been impossible to guess, but the illustrations in this book show that they can be very great. Most strikingly, each stage no longer improves smoothness beyond the earlier stages, but rather adds meaningful detail beyond anything brought in by earlier stages. Thus, the limit to which our present process converges includes details of every size, down to infinitesimally small detail.

Though it is hard to believe, Fatou and Julia had discovered this property of the limit without any graphics. This may be one reason why their results went on to hover for sixty years in an intellectual no-man's land. A few mathematicians viewed them as great achievements, but their study hardly moved forward. A new tool that I introduced into supposedly "pure" mathematics—namely, computer graphics—gave new life to this study, and my observations brought to it many new practitioners worldwide.

Julia's powerful geometric intuition only could tell him that some curves later called Julia sets bear a general resemblance to a shape called a snowflake curve. That curve includes details of every conceivable size down to the infinitesimal, yet can be drawn with ease, which is precisely why the Swede Helge von Koch defined it in 1904. Figure X of Chapter C shows how the Koch curve is obtained. Analogous procedures can lead to either a branched curve or a totally disconnected collection of points called Cantor dust.

Furthermore, it is obvious (and therefore need not be illustrated) that iterative deletion can equally well serve to obtain a straight line, e.g., by erasing all the small squares other than the squares that intersect the main diagonal. In other words, there is much parallelism between deletion of directly specified triangles and squares, and deletion of pieces that are specified indirectly via the squaring operation.

One difference is that the former works to rules that take many lines to write down fully, while to specify squaring takes only a very short formula, hence the surprising conclusion that the operation of deletion of triangles and squares is complicated from an algorithmic viewpoint. On the other hand, ordinary draftsmen could do nothing with squaring, while they find Koch curves, Cantor sets, and the other diagrams on the same page easy to draw. Hence snowflakes are the easier shapes from the practical viewpoint, which is why they are often found in books of mathematical games.

But being widely known does not mean that they mattered much. They served only the very limited purpose for which they had been introduced: To act as counterexamples.

Let us elaborate. Physics, mechanics, and mathematics used to belong to a single discipline, called natural philosophy, and the emphasis of dynamics on smooth behavior was paralleled by the emphasis of mathematics on smooth functions and curves. At the same time that Poincaré was envisioning nonsmooth dynamics, mathematicians became increasingly aware of nonsmooth functions and curves. Partly to respond to criticism of rigorous proofs as nitpicking and roundabout ways to reach the obvious, mathematicians looked for instances in which the seemingly obvious was in fact drastically incorrect. Thus, Koch curves were contrived to show that rigor was needed. On the other hand, they were viewed as having limited intrinsic interest and also as having nothing in common except for being "pathological" counterexamples.

Quite to the contrary, I discovered that some of those would-be counterexamples have in common an unheralded property called self-similarity, and, in addition, are of great and direct practical use, for example as models (or rather "cartoons") of mountains or coastlines. I called these shapes fractals, and constructed around them a new fractal geometry of nature. Thus, the snowflake is flesh whose skin is a fractal curve, the infinite tree is a fractal curve not bounding any flesh, and the Cantor set is a fractal dust.

Now we can turn back to the transformation $z \to z^2 + c$ and define filled-in Julia sets as consisting of the points that the repeated quadratic transformation fails to remove to infinity. The special value $c = 0$ leads to a disk (filled-in circle), and the special value $c = 2$ leads to an interval. But all other values of c generate fractal sets.

A remarkable observation due to Pierre Fatou and Gaston Julia is that a Julia set's topology is deeply dependent on the value of c. It can be like a skin around flesh, i.e., a curve that surrounds a domain with a positive area. It can be a skin without flesh, i.e., a curve (in general, a heavily branched one) that surrounds nothing, or it can be a disconnected dust. All these diverse geometric possibilities did not have to be called upon deliberately, because all had always been latently present in $z^2 + c$, that is, in the simplest function other than $az + b$.

Clearly, it is imperative to classify the value of c according to some basic characteristics of the Julia set. Fortunately, Fatou and Julia discovered a powerful criterion to determine whether the Julia set is connected. It suffices to follow the transforms of a special point called "critical." In the quadratic case, it is the point $z = 0$. If these transforms go to infinity, the Julia set is a totally disconnected dust; if these transforms remain bounded, the Julia set is connected and the set to which c belongs has been called the Mandelbrot set M.

This set's definition is as simple as can be. Therefore, even in 1980, the dark ages of computing, drawing M made small demands in terms of computer power. (Today, every child knows how to draw M.) But the original interpretation of the drawings demanded skill and experience in experimental mathematics. I drew many pictures of M for a large number of values of c, covering the interesting portions of the plane quite tightly, and made a mass of surprising inferences that led to rigorous mathematics, making the topic popular again both for the specialist and the very important common man, woman, or child.

The story of how precisely the Mandelbrot set came to be discovered is this book's main topic.

APPENDIX: THE OPERATION $z^2 - c$ IN COMPLEX NUMBERS

Being available, this appendix was reproduced, though it is hard to guess whether it will help even one reader benefit from the contents of this book. A complex number is a convenient way of denoting the position of a point P in the plane with respect to an origin 0 and two perpendicular coordinate axes. The position of P can be specified by the two distances x and y measured along two orthogonal directions. If so, x and y are called "Cartesian coordinates." One writes $z = x + iy$, and one calls x the real part and y the imaginary part of z. Of course, the computer knows nothing about complex numbers as such. To perform the transformation $zz - c$, where $c = c' + ic''$, the computer replaces the initial numbers x by $x^2 - y^2 - c'$ and y by $2xy - c''$. Many available programs can be used without the user's knowing anything more than the computer does. However, a few words of elaboration may be worthwhile. The position of P can also be specified by "polar coordinates," which are the distance from the origin to the point P (called the modulus) and the angle θ between the x-axis and the line that joins the origin to P. In these coordinates, one writes $z = re^{i\theta}$. To add or subtract complex numbers, that is, to obtain $z' + z''$ or $z' - z''$, one adds or subtracts separately the real and imaginary parts of z' and z''. The transformation zz consists either in replacing x by $x^2 - y^2$ and y by $2xy$, or in squaring the modulus, $r \rightarrow r^2$ and in doubling the angle, $\theta \rightarrow 2\theta$. When P is the point of Cartesian coordinates 0 and 1, one has $z = i$; hence $r = 1$ and $\theta = 90°$; in this case, z^2 has modulus 1 (again) and angle 180°; hence $z^2 = -1$. The operation "square root of -1 cannot be performed with ordinary "real" numbers, and complex numbers were originally introduced to provide that operation with meaning.

First publication

C24

The Fatou and Julia stories

THE THEORY OF ITERATION OF RATIONAL FUNCTIONS goes back to the mid-nineteenth century and perhaps even to Abel. During its first classic period, it indissolubly linked the names of Pierre Fatou and Gaston Julia. It also generated great controversy accompanied by hasty anecdotes and schematic or fanciful stories. The account in Alexander 1994 is excellent: scholarly and balanced. But it is necessarily based on secondary published sources. I was closer. In fact, Julia was my teacher of differential geometry at École Polytechnique. More important were the stories told about him and Fatou by my uncle Szolem Mandelbrojt (1899–1983) and other persons I knew well. Combining what I heard and what it made me read, this chapter discusses Fatou, then Julia, and finally their interaction. The last section deals with the aftermath.

1. PIERRE FATOU (1878–1929)

Fifty years after his death, Pierre Fatou resides in the mathematical Pantheon, but he had been a shabbily treated outsider and (strictly speaking) only an amateur mathematician.

Fatou was born in Louest, in Brittany, on February 28, 1878 and died in 1929 in Pornichet in Brittany. He attended École Normale Supérieure from 1898 to 1901, when he became "agrégé" (a degree somewhat comparable to an Ed. D. in mathematics). At this point, he joined the Paris Observatory, where he remained as junior astronomer until 1928 ("stagiaire" = intern then "aide" = assistant, and later "adjoint"= associate). His assignments were specific, and they often changed, the last one being concerned with measurements of double stars. Astronomers marked a dutiful employee's death by dutiful obituaries.

His name survives because he also "moonlighted" as a mathematician. He received his Doctorate ès Sciences in February 1907, for a thesis highly praised by Paul Painlevé (Gispert 1991, pp. 387–388) and for the rest of his life tackled topics in the mathematical mainstream of his day.

History records a number of composers and poets who earned their living in totally unrelated ways; For example, the American composer Charles Ives was an insurance executive. In the same vein, French diplomacy in the 1930s was run (disastrously) by Alexis Léger de Saint-Léger. He wrote hermetic poetry on the side under the pseudonym of Saint-John Perse, and eventually won the Nobel Prize in Literature. Similarly, nineteenth-century mathematicians included imprisoned generals (Poncelet), police officials (Fourier), and high-school teachers (Weierstrass).

Fatou may have been the last of that breed. Being a very private person, he did not record his reason for staying out of the mainstream, but S. Mandelbrojt reported that early on, Fatou retained the stigma of having been branded by French Academia as something of a black sheep or renegade. This may have happened after his Ph.D., when he refused a first job in the provinces. At that time, this was a near-absolute prerequisite to a later promotion to Paris. Moreover, it was not necessarily onerous: For example, before Mandelbrojt was appointed to Clermont-Ferrand in 1929, the advanced calculus course was not actually given — for lack of students — but the professor was assigned no other duty.

A search for other personal material about Fatou yielded one tidbit. Henri Cartan recalled that Fatou was very fond of music and particularly admired Prokofiev. Cartan (with his friends, all in their twenties) and Fatou (alone and in his late forties) met in the cheap seats of concert halls.

Only the bare facts of Fatou's last few years are available, and they have become hard to interpret. In 1927, he presided over the Société Mathématique de France, and in July 1928 (aged 50!) he was promoted by the Observatoire to the rank of full astronomer. In 1929, he wrote a *Notice* on his work, about four-fifths of which concerned mathematics. French Academic *Notices* were and perhaps continue to be written to apply for a specific position, but this particular *Notice* does not indicate its purpose. The Académie des Sciences never elected him, not even to its lower rank of *Correspondant*, so perhaps he was seeking this post.

One would like to think that he was about to be welcomed into the mainstream, but there is a sharply discordant piece of evidence. Check on

Appell & Goursat 1929–1930, a book in two volumes, each over 500 pages long. A first edition of 1895 had a preface by C. Hermite. Fatou prepared Volume I of the second edition, and the credit he received was a reasonable 10% of the working area of the cover.

He received the same "billing" for Volume II, but the situation there was altogether different. The overall scope had broadened, Paul Appell (1855–1930) had died, and the one-page foreword was written by Edouard Goursat (1853–1936). It contains the following words: "When the publisher asked for a new edition of our *Theory of Algebraic Functions,...* it seemed useful to add at least a summary of the theory of Fuchsian functions. The recently deceased Fatou had kindly undertaken this task. However, instead of the few chapters we were expecting from him, a treatise on automorphic functions was left to us. We are happy to have provoked the publication of a masterful book that the excessive modesty of the author may have prevented him from writing.... Reading it could only increase the grief inspired in all who have known and appreciated this man, by the premature death of Fatou. Science was still expecting much from him."

Goursat also mentions that, "the manuscript was entirely written by Fatou, who finished it a few months before his death." Two questions remain unanswered. Why did the President of the French Mathematical Society accept an assignment fit for a junior person? And why did his text on automorphic functions fail to be published with him as sole author? Appell contributed nothing to it, and Goursat contributed only a foreword.

Fatou also wrote on Taylor series. Classical analysts like my uncle labored mightily to extend Fatou's work. Its day may come back, but today it is known to few persons. Fatou also wrote on the Lebesgue integral, where he is hailed for a *lemma.*

A lemma is, of course, a proposition that is not perceived as of great value by itself, but only as a step toward some much deeper theorem. *Fatou's lemma* recalls the terms *Abel's lemma* or *Schwarz's lemma.* One may read a great deal of mathematics while believing that these great scholars authored no deep theorem to match. Why is this so? Perhaps one is never a good judge of one's own work, and there may be a tendency to reserve *theorem* for the fruit of one's hardest work, while downgrading as *lemmas* the results that have been easier to obtain, even when they are of wider importance. Or could it be that posterity chooses sometimes to honor the greatest among scholars by giving their names to their most widely used results, which are also the simplest?

2. GASTON JULIA (1893–1978)

Julia's life story is well documented in several accessible references. For example, a long obituary, Garnier 1978, appeared in the *Comptes rendus de l'Académie des Sciences* (Vie Académique) **286**, June 12, 1978, 126–133.

Therefore, this section will be very brief and center on my own recollections. Gaston Julia was born in February 1893, in Sidi bel Abbes, West Algeria, where his family had moved from Spain. In a Catholic parochial school, his talents were promptly recognized. After graduating from the Lycée d'Oran, he received a scholarship for study in Paris. At age 18, he was admitted first in his class to both École Polytechnique and École Normale Supérieure. He went to Normale from 1911 to 1914, when World War I broke out. After a brief training period, he was sent to the trenches on January 25, 1915, as a second lieutenant of infantry. All the other officers in his company died that day; he survived, but with multiple wounds, including a deep wound to the face. For the rest of his life, he lectured with a leather mask over his nose and cheek. Painful operations (many done under incomplete anesthesia) continued throughout his life. In 1916, while in the hospital, he prepared his Doctorat ès Sciences, and his Ph.D. thesis received the (highly rated) Prix Bordin. His next research topic was the theory of iteration of rational functions.

Around 1920, Julia was one of very few young and active mathematicians in France. Life in the trenches, wounds, and other operations had deeply affected many veterans but not Julia's intellect. He was a brilliant man and also a national symbol in several ways. Being a native of French Algeria of Spanish background counted, because France prided itself on its "melting pot" tradition. And the war had made him a "gueule cassée," that is, a "broken muzzle." He moved up the academic ranks very rapidly, becoming soon a professor at the Faculté des Sciences de Paris (the "Sorbonne"), and at age 41 joined the Académie des Sciences. Of those elected during that period, only one was younger: the aristocratic pioneer of quantum theory, Prince Louis de Broglie. During World War II, Julia lectured in Germany, which later was criticized. The last years of his life were spent in the old Hospice des Invalides (near Napoléon's tomb), and his funeral was held very formally in the Eglise des Invalides.

In the fall of 1945, I was a first-year student at École Polytechnique and Julia was the professor of (basic differential) geometry. I first saw him from the very back of a large auditorium. A few times during the term, he and the physicist Louis Leprince-Ringuet deviated from the norm by inviting interested students to chat. This was the nearest my

Polytechnique professors came to having office hours. He was also a professor of higher analysis at the Faculté des Sciences de l'Université de Paris, where he was expected to teach an advanced course.

He was a very vigorous man, with a strong and well-articulated voice. His temperamental style, which we used to call "Mediterranean," stood in sharp contrast to the grayness of the professor of mathematical analysis, Paul Lévy. Unfortunately, Julia was largely reading stale course notes. The École distributed them at regular intervals in the form of *feuilles*, that is, large (*quarto*) signatures, to be bound at year's end as mementos.

He was also fond of peppering his course with "maxims," several of which stick in my memory: "Those among you who go into scientific research will often find that you have been scooped. If the person who scoops you reaps no glory, little is lost. If that person does reap glory, you should be even prouder of your work, and next time you will be luckier." The class repeatedly heard him say that "To simplify, you should 'complexify.' That is, when you have a complicated problem and wish to simplify it, it is a good idea to replace all reals by complex numbers." Needless to say, this last maxim was present in my mind in the late 1970s when the real function $x^2 + c$ was of wide concern among my physicist friends and I turned to the Fatou–Julia theory of the iteration of the complex function $z^2 + c$.

Chapter C1 reports that in the late 1940s, Paul Montel and my uncle greatly admired Julia's early work on iteration and other areas of mathematical analysis. But this admiration was shared by few other mathematicians. Analysts felt beleaguered at that time (see Chapter 25), disrespected by the reigning "Bourbaki" movement. Paradoxically, Bourbaki had "rented" Julia's seminar during the thirties, but their taste for grand abstract structures was as far removed as possible from Julia's meticulous study of a plethora of special and subtle examples.

3. THE FATOU–JULIA STORY

A scientist's life is never the stuff of legend except when that scientist clashes with either the People or the State, is a healer, or is Galileo meeting the Inquisition, or Oppenheimer meeting Teller. To the man of theater, science by itself is not the stuff of drama, and the artist does not recognize the beauty of mathematics as an acceptable form of beauty (with, it must be interjected, the possible exception of fractals).

In particular, arguments about scientific priority are inevitably taken by outside observers to be pitiful family squabbles. As an exception, the academic Paris of my youth often alluded to the priority argument that raged in 1917 and 1918 between Pierre Fatou and Gaston Julia. Both had implemented one of the bright mathematical innovations of 1912: Montel's theory of "normal families of functions".

The Fatou–Julia story was settled by Reason of State in the midst of War. A tribunal was designated, judges decided in favor of Julia, gave a smaller "consolation" prize to Fatou, and their decision was printed. Section 2 suggests that there was a continuing and widespread feeling that this award was at least in part meant to alleviate the deep physical suffering of someone who had previously been marked as a winner. But Section 1 also suggests that the loser was already familiar with loss.

The Fatou versus Julia encounter will be described through a few documents in the record collected by my wife, Aliette Mandelbrot. The originals are completely open and available but known to few. The events to be described may perhaps help Fatou & Julia be remembered as men of flesh and blood, and not as cold marble statues. Needless to say, given the horrible context, I seek neither heroes nor persons to blame.

To appreciate the following quotations, one must know that a "sealed envelope" was a letter that an author could ask the Académie des Sciences to register and file, to be opened either on request or after a prescribed number of years. A "sealed envelope" puts an idea on record without revealing it to perceived competitors. Old "sealed envelopes" are continually being opened, and some add to the history of sciences. Julia 1968 is a reprint of most of Julia's *Works*.

Note on rational substitutions, by Mr. Gaston Julia, presented to the Academy on December 24, 1917, reprinted in Julia 1968, **I**, pp. 105–107. [Remark: There is no indication that this Note was presented by a Member of the Academy.]

"I just read with interest the note Mr. Fatou published in the *Proceedings* of the meeting of December 17, 1917. Its main results I have myself consigned earlier in four sealed envelopes filed at the Academy Office on June 4, 1917 (8401), August 27, 1917 (8431), September 17, 1917 (8438), and December 10, 1917 (8466). By unsealing these four sealed envelopes, the Academy will observe that results identical to Mr. Fatou's (aside from notation and examples) are recorded there with brief indications of the methods of proof. By a curious coincidence, one of my methods happens to make use of Mr. Montel's results on normal families of analytic functions, which Mr. Fatou also uses. The Academy will decide, as to methods as well as to results, who should be given priority...

"I decided to file these 'sealed envelopes' because in his Note of May 21, 1917, Mr. Fatou had reported to the Academy some of the results I had achieved... ."

Report on the preceding Note, by Georges Humbert. Reprinted in Julia 1968, **I**, p. 107.

"The Academy has asked me to examine four sealed envelopes by Mr. Julia, opened during the meeting of December 24, and a Note in which the same author brings up a question of priority... [in reference] to Mr. Fatou's [Note of December 17, 1917]... . The comparison of the two Notes and the four sealed envelopes shows that Mr. Julia's assertion is well founded. This sealed envelopes include, indeed, among others, the results for which he claims priority... ."

Report by Emile Picard and Georges Humbert on the grant of a Grand Prize in Mathematical Science to Gaston Julia. (Proceedings of the Academy meeting of December 2, 1918), not reprinted in Julia 1968.

"Competing for this Prize, three memoirs were filed at the Academy Office. The Committee has restricted its attention to those by Mr. Lattès, Professor at the University of Toulouse, and by Mr. Julia, lieutenant in the 34th Infantry Regiment.

"...As his research proceeded, Mr. Julia filed sealed envelopes at the Academy. After the last had been filed, in December 1917, a known geometer, Mr. Fatou, to whom the theory of iteration already owes interesting advances in a new direction, recorded in our *Proceeding* the bulk ["la plus grande partie"] of the same results. He too had obtained them by using the properties of the normal families of Mr. Montel. It is not the first time, in the history of Science, that one or two significant scholars arrive together, by the same path, at the same discovery... ."

It is hard not to imagine that the Academicians of 1917 were thinking of János Bolyai (1802–1860). When Bolyai discovered his form of non-Euclidean geometry, his father Wolfgang urged him to publish fast, "because it seems to be true that many things have, as it were, an epoch in which they are discovered in several places simultaneously, just as the violets appear on all sides in springtime." Indeed, János' result had already been published by Lobachevsky a few years before, and had long been known to Gauss, who had chosen "not to allow it to become known during [his] lifetime."

A few dates deserve elaboration. December 24, 1917, was Christmas Eve of the year of the battle of Verdun. Since World War I ended on November 11, 1918, the report published on December 2, 1918, was written while the war was barely over, or perhaps still raging.

That is, the originals state (with flourishes) that the Academy met in wartime on a Christmas Eve to compare results in *already* published papers by Fatou with results that Julia had *not* published but filed in sealed envelopes. Julia was not referenced by Fatou on May 21, 1917, for the good reason that Fatou was resuming the work started in Fatou 1906, and Julia had not yet done anything along these lines. One may conjecture that Julia's interest may have been to avoid revealing some technique that he wished to preserve for himself for a while longer.

Last but not least, several questions beg to be answered. In our present scientific culture, as soon as a mathematical topic becomes interesting, a swarm of locusts seems to drop all other activities and converge to the new cluster of activity and share in the pleasure of exploring it. So how did Julia and Fatou manage to "clean up" their common topic so thoroughly that not much else could be done for so long? A general answer is that the world of science was far less crowded than today. A more specific answer is that World War I put intellectual activities at a near standstill, leaving mathematics in the hands of Fatou, a cripple, and Julia, a war casualty.

3. FATOU AND JULIA AFTER 1917

One crude token of nonrecognition is the *World Who's Who in Science from Antiquity to the Present*, published in 1968. The French mathematicians listed include the above-mentioned Appell and Goursat but also the ephemeral R. Garnier, but neither Fatou nor Julia.

Julia, still alive, was not eligible for the far more scholarly *Dictionary of Scientific Biography*, Gillispie 1970–1976. Fatou received two full columns, but iteration received one line and no mention in the bibliography.

Additional evidence and my own recollections are brought together in Chapter C25. Be that as it may, while beautiful mathematics may wax and wane with the winds of fashion and the availability of tools, it never dies.

First publication

C25

Mathematical analysis while in the wilderness

HAVING A BROAD INTENDED READERSHIP, this chapter should begin by a very rough characterization. Mathematical analysis is "like calculus but far more advanced." An exact definition does not exist. This is not surprising, because truly important notions are left undefined, even in mathematics! This is argued in an appendix to the preface of M 1999N and in an overview chapter in M 2002H.

During the glory days of the often mentioned Nicolas Bourbaki, the 1950s to the 1970s, mathematical analysts had good reason to feel beleaguered. Under most definitions, my uncle, Szolem Mandelbrojt, was a leading mathematical analyst. Overcoming adverse conditions, he was one of a few persons whose work and students kept the flame of analysis alive.

This chapter samples a few readily checked items that help recapture the mood that ruled mathematics, first in the 1920s, and later in the 1950s and 1960s. The last section is an amusing text on experimental mathematics.

Hadamard, writing in 1934, as witness to an earlier period in mathematics. Hadamard has been repeatedly mentioned. This truly great man has at long last received a scientific biography (Shaposhnikova & Mazya 1995). He gained his spurs by proving the prime number theorem, which for nearly a century had been a challenging conjecture that Gauss conceived on the basis of explicit calculations. Later, he contributed massively to the theory of the propagation of waves. In particular, in an achievement that deserves to be known and hailed by physicists, he showed that waves propagate differently in spaces of even and odd embedding dimension. In fact, for much of his life, Hadamard was Professor of *Mechanics!* This title was not interpreted narrowly, since his main teaching consisted in directing a twice-weekly seminar in which every aspect of mathematics was welcome.

To all those interested in history, I recommend particularly Hadamard 1934. From this eulogy of Paul Painlevé (1863–1933), unaccountably not included in the collected *Oeuvres*, only a few excerpts must suffice.

"Around the years 1880–1890, the great problem mathematics had faced since the invention of calculus was the integration of differential equations. Without this integration, we have no language capable of describing change. However, aside from a few cases that are simple and quickly exhausted, the difficulty of integration proves to be insurmountable, increasingly so as we become better acquainted with the problem. A grand discovery led the march of science during the second half of the nineteenth century and promised important progress. This was the theory of analytic functions, which brought light to the notion of function, that is, the notion of change itself... . The precision and harmony of the study of those functions could not have been expected as long as one remained on the real line... . But the third quarter of the nineteenth century did not bring to the theory of differential equations anywhere near the marvelous progress it brought to the functions of complex variables... .

"But the face of science was changing. Poincaré had arrived on the scene and in both directions brought powerful new advances. As fulgurant discoveries ... were coming forth, I often saw, around me, admiration being combined with a kind of concern. Most of those marvelous results did not seem to be followed up. A serious drawback, indeed, if it had been confirmed. But we understand now that after such long jumps, science must, so to speak, get back its wind. To develop Poincaré, a necessary first step was for the mind to be filled with his spirit."

The 1950s and 1960s. The title of the multivolume work by Nicolas Bourbaki is *Élements de mathématique*. The French singular *mathématique* was an unusual counterpart of the English *mathematik*, which survives in the names of ancient chairs. Until the 1950s, there was a self-translating subtitle: *Première partie: les structures fondamentales de l'analyse*. Then the subtitle vanished with its promise of a *Deuxième partie*, and *structure* remained for its own sake.

A sense of how analysis lived in the Bourbaki environment is given by the following statements by authors who directly affected my work.

Kahane and Salem 1962. Jean-Pierre Kahane was my uncle's student, and the major monograph he wrote with Raphael Salem had a major effect on my scientific life, as exemplified by Kahane & M 1974{N11}. Until that monograph I thought that Hausdorff dimension was part of ancient, not current, esoterica.

Kahane & Salem opens with a quote from *Odile,* by Raymond Queneau, a renowned literary figure who fancied mathematics (and once wrote to invite me to visit him): "It is not to architecture, to buildings, that geometry and analysis should be compared, but to botany, to geography, even to the physical sciences. The task is to describe a world, to discover it and neither to construct nor to invent it, since it exists outside the human mind and independently of it."

The Preface that follows elaborates on this quotation: "A few decades ago, this book would not have needed this Preface, which is written as an apology. Today, at a time when most mathematicians — and the best ones — are mostly interested in questions of structure, this book may appear obsolete and, in a way, may look like a collection of dried leaves. We must therefore explain that our goals are in no way reactionary. We know the beauty of the great modern theorems... but believe that, without disdaining the architecture which dominates the object of mathematics, it is permissible to show interest for these objects for their own sake. As isolated as they may appear, they often hide in themselves properties that reward attention by posing fascinating problems. Some of our friends call this approach "refined" mathematics, and we have often wondered whether this term was meant to express appreciation or disdain."

Magnus 1974. This major textbook has already been mentioned in Chapter C2 as having influenced my work on Kleinian groups. It opens with this quotation from C.L. Siegel: "The mathematical universe is inhabited not only by important species but also by interesting individuals."

Mandelbrojt 1952. This piece, arguably the only written statement of my uncle's philosophical stance, was reproduced photographically as a supplement of Mandelbrojt 1985: "I must say that my very strong wish to do mathematics originated in the will to understand the outside world... in the highest sense... . [After someone proves that there is an infinity of] twin primes like 3-5, 11-13, or 17-19, I shall feel as fulfilled as when I finally understand the expansion of the Universe... . Generality is a great virtue in mathematics, and some scorn is directed to particular cases. But I cannot view generality as a god... Generality is beautiful when it explains ... . *There is an optimum* to generality ... . To generalize merely because of an attraction to generality or abstraction risks entering a formal world... . I would not like to live in the world of formal logic envisioned by some of my colleagues... . No divine law that I know forces us to abandon a being I view as complete for one that has the same virtues but is drier and more formal. It is said that the goal of mathematics is to study the relation between things and not the things themselves. I approve, but [as] I see it, a mathematician lives a double life: ... in a world of intuitive ideas and ... a world of difficulties that bring so many joys and without which his life would be too vague or too easy."

The preceding text criticized Bourbaki firmly but discreetly. In private, Szolem was more outspoken, and it is clear that his views deeply influenced me.

Lucas 1891. The topic of the next and the previous quotations is not mathematical analysis but experimental mathematics. But I wanted to include it in this book and no other place would have been better.

When I noticed the preface of Lucas 1891, it begged to be copied for some future use. At that time, experimental mathematics had every reason to feel unloved, in fact, beleaguered. Mercifully, the situation has much improved. Little knowledge of French is needed to skim this text. Therefore, translating it would be pointless. Edouard Lucas (1842–1891) was a professor of "Mathématiques Spéciales," his last position being at the Lycée Saint-Louis, in Paris.

"Comme toutes les sciences, l'Arithmétique résulte de l'observation; elle progresse par l'étude des phénomènes numériques donnés par des calculs antérieurs, ou fabriqués, pour ainsi dire, par l'expérimentation; mais elle n'exige aucun laboratoire et possède seule le privilège de convertir ses inductions en théorèmes déductifs. Comme en Chimie, par exemple, on prépare les nombres au moyen du calcul; par la divisibilité, on décompose ceux-ci en éléments simples, les facteurs premiers; par la théorie des résidus potentiels, on détermine leur aspect et, en quelque sorte, leurs réactions mutuelles; enfin par la juxtaposition des nombres triangulaires, carrés, polygonaux, cubiques, etc., la théorie des formes numériques rappelle l'étude des systèmes cristallins. C'est par l'observation du dernier chiffre dans les puissances successives des nombres entiers que Fermat, notre *Divus Arithmeticus*, créa l'Arithmétique Supérieure, en donnant l'énoncé d'un théorème fondamental; c'est par la méthode expérimentale, en recherchant la démonstration de cette proposition, que la théorie des racines primitives fut imaginée par Euler; c'est par l'emploi immédiat de ces racines primitives que Gauss obtint son célèbre théorème sur la division de la circonférence en parties égales, et celui-ci fut le point de départ des profondes recherches d'Abel et de Galois, de MM. Kummer, Hermite et Kronecker, dans l'Algèbre supérieure. ...

"La théorie des suites récurrentes est une mine inépuisable qui renferme toutes les propriétés des nombres; en calculant les termes successifs de telles suites, en décomposant ceux-ci en facteurs, en recherchant par l'expérimentation les lois de l'*apparition* et de la *reproduction* des nombres premiers, on fera progresser d'une manière systématique l'étude des propriétés des nombres et de leurs applications dans toutes les branches des Mathématiques."

The flow and ebb of history. The depth of Bourbaki's past stranglehold on many minds is fast becoming a distant memory. Historians will have to unscramble the many forces that contributed to its end and assess the relative contributions of my work. Be that as it may, the defensiveness of the preceding excerpts astonishes those who did not witness that era but confirms what was said earlier.

During the 1950s and 1960s, the Fatou–Julia theory was utterly unfashionable, in the wilderness, because it consisted of many special examples and few "great modern theorems." The terms "dried leaves," "isolated," "interesting individual," or "particular case" fitted it perfectly. This is the reason why it fell altogether outside of the mathematical mainstream and was, for all practical purposes, abandoned. Do not forget that for Bourbaki, Poincaré was the devil incarnate, who had left behind a mess of unproven assertions and loose ends. They boasted that they had cleaned up that mess. Of course, Poincaré has long been a source of concern to French mathematicians. In the 1880s, Hermite kept writing to Mittag-Leffler to complain that young Poincaré never completed a proof. For students of chaos and fractals, Poincaré is, of course, God on Earth.

Was a swing from Poincaré to Bourbaki and back normal and preordained, a matter of healthy overall development? There is a widely held view of history as a harmonic pendulum swinging back and forth, from boom to bust to boom, from anarchy to tyranny to anarchy. Such labels chop the history of mathematics into manageable chunks, to echo Lord Keynes's opinion that business cycles help divide books of economic history into manageably thin tomes. This opinion was sarcastic because Keynes thought that business cycles have no predictive value, and I extend his opinion to the cyclic view of how mathematics develops. One of the best effects of chaos theory and of the current revival of Poincaré might be that the crude pendulum model of history may at long last be put to a final rest.

CB

Cumulative bibliography, including copyright credits

This bibliography's overlapping parts have different motivations and very unequal lengths.

The shorter part consolidates the lists of references of the reprints and of the new chapters. The originals of the reprints tended to have few references, most of them either generic or ancient. The reason was a lack of items to refer to. Indeed, the topics I dealt with were themselves ancient but their study had long been interrupted. Extensive activity in developing those old works of mine came too late to be referenced. To propose a balanced reading list usable in 2003, I feel neither duty-bound nor competent.

The longer part in the middle of this bibliography lists most of my publications, together with copyright credits for those reproduced in this book. Such a list used to be a must *in volumes of* Selected Papers. *The internet – a new high-tech form of samizdat – may have transformed it into an obsolete symptom of* Vanitas Vanitarum. *The publications reprinted in a* Selecta *volume are preceded by an annotation of the form* ***N16**, *which refers to Volume N, Chapter 16. The letter code is E=M1997E, N=M1999N, H=M2002H, FE=M1997FE.*

The sources of the references in this list are very diverse and some are known to few readers. Therefore available variants are included and the only abbreviations used are J. for Journal, Proc. for Proceeding, Tr. for Transactions, and Z. for Zeitschrift.

AHARONY, A. & FEDER, J. Eds. 1989. *Fractals in Physics.* Amsterdam: North-Holland, *Physica*: **D 38**, 1-398.

ALEXANDER, D.S. *A History of Complex Dynamics from Schröder to Fatou and Julia,* Braunschweig/Wiesbaden: Vieweg, 1994.

APOSTEL, L., MANDELBROT, B. & MORF, A. 1957. *Logique, langage et théorie de l'information.* Paris: Presses Universitaires de France.

APPELL, P. & GOURSAT, E. 1895, deuxième édition, 1929-1930. *Théorie des fonctions algébriques.* Paris: Gauthier-Villars.

ASIKAINEN, J. AHARONY, A., RAUSCH, E., HOVI, J.P., & MANDELBROT, B.B., 2003. Fractal properties of critical Potts clusters. *European Physical Journal*: **B34**, 479- .

BARRAL, J., COPPENS, M-O. & MANDELBROT, B.B. 2003. Multiperiodic multifractal martingale measures. *Journal des mathématiques pures et appliquées,* in the press.

BARRAL, J. & MANDELBROT, B.B. 2002. Multifractal products of cylindrical pulses. *Probability Theory and Related Fields*: **124**, 409-430.

BARRAL, J. & MANDELBROT, B.B. 2004. Multifractal products of independent random functions: *Fractals.* Ed. Michael L. Lapidus, Providence, RI: American Mathematical Society.

BARTON, C. & LAPOINTE, P. Eds. 1994. *Fractal Geometry and its Use in the Earth Sciences.* New York: Plenum.

BARTON, C. & LAPOINTE, P. Eds. 1995. *Fractals in Petroleum Geology and Earth Processes.* New York: Plenum.

BEARDON, A. 1983. *The Geometry of Discrete Groups,* New York: Springer-Verlag.

BEDFORD, T. J. 1984. *Crinkly Curves, Markov Partitions and Dimension.* Unpublished Ph.D. Thesis, Warwick University, U.K..

N6 BERGER, J.M. & MANDELBROT, B.B., 1963. A new model for the clustering of errors on telephone circuits. *IBM Journal of Research and Development*: **7**, 224-236.

BIDAUX, R., BOCCARA, N. SARMA, G., SÈZE, L., DE GENNES, P. G., PARODI, O., 1973. Statistical properties of focal conic textures in smectic liquid crystals. *Le J. de Physique* **34**, 661-672.

BLANCHARD, P. 1984. Complex analytic dynamics on the Riemann sphere. *Bull. American Mathematical Society:* **11**(1), 85-141.

BLUMENFELD, R., & AHARONY, A. 1989. Breakdown of multifractal behavior in diffusion-limited aggregates. *Physical Review Letters*: **62**, 2977-2980.

BLUMENFELD, R. & MANDELBROT, B.B., 1997. Lévy dusts, Mittag-Leffler statistics, mass fractal lacunarity, and perceived dimension. *Physical Review*: **E 56**. 112-118.

BOURBAKI, N. 1960. *Eléments d'histoire des mathématiques.* Paris: Hermann.

BOURGUIGNON, J. P. 1999. An interview concerning the "World Mathematical year 2000." wmy2000.math.jussieu.fr/wmy_6.html.

BOX, G. E. & JENKINS, G. M. 1970. *Time Series Analysis, Forecasting and Control.* San Francisco: Holden-Day.

BOYD, D. W. 1973a. The residual set dimension of the Apollonian packing. *Mathematika*: **20**, 170-174.

BOYD, D. W. 1973b. Improved bounds for the disk packing constant. *Aequationes Mathematicae*: **9**, 99-106.

BRAGG, W. H. 1934. Liquid crystals. *Nature*: **133**, 445-456.

BROLIN, H, 1965. Invariant sets under iteration of rational functions. *Arkiv för Matematik*: **6**, 103-144.

BROOKS, R. 1989. The Mandelbrot set (Letter to the Editor). *The Mathematical Intelligencer*: **12(1)**, 3.

BROOKS, R. & METELSKI, J. P. 1981. The dynamics of 2-generator subgroups of PSL (2,C), *Riemann Surfaces and Related Topics,* Eds. I. Kra and B. Maskit, Princeton University Press.

CALVET, L., FISHER, A., & MANDELBROT, B.B. 1997. *Large Deviations and the Distribution of Price Changes.* See under Mandelbrot, Calvet & Fisher 1997.

CANTOR, G. 1932. *Gesammelte Abhandlungen mathematischen und philosophischen Inhalts.* Ed. E. Zermelo. Berlin: Teubner. Olms reprint.

CARLESON, L. & JONES, P.W. 1992. On coefficient problems of univalent functions and conformal dimension, *Duke Mathematical J.*: **66**, 169-206.

CARPENTER, L.C., FOURNIER, A., and FUSSEL, D., Display of fractal curves and surfaces, to appear, *Communications of the Association for Computing Machinery.*

CHOUCHAN, M. 1995. *Nicolas Bourbaki: Faits et légendes.* Argenteuil: Éditions du Choix.

CIOCZEK-GEORGES, R. & MANDELBROT, B. B. 1995. A class of micropulses and antipersistent fractional Brownian motion. *Stochastic Processes and their Applications*: **60**, 1-18.

CIOCZEK-GEORGES, R. & MANDELBROT, B. B. 1996. Alternative micropulses and fractional Brownian motion (with Renata Cioczek-Georges). *Stochastic Processes and their Applications*: **64**, 143-152.

CIOCZEK-GEORGES, R. & MANDELBROT, B. B. 1996. Stable fractal sums of pulses: the general case.

CIOCZEK-GEORGES, R., MANDELBROT, B. B., SAMORODNITSKY, G., & TAQQU, M. S. 1995. Stable fractal sums of pulses: the cylindrical case. *Bernoulli*: **1**, 201-216.

COLLET, P. and ECKMAN, J. P. 1980. *Iterated Maps on the Interval as Dynamical Systems.* Boston: Birkhauser.

COPPENS, M-O. & MANDELBROT, B.B. 1999. Easy and natural generation of multifractals; multiplying harmonics of periodic functions. *Fractals in Engineering*. Eds. J. Lévy-Véhel, E. Lutton, & C. Tricot. New York. Springer, 113-122.

COXETER, H. S. M., 1979. The non-Euclidean symmetry of Escher's picture "Circle Limit III." *Leonardo* 12, 19-25.

CVITANOVIC, P. & MYRHEIM, J. 1982. Universal for period *n* -triplings in complex mappings, *Physics Letters:* **94A**, 329-333.

***H29** DAMERAU, F.J. & MANDELBROT, B.B. 1973. Tests of the degree of word clustering in samples of written English. *Linguistics:* **102,** 1973, 58-75.

DENJOY, A. 1932. Sur quelques points de la theorie des fonctions: *Comptes rendus* (Paris): **194**, 44-46.

DENJOY, A. 1938. Sur une fonction réelle de Minkowski, *Journal des Mathématiques Pures et Appliquées*: **62**, 105-151.

DENJOY, A. 1955. *Articles et mémoires*, Paris, Gauthier-Villars, 1955, **2**, 925-971.

DEVANEY, R., & KEEN, L. Eds. 1989. *Chaos and Fractals. The Mathematics Behind the Computer Graphics,* Providence, RI: American Mathematical Society.

DEWDNEY, A.K. 1985. A computer microscope zooms in for a look at the most complex object of mathematics. *Scientific American*, August, cover and 16-21.

DIEUDONNÉ, J. 1975. L'abstraction et l'intuition mathématique, *Dialectica*: **29**, 39-54.

DOUADY, A. 1982. Systems dynamiques holomorphes. *Séminaire Boubaki. Astérisque*: **599**.

DOUADY, A. 1986. Julia sets and the Mandelbrot set, *The Beauty of Fractals,* by Heinz-Otto Peitgen & Peter H. Richter, New York: Springer, 161-173.

DOUADY, A. and HUBBARD, J. 1982. Iteration des polynomes quadratiques complexes. *Comptes rendus* (Paris): **294**-I, 123-126.

DOUADY, A. and HUBBARD, J. 1984. *Étude dynamique des polynomes complexes. I, II,* Publications Mathématiques d'Orsay, Université de Paris-Sud, Orsay, France.

DOUADY, A. and HUBBARD, J. 1985. On the dynamics of polynomial-like mappings. *Ann. Scientifiques de l'École Normale Supérieure:* **18**, 287-343.

DUBUC, S. 1982. Theoretical and numerical study of the Dekarlin-McGregor function. *J. Analyse Mathématique:* **42**, 15-37.

EVERTSZ, C. J. G., JONES, P. W., & MANDELBROT, B. B. 1991. Behavior of the harmonic measure at the bottom of fjords. *J. Physics*: **A24**, 1889-1901.

EVERTSZ, C. J. G. & MANDELBROT, B. B. 1991n. Steady state noises in diffusion limited fractal growth. *Europhysics Letters*: **15**, 245-250.

EVERTSZ, C. J. G. & MANDELBROT, B. B. 1992a. Multifractal measures. Appendix in *Chaos and Fractals: New Frontiers in Science,* by H.-O. Peitgen, H. Jürgens & D. Saupe. New York: Springer, 849-881.

EVERTSZ, C. J. G. & MANDELBROT, B. B. 1992b. Self-similarity of the harmonic measure on DLA. *Physica*: **A185**, 77-86.

EVERTSZ, C. J. G., MANDELBROT, B. B., & NORMANT, F. 1991f. Fractal aggregates, and the current lines of their electrostatic potentials. *Physica*: **A177**, 589-592.

EVERTSZ, C. J. G., MANDELBROT, B. B., & NORMANT, F. 1992t. Harmonic measure around linearly self-similar trees. *J. Physics*: **A25**, 1781-1797.

EVERTSZ, C. J. G., MANDELBROT, B. B., NORMANT, F., & WOOG, L. 1992. Variability of the form and the harmonic measure for small off-off lattice diffusion limited aggregates. *J. Physics*: **A45**, 5798-5804 & 8985-8986.

EVERTSZ, C. J. G., MANDELBROT, B. B., & WOOG, L. 1992. Variability of the form and of the harmonic measure for small off-off-lattice diffusion limited aggregates. *Physical Review*: **A 45**, 5798-5804.

FATOU, P. 1906. Sur les solutions uniformes de certaines équations fonctionnelles. *Comptes rendus* (Paris): **143**, 546-548.

FATOU, P. 1919-1920. Sur les solutions équations fonctionnelles. *Bull. Société Mathématique de France*: **47**, 161-271; **48**, 33-94, & **48**, 208-314.

FEDER, J. 1988. *Fractals*. New York: Plenum.

FEIGENBAUM, M. J. 1978. Quantitative universality for a class of nonlinear transformations. *J. of Statistical Physics*: **19**, 25-52.

FEIGENBAUM, M. J. 1981. Universal behavior in nonlinear systems. *Los Alamos Science*: **1**, 4-27.

FISHER, A., CALVET, L., & MANDELBROT, B. B. 1997. The Multifractality of the Deutshmark / US Dollar Exchange Rates. See under Mandelbrot, Calvet & Fisher 1997.

FRAME, M.L. & MANDELBROT, B.B. 2002. *Fractals, Graphics, and Mathematics Education*, Washington, DC: Mathematical Association of America & Cambridge: The University Press.

FRAME, M.L. & MANDELBROT, B.B. 2003. *Panorama of Fractals and their Uses*. Scheduled to become a book. At this point, a version under construction can be found at the following internet address: http://math.yale.edu/panorama/panorama.htu/.

FRICKE, R & KLEIN, F. 1897. *Vorlesungen über die Theorie der automorphen Functionen*. Leipzig: Teubner (Johnson reprint).

***N2** FRISCH, U. & PARISI, G. 1985. Fully developed turbulence and intermittency, in *Turbulence and Predictability in Geophysical Fluid Dynamics and Climate Dynamics*. International School of Physics "Enrico Fermi." Course 88, Eds. M. Ghil et. al. Amsterdam: North-Holland, 84-88.

GEFEN, Y., AHARONY, A., & MANDELBROT, B. B. 1983. Phase transitions on fractals: I. Quasi-linear lattices. *J. of Physics*: **A 16**, 1267-1278.

GEFEN, Y., AHARONY, A., & MANDELBROT, B. B. 1984. Phase transitions on Fractals: III. Infinitely ramified lattices. *J. of Physics*: **A 17**, 1277-1289.

GEFEN, Y., AHARONY, A., MANDELBROT, B. B. & KIRKPATRICK, S. 1981. Solvable fractal family, and its possible relation to the backbone at percolation. *Physical Review Letters*: **47**, 1771-1774.

GEFEN, Y., AHARONY, A., MANDELBROT, B. B. & SHAPIR, Y. 1984. Phase transitions on fractals: II. Sierpinski gaskets. *J. of Physics*: **A 17**, 435-444.

GEFEN, Y., MANDELBROT, B. B., & AHARONY, A. 1980. Critical phenomena on fractals. *Physical Review Letters*: **45**, 855-858.

GEFEN, Y., MEIR, Y., MANDELBROT, B. B. & AHARONY, A. 1983. Geometric implementation of hypercubic lattices with noninteger dimensionality, using low lacunarity fractal lattices. *Physical Review Letters*: **50**, 145-148.

GERNSTEIN, G. L. & MANDELBROT, B. B. 1964. Random walk models for the spike activity of a single neuron. *The Biophysical J.*: **4**, 41-68.

GILLESPIE, C.C. (Ed.) 1970-6. *Dictionary of Scientific Biography*. New York: Scribner's.

GISPERT, H. 1991. La France mathématique, 1890-1914. *Cahiers d'Histoire et de Philosophie des Sciences*: **34**.

GIVEN, J. A. & MANDELBROT, B. B. 1983. Diffusion on fractal lattices and the fractal Einstein relation. *J. Physics*: **A 16**, L565-L569.

GNEDENKO, B.V. 1967. *The Theory of Probability*. New York: Chelsea.

GNEDENKO, B.V. & KOLMOGOROV, A.N. 1968. *Limit Distributions for Sums of Independent Random Variables*. Reading, MA: Addison-Wesley.

GROSSMAN, S. & THOMAE, S. 1977. Invariant distributions and stationary correlation functions of one-dimensional discrete processes. *Z. für Naturforschung*: **32A**, 1353-1363.

GUCKENHEIMER, J. & MCGEHEE, R. 1984. *A Proof of the Mandelbrot N^2 Conjecture*, http://www.math.umn.edu/~mcgehee/publications/Mandelbrot N2/Mandelbrot N2.pdf. See Report No. **15**, Djursholm (Sweden): Institut Mittag-Leffler.

GUREL, O. & RÖSSLER, O. E. Eds., 1979. Bifurcation theory and applications in scientific disciplines, *Annals of the New York Academy of Sciences*: **316**, 1-708.

GUTZWILLER, M. C., 1980. Classical quantization of a Hamiltonian with ergodic behavior. *Physical Review Letters*: **45**, 150-153.

GUTZWILLER, M. C. 1983. Stochastic behavior in quantum scattering. *Physica* (Amsterdam): **7D**, 341-355.

*C GUTZWILLER, M. C. & MANDELBROT, B. B. 1988. Invariant multifractal measures in chaotic Hamiltonian systems, and related structures. *Physical Review Letters*: **60**, 673-676.

HADAMARD, J. 1912. L'oeuvre mathématique de Poincaré. *Acta Mathematica*: **38**, 203-287. Reprints in Poincaré 1916-, XI, 152-242 and Hadamard 1968, 4, 1921-2005.

HADAMARD, J. 1934. L'oeuvre scientifique de Paul Painlevé. *Revue de Métaphysique et de Morale:* **XLI (3)**, 289-325.

HADAMARD, J. 1968. *Oeuvres de Jacques Hadamard.* Paris: Editions de CNRS.

HALSEY, T. C., JENSEN, M. H., KADANOFF, L. P., PROCACCIA, I. & SHRAIMAN, B. I. 1986. Fractal measures and their singularities: the characterization of strange sets. *Physical Review:* **A33**, 1141-1151. IMPORTANT ERRATA: *Physical Review*: **A34**, 1986, 1601.

HARDY, G. H. & WRIGHT, E. M. 1960. *An Introduction to the Theory of Numbers* (4th edition) Oxford: Clarendon Press.

HELLEMAN, R. H. G. Ed. 1980. Nonlinear dynamics. *Annals of the New York Academy of Sciences*: **357**, 1-507.

HERMANN, R. 1991. Fractal Theory. *Mathematical Intelligencer*: **13**, 4.

HEWITT, E. 1990. So far, so good: my life up to now. *Mathematical Intelligencer*: **12 (4)**, pp. 58-63.

HOKUSAI, K. 1834. *Fugaku Hyakkei* or *One Hundred Views of Mt. Fuji.* The Original in three small volumes is a priceless library treasure. *Reprint with commentaries.* New York: George Braziller, 1988.

HOFSTADTER, D. R. 1981. Strange attractors: mathematical patterns delicately poised between order and chaos. *Scientific American*: **245**, (November issue), 16-29.

HOVI, J.-P., AHARONY, A., STAUFFER, D., & MANDELBROT, B.B. 1996. Gap independence and lacunarity in percolation clusters. *Physical Review Letters*: **77**, 877-890.

HUGHES, B.D. 1995-1996. *Random walks and random environments.* Oxford: Clarendon Press.

HURD, A. J., MANDELBROT, B. B. & WEITZ, D. A. (eds) 1987. *Fractal Aspects of Materials: Disordered Systems.* Extended Abstracts of a MRS Symposium, Boston. Pittsburgh PA: Materials Research Society.

JAFFARD, S. & MANDELBROT, B. B. 1996. Local regularity of nonsmooth wavelet expansions and application to the Polyà function. *Advances in Mathematics*: **120**, 265-282.

JULIA, G. 1918. Mémoire sur l'iteration des fonctions rationnelles. *J. de Mathématiques Pures et Appliquées:* **4** 47-245. Reprinted (with related texts) in Julia 1968, 121-319.

JULIA, G. 1968. *Oeuvres de Gaston Julia,* Paris: Gauthier-Villars.

***N11** KAHANE, J. P. & MANDELBROT, B. B. 1965. Ensembles de multiplicité aléatoires. *Comptes rendus* (Paris): **261**, 3931-3933.

KAHANE, J. P. & SALEM, R. 1963. *Ensembles parfaits et séries trigonométriques.* Paris: Hermann.

KAKUTANI, S. 1944. 2-dimensional Brownian motion and harmonic functions. *Proc. Imperial Academy of Sciences* (Tokyo): **20**, 706-714.

KINNEY, J.R. 1960. Note on a singular function of Minkowski. *Proc. American Mathematical Society*: **11**, 788-794.

KRANTZ, S.G. 1989. Fractal geometry. *Mathematical Intelligencer*: **11(4)**, 12-16.

KRANTZ, S.G. 1990. Mathematical anecdotes. *Mathematical Intelligencer*: **12(3)**, 32-39.

KRANTZ, S.G. 1991. Mathematical anecdotes. *Mathematical Intelligencer*: **13(4)**, 5.

LAIBOWITZ, R. B., MANDELBROT, B. B. & PASSOJA, D. E. Eds. 1985. *Fractal Aspects of Materials.* Extended Abstracts of a MRS Symposium, Boston. Pittsburgh PA: Materials Research Society.

LAM, C.-H., KAUFMAN, H. & MANDELBROT, B. B. 1994. Orientation of particle attachment and local isotropy in diffusion limited aggregates (DLA). *J. Physics*: **A28**, L 213-217.

LATTÈS, S. 1918a. Sur l'itération des substitutions rationnelles et les fonctions de Poincaré. *Comptes rendus* (Paris): **166**, 26-28.

LATTÈS, S. 1918b. Sur l'itération des substitutions rationnelles à deux variables. *Comptes rendus* (Paris): **166**, 151-153.

LATTÈS, S. 1918c. Sur l'itération des fractions irrationnelles. *Comptes rendus* (Paris): **166**, 486-488.

LEE, J. & STANLEY, H.E. 1988. Phase transition in the multifractal spectrum of diffusion-limited aggregates. *Physical Review Letters*: **61**, 2945-2948.

LEE, J., ALSTROM, P. & STANLEY, H.E. 1989. Scaling of the minimum growth probability for the "typical" diffusion-limited aggregation configuration. *Physical Review*: **A39**, 6545-6556; *Physical Review Letters Comm.*: **62**, 3013-3013.

LEIBNIZ, G. W. 1849-. *Mathematische Schriften.* Ed. C.I. Gerhardt. Halle: H.W. Schmidt (Olms reprint).

LÉVY, P., MANDELBROJT, S., MALGRANGE, B. & MALLIAVIN, P. 1967. *La vie et l'oeuvre de J. Hadamard.* Genève: Monographies de l'Enseignement Mathématique: **16**.

LUCAS, E. 1891. *Théorie des nombres.* Paris: Gauthier-Villars.

MAGNUS, W. 1974. *Noneuclidean Tesselations and their Groups.* New York: Academic Press.

MAKAROV, N.G. 1998. Fine structure of harmonic measure. *Algebra i Analiz*: **10**, 1-62; translation in *St. Petersburg Mathematical J.*: **10(2)**, 217-268.

MANDELBROJT, S. 1952. Pourquoi je fais des mathématiques. *Revue de Métaphysique et de Morale*: **57(4)**, 442-429.

MANDELBROJT, S. 1985. Souvenirs à bâtons rompus. *Publications du Séminaire d'Histoire des Mathématiques de l'Université de Paris*: **6**, 1-46.

MANDELBROT, B. B. 1951. Adaptation d'un message à la ligne de transmission, I & II. *Comptes rendus* (Paris): **232**, 1638-1740 & 2003-2005.

MANDELBROT, B. B. 1953i. An informational theory of the statistical structure of language, in *Communication Theory.* Ed. W. Jackson. London: Butterworth, 486-504.

MANDELBROT, B. B. 1953t. Contribution à la théorie mathématique des jeux de communication (Ph.D. Thesis). *Publications de l'Institut de Statistique de l'Université de Paris*: **2**, 1-124.

MANDELBROT, B. B. 1954w. Structure formelle des textes et communication (deux études). *Word* **10**, 1-27. Corrections. *Word*: **11**, 424. Translations into English, Czech and Italian.

MANDELBROT, B. B. 1955b. On recurrent noise limiting coding. *Information Networks, the Brooklyn Polytechnic Institute Symposium*, 205-221. Ed. E. Weber. New York: Interscience. Translation into Russian.

MANDELBROT, B.B. 1955t. Théorie de la précorrection des erreurs de transmission. *Annales des Télécommunications:* **10**, 122-134.

MANDELBROT, B. B. 1956c. La distribution de Willis-Yule, relative au nombre d'espèces dans les genres taxonomiques. *Comptes rendus* (Paris) **242**, 2223-2225.

MANDELBROT, B. B. 1956g. Memorandum. University of Geneva: Mathematics Institute.

MANDELBROT, B. B. 1956m. A purely phenomenological theory of statistical thermodynamics: canonical ensembles. *IRE Tr. on Information Theory*: **112**, 190-203.

MANDELBROT, B. B. 1956t. Exhaustivité de l'énergie d'un système, pour l'estimation de sa température. *Comptes rendus* (Paris): **243**, 1835-1837.

MANDELBROT, B.B. 1956w. On the language of taxonomy: an outline of a thermo-statistical theory of systems of categories, with Willis (natural) structure. *Information Theory, the Third London Symposium.* Ed. C. Cherry. London: Butterworth; New York: Academic, 1956, 135-145.

MANDELBROT, B. B. 1957b. Note on a law of J. Berry and on insistence stress. *Information and Control* : **1**, 76-81.

MANDELBROT, B. B. 1957p. *Linguistique statistique macroscopique.* In Apostel, Mandelbrot and Morf, 1-80.

MANDELBROT, B. B. 1957r. *Application of Thermodynamical Methods in Communication Theory and Econometrics.* Memorandum of the University of Lille Mathematics Institute.

MANDELBROT, B. 1957t. *Application of thermodynamical methods in communication theory and in econometrics.* Institut Mathématique de l'Université de Lille.

MANDELBROT, B. B. 1958p. Les lois statistiques macroscopiques du comportement (rôle de la loi de Gauss et des lois de Paul Lévy). *Psychologie Française*: **3**, 237-249.

MANDELBROT, B. B. 1959g. Ensembles grand canoniques de Gibbs; justification de leur unicité basée sur la divisibilitié infinie de leur énergie aléatoire. *Comptes rendus* (Paris): **249**, 1464-1466.

***E10** MANDELBROT, B. B. 1959p. Variables et processus stochastiques de Pareto-Lévy et la répartition des revenus, I & II. *Comptes rendus* (Paris): **249**, 613-615 & 2153-2155.

MANDELBROT, B. B. 1959s. A note on a class of skew distribution functions. Analysis and critique of a paper by H. A. Simon. *Information and Control*: **2**, 90-99.

***E10** MANDELBROT, B. B. 1960i. The Pareto-Lévy law and the distribution of income. *International Economic Review:* **1**, 79-106.

MANDELBROT, B. B. 1961b. On the theory of word frequencies and on related Markovian models of discourse. *Structures of language and its mathematical aspects.* Ed. R. Jakobson. 120-219. New York: American Mathematical Society

***E11** MANDELBROT, B. B. 1961e. Stable Paretian random functions and the multiplicative variation of income. *Econometrica*: **29**, 517-543.

MANDELBROT, B. B. 1961s. Final note on a class of skew distribution functions (with a postscript). *Information and Control*: **4**, 198-216 & 300-304.

***E1 FE** MANDELBROT, B.B. 1962c. Sur certains prix spéculatifs: faits empiriques et modèle basé sur les processus stables additifs de Paul Lévy. *Comptes rendus* (Paris): **254**, 3968-3970.

***E14,15** MANDELBROT, B.B. 1962i. *The Variation of Certain Speculative Prices.* IBM External Research Report: **NC-87**, March, 1962.

MANDELBROT, B. B. 1962n. Statistics of natural resources and the law of Pareto. IBM External Research Note: **NC-146**, June 29, 1962. See Mandelbrot 1995b.

E12 MANDELBROT, B. B. 1962q. Paretian distributions and income maximization. *Quarterly J. of Economics of Harvard University*: **76**, 57-85.

MANDELBROT, B. B. 1962t. The role of sufficiency and estimation in thermodynamics. *The Annals of Mathematical Statistics*: **33**, 1021-1038.

MANDELBROT, B. B. 1963. *Towards a Revival of the Statistical Law of Pareto.* IBM Research Report: **NC-227**, March, 1963.

***E14** MANDELBROT, B. B. 1963b. The variation of certain speculative prices (Abstract). *Econometrica*: **31**, 757-758.

***E14** MANDELBROT, B. B. 1963b. The variation of certain speculative prices. *J. of Business* (Chicago): **36**, 394-419. Reprinted in Cootner 1964: 297-337.

***E3** MANDELBROT, B. B. 1963e. New methods in statistical economics. *J. of Political Economy*: **71**, 421-440. Reprint in *Bulletin of the International Statistical Institute,* Ottawa Session: **40** (2), 669-720.

***E8** MANDELBROT, B. B. 1963g. *A Survey of Growth and Diffusion Models of the Law of Pareto.* IBM External Research Note: **NC-253**.

***E10** MANDELBROT, B. B. 1963i. The stable Paretian income distribution when the apparent exponent is near two. *International Economic Review*: **4**, 111-115.

MANDELBROT, B. B. 1963j. *The Stable Paretian Income Distribution when the Apparent Exponent is Near Two.* Memorandum appended to a joint reprint of M 1960i and M 1963i.

MANDELBROT, B. B. 1963o. *Oligopoly, Mergers, and the Paretian Size Distribution of Firms.* IBM External Research Note: **NC-246**, March 1963.

MANDELBROT, B.B. 1964h. *Self-Similar Random Processes and the Range,* IBM External Research Report: **RC-1163.**

MANDELBROT, B.B. 1964i. *Self-similar Random Processes: Extrapolation, Intepolation and Decay of Perturbations,* IBM External Research Report: **RC-1241.**

MANDELBROT, B. B. 1964j. The epistemology of chance in certain newer sciences. Read at *The Jerusalem International Congress on Logic, Methodology and the Philosophy of Science* (unpublished).

***E8** MANDELBROT, B. B. 1964o. Random walks, fire damage amount, and other Paretian risk phenomena. *Operations Research*: **12**, 582-585.

MANDELBROT, B. B. 1964s. *Self-similar random processes and the range.* IBM External Research Report: **RC-1163,** April 13.

MANDELBROT, B. B. 1964t. Derivation of statistical thermodynamics from purely phenomenological principles. *J. of Mathematical Physics*: **5**, 164-171.

***N5** MANDELBROT, B. B. 1964w. *Self-similar Turbulence and Non-Wienerian Conditioned Spectra.* IBM Research Report: **RC-134**.

***N8** MANDELBROT, B. B. 1965b. Time-varying channels, *1/f* noises and the infrared catastrophe, or: why does the low frequency energy sometimes seem infinite. *IEEE Communication Convention.* Boulder, CO.

***N7** MANDELBROT, B. B. 1965c. Self-similar error clusters in communications systems and the concept of conditional stationarity. *IEEE Transaction on Communications Technology*: **13**, 71-90.

***H9** MANDELBROT, B. B. 1965h. Une classe de processus stochastiques homothétiques à soi; application à la loi climatologique de H. E. Hurst. *Comptes rendus* (Paris): **260**, 3274-3277.

MANDELBROT, B. B. 1965m. Very long-tailed probability distributions and the empirical distribution of city sizes. *Mathematical Explorations in Behavioral Science* (Cambria Pines CA, 1964). Eds. F. Massarik & P. Ratoosh. Homewood, Ill.: R. D. Irwin, 322-332.

MANDELBROT, B. B. 1965s. Leo Szilard and unique decipherability. *IEEE Tr. on Information Theory* **IT-11**, 455-456.

***FE** MANDELBROT, B. B. 1965z. Information theory and psycholinguistics. *Scientific Psychology: Principles and Approaches* Eds. B. B. Wolman & E. N. Nagel. New York: Basic Books 550-562.. Reprint in *Language, Selected Readings.* Eds. R. C. Oldfield & J. C. Marshall. London: Penguin. Reprint with appendices, *Readings in Mathematical Social Science.* Eds. P. Lazarfeld and N. Henry. Chicago, Ill.: Science Research Associates (1966: hardcover). Cambridge, MA: M.I.T. Press (1968: paperback). Russian translation.

***E19** MANDELBROT, B. B. 1966b. Forecasts of future prices, unbiased markets, and "martingale" models. *J. of Business* (Chicago): **39**, 242-255. Important errata in a subsequent issue of the same journal.

***FE** MANDELBROT, B. B. 1966r. Nouveaux modèles de la variation des prix (cycles lents et changements instantanés). *Cahiers du Séminaire d'Économétrie*: **9**, 1966, 53-66.

***N10** MANDELBROT, B. B. 1967b. Sporadic random functions and conditional spectral analysis; self-similar examples and limits. *Proc. of the Fifth Berkeley Symposium on Mathematical Statistics and Probability*: **3**, 155-179. Eds. L. LeCam & J. Neyman. Berkeley: University of California Press.

***N9** MANDELBROT, B. B. 1967i. Some noises with $1/f$ spectrum, a bridge between direct current and white noise. *Institute of Electrical and Electronic Engineers Tr. on Information Theory*: **13**, 289-298.

***E15** MANDELBROT, B. B. 1967j. The variation of some other speculative prices. *J. of Business* (Chicago): **40**, 393-413.

***N12** MANDELBROT, B. B. 1967k. Sporadic turbulence. *Proc. International Symposium on Boundary Layers and Turbulence, including Geophysical Applications (Kyoto, 1966). Supplement to The Physics of Fluids*: **10,** Sept. 1967, S302-303.

MANDELBROT, B. B. 1967p. Sur l'épistémologie du hasard dans les sciences sociales: invariance des lois et vérification des hypothèses, *Encyclopédie de la Pléiade: Logique et Connaissance Scientifique.* Ed. J. Piaget. 1097-1113. Paris: Gallimard.

MANDELBROT, B. B. 1967s. How long is the coast of Britain? Statistical self-similarity and fractional dimension. *Science*: **155**, 636-638.

MANDELBROT, B. B. 1968i. Some aspects of the random walk model of stock market prices: comment. *International Economic Review*: **9**, 258.

MANDELBROT, B. B. 1968p. Les constantes chiffrées du discours. *Encyclopédie de la Pléiade: Linguistique*, Ed. J. Martinet, Paris: Gallimard, 46-56.

***N13** MANDELBROT, B. B. 1969b. On intermittent free turbulence: Abstract, followed by unpublished draft. *Proc. of the Symposium on Turbulence of Fluids and Plasmas.* (Polytechnic Institute of Brooklyn). New York: Interscience.

***H30** MANDELBROT, B. B. 1969e. Long-run linearity, locally Gaussian processes, H-spectra and infinite variances. *International Economic Review*: **10**, 82-111.

MANDELBROT, B. B. 1970a. Long-run interdependence in price records and other economic time series. *Econometrica*: **38**, 122-123.

***H30** MANDELBROT, B. B. 1970e. Statistical dependence in prices and interest rates. *Papers of the Second World Congress of the Econometric Society*, Cambridge, England (8-14 Sept. 1970).

MANDELBROT, B. B. 1970n. Analysis of long-run dependence in time series: the R/S technique. *Fiftieth Annual Report of the National Bureau of Economic Research*, 107-108.

MANDELBROT, B. B. 1970p. On negative temperature for discourse. Discussion of a paper by Prof. N. F. Ramsey. *Critical Review of Thermodynamics*, 230-232. Ed. E. B. Stuart et al. Baltimore, MD: Mono Book.

MANDELBROT, B. B. 1970y. *Statistical Self-Similarity and Non-Laplacian Chance.* Unpublished Trumbull Lectures, Yale University.

MANDELBROT, B. B. 1971. Comments on "Application of linear random models to four annual streamflow series." *Water Resources Research*: **7**, 1360-1362.

***E20** MANDELBROT, B. B. 1971e. When can a price be arbitraged efficiently? A limit to the validity of the random walk and martingale models. *Review of Economics and Statistics*: **53**, 225-236.

***H15** MANDELBROT, B. B. 1971f. A fast fractional Gaussian noise generator. *Water Resources Research:* **7**, 543-553.

MANDELBROT, B. B. 1971g. *The conditional cosmographic principle and the fractional dimension of the universe.* (Submitted unsuccessfully to several periodicals; first published as part of Mandelbrot 1975O.)

***H30** MANDELBROT, B. B. 1971n. Statistical dependence in prices and interest rates. *Fifty-first Annual Report of the National Bureau of Economic Research*: 141-142.

MANDELBROT, B. B. 1971q. Analysis of long-run dependence in economics: the R/S technique. *Econometrica*: **39** (July Supplement): 68-69.

***E14** MANDELBROT, B. B. 1972b. Correction of an error in "The variation of certain speculative prices (1963)." *J. of Business:* **40**, 542-543.

***H30** MANDELBROT, B. B. 1972c. Statistical methodology for nonperiodic cycles: from the covariance to *R/S* analysis. *Annals of Economic and Social Measurement*: **1**, 259-290.

MANDELBROT, B. B. 1972d. On Dvoretzky coverings for the circle. *Z. für Wahrscheinlichkeitstheorie*: **22**, 158-160.

***N14** MANDELBROT, B. B. 1972j. Possible refinement of the lognormal hypothesis concerning the distribution of energy dissipation in intermittent turbulence. *Statistical Models and Turbulence.* Eds. M. Rosenblatt & C. Van Atta. Lecture Notes in Physics: **12**, New York: Springer, 333-351.

***H16** MANDELBROT, B. B. 1972w. Broken line process derived as an approximation to fractional noise. *Water Resources Research*: **8**, 1354-1356.

MANDELBROT, B. B. 1972z. Renewal sets and random cutouts. *Z. für Wahrscheinlichkeitstheorie*: **22**, 145-157.

***E21** MANDELBROT, B. B. 1973c. Comments on "A subordinated stochastic process model with finite variance for speculative prices," by Peter K. Clark. *Econometrica*: **41**, 157-160.

FE MANDELBROT, B. B. 1973f. Formes nouvelles du hasard dans les sciences. *Economie Appliquée*: **26**, 307-319.

FE MANDELBROT, B. B. 1973j. Le problème de la réalité des cycles lents, et le syndrome de Joseph. *Economie Appliquée*: **26**, 349-365.

FE MANDELBROT, B. B. 1973v. Le syndrome de la variance infinie, et ses rapports avec la discontinuité des prix. *Economie Appliquée*: **26**, 321-348.

*N16 MANDELBROT, B. B. 1974c. Multiplications aléatoires itérées et distributions invariantes par moyenne pondérée. *Comptes rendus* (Paris): **278A**, 289-292 & 355-358.

*E8 MANDELBROT, B. B. 1974d. A population birth and mutation process, I: Explicit distributions for the number of mutants in an old culture of bacteria. *J. of Applied Probability*: **11**, 437-444. (Part II distributed privately).

*N15 MANDELBROT, B. B. 1974f. Intermittent turbulence in self-similar cascades: divergence of high moments and dimension of the carrier. *J. of Fluid Mechanics*: **62**, 331-358.

*H17 MANDELBROT, B. B. 1975b. Fonctions aléatoires pluritemporelles: approximation poissonienne du cas brownien et généralisations. *Comptes rendus* (Paris): **280A**, 1075-1078.

*H18 MANDELBROT, B. B. 1975f. On the geometry of homogeneous turbulence, with stress on the fractal dimension of the iso-surfaces of scalars. *J. of Fluid Mechanics*: **72**, 401-416.

MANDELBROT, B. B. 1975m. Hasards et tourbillons: quatre contes à clef. *Annales des Mines* (November): 61-66.

MANDELBROT, B. B. 1975O, 1984O, 1989O, 1995O. *Les objets fractals: forme, hasard et dimension.* Paris: Flammarion.

MANDELBROT, B. B. 1975u. Sur un modèle décomposable d'univers hiérarchisé: déduction des corrélations galactiques sur la sphère céleste. *Comptes rendus* (Paris): **280A**, 1551-1554.

*H19 MANDELBROT, B. B. 1975w. Stochastic models for the Earth's relief, the shape and the fractal dimension of the coastlines, and the number-area rule for islands. *Proc. of the National Academy of Sciences USA*: **72**, 3825-3828.

*H26 MANDELBROT, B. B. 1975z. Limit theorems on the self-normalized range for weakly and strongly dependent processes. *Z. für Wahrscheinlichkeitstheorie*: **31**, 271-285.

*N19 MANDELBROT, B. B. 1976c. Géométrie fractale de la turbulence. Dimension de Hausdorff, dispersion et nature des singularités du mouvement des fluides. *Comptes rendus* (Paris): **282A**, 119-120.

*N18 MANDELBROT, B. B. 1976o. Intermittent turbulence and fractal dimension: kurtosis and the spectral exponent $5/3 + B$. In *Turbulence and Navier Stokes Equations.* Ed. R. Temam. (Lecture Notes in Mathematics **565**.) New York: Springer, 121-145.

MANDELBROT, B. B. 1977b. Fractals and turbulence: attractors and dispersion. *Turbulence Seminar Berkeley 1976/1977.* Eds. P. Bernard & T. Ratiu. *Lecture Notes in Mathematics*: **615** 83-93. New York: Springer. Russian translation.

MANDELBROT, B. B. 1977F. *Fractals: Form, Chance, and Dimension.* San Francisco: W. H. Freeman.

MANDELBROT, B. B. 1977h. Geometric facets of statistical physics: scaling and fractals. *Statistical Physics 13*, International IUPAP Conference, 1977. Eds. D. Cabib et al. *Annals of the Israel Physical Society*: **2**, 225-233.

MANDELBROT, B. B. 1977l. Physical objects with fractional dimension: seacoasts, galaxy clusters, turbulence and soap. *The Institute of Mathematics and its Applications* (Great Britain) Bulletin: **13**, 189-196. Also in *Fluid Dynamics – les Houches.* 1973. Eds. R. Balian & J. L. Peube. New York: Gordon & Breach, 557-578.

MANDELBROT, B. B. 1978b. The fractal geometry of trees and other natural phenomena. *Buffon Bicentenary Symposium on Geometrical Probability*, Eds. R. Miles & J. Serra. *Lecture Notes in Biomathematics*: **23** 235-249. New York: Springer.

MANDELBROT, B. B. 1978c. Colliers aléatoires et une alternative aux promenades au hasard sans boucle: les cordonnets discrets et fractals. *Comptes rendus* (Paris): **286A**, 933-936.

*N5 MANDELBROT, B. B. 1978h. Geometric facets of statistical physics: scaling and fractals. *Annals of the Israel Physical Society*: **2** (1), 225-233.

MANDELBROT, B. B. 1978r. Les objets fractals. *La Recherche*: **9**, 1-13.

MANDELBROT, B. B. 1979n. Comment on bifurcation theory and fractals. *Bifurcation Theory and Applications*, Eds. Gurel & O. Rössler. *Annals of the New York Academy of Sciences*: **316**, 463-464.

MANDELBROT, B. B. 1979u. Corrélations et texture dans un nouveau modèle d'Univers hiérarchisé, basé sur les ensembles trémas. *Comptes rendus* (Paris): **288A**, 81-83.

MANDELBROT, B. B. 1980b. Fractals and geometry with many scales of length. *Encyclopedia Britannica 1981 Yearbook of Science and the Future*: 168-181.

***C3** MANDELBROT, B. B. 1980n. Fractal aspects of the iteration of $z \to \lambda z(1-z)$ for complex λ and z. *Non Linear Dynamics*, Ed. R. H. G. Helleman. *Annals of the New York Academy of Sciences*: **357**, 249-259.

MANDELBROT, B. B. 1981s. Scalebound or scaling shapes: a useful distinction in the visual arts and in the natural sciences. *Leonardo:* **14**, 45-47.

***E14** MANDELBROT, B. B. 1982c. The variation of certain speculative prices. *Current Contents*: **14**, 20.

MANDELBROT, B.B. 1982f. Comments on computer rendering of fractal stochastic models. *Communications of the Association for Computing Machinery:* **25,** cover and pp. 581-584.

***C4, 16** MANDELBROT, B. B. 1982F *{FGN}*. *The Fractal Geometry of Nature.* New York: W. H. Freeman.

MANDELBROT, B. 1982n. Des monstres de Cantor et Peano à la géométrie fractale de la nature. *Penser les mathématiques,* textes réunis par J. Dieudonné, M. Loi & R. Thom, Paris: Editions du Seuil, 226-251.

***C23** MANDELBROT, B. B. 1982s. The inexhaustible function z squared plus c. Invited by *Scientific American,* never published there, but adapted into this book's Chapter C23.

MANDELBROT, B. B. 1982t. The many faces of scaling: fractals, geometry of nature and economics. *Self-Organization and Dissipative Structures.* Eds. W. C. Schieve & P. M. Allen: 91-109.

MANDELBROT, B. B. 1983d. Les fractales, les monstres et la beauté. *Le Débat* : **24**, 54-72.

MANDELBROT, B. 1983. In *Percolation Structures and Processes.* Eds. G. Deutscher, R. Zallen, & J. Adler. *Annals of the Israel Physical Society*: **5**, 59-80.

***C18** MANDELBROT, B. 1983m. Self-inverse fractals osculated by sigma discs, and the limit sets of inversion groups. *Mathematical Intelligencer:* **5** (Spring), **(2)**, Front and back covers and pp. 9-17.

***C5** MANDELBROT, B. B. 1983p. On the quadratic mapping $z \to z^2 - \mu$ for complex μ and z : the fractal structure of its *M*-set, and scaling. *Physica*: **D7**, 224-239.

MANDELBROT, B. B. 1984. Some "facts" that evaporate upon examination. *Mathematical Intelligencer*: **11** (Fall) 17-19.

MANDELBROT, B. B. 1984c. *Fractal Sums of Pulses, and New Random Variables and Functions.* (Memorandum superseded by M 1995n and four papers coauthored by R. Cioczek-Georges.)

MANDELBROT, B. B. 1984d. Profile by Monte Davis. *Omni* (New York) (February issue) 65.

MANDELBROT, B. B. 1984e. Fractals and physics: squig clusters, diffusions, fractal measures and the unicity of fractal dimensionality. *J. of Statistical Physics:* **34**, 1984, 895-910.

MANDELBROT, B. B. 1984f. Squig sheets and some other squig fractal constructions. *J. of Statistical Physics:* **36**, 519-539.

MANDELBROT, B. B. 1984j. *Fraktal Kikagaku.* Traduction japonaise de Mandelbrot 1982F*{FGN}*, par H. Hironaka. Tokyo: Nikkei Science.

***C13** MANDELBROT, B. B. 1984k. On the dynamics of iterated maps, VIII: The map $z \to \lambda(z+1/z)$, from linear to planar chaos, and the measurement of chaos. *Chaos and Statistical Mechanics.* (Kyoto Summer Institute.) Ed. Y. Kuramoto, New York: Springer, 32-41.

MANDELBROT, B. B. 1984r. Comment on the equivalence between fracton/spectral dimensionality and the dimensionality of recurrence. *J. of Statistical Physics:* **36**, 543-545.

MANDELBROT, B. 1984s. Les images fractales: un art pour l'amour de la science et ses applications. *Sciences et Techniques:* Mai 1984, 16-19, 34-35, & 65.

***N5** MANDELBROT, B. B. 1984w. On fractal geometry and a few of the mathematical questions it had raised. *Proc. of the Twelfth International Congress of Mathematicians* (Warsaw, 1983). Ed. Z. Ciesielski. Warsaw: PWN & Amsterdam: North-Holland, 1984, 1661-1675.

MANDELBROT, B. B. 1985b. Interview by Anthony Barcellos. *Mathematical People.* Eds. D. J. Albers and G. L. Alexanderson, Boston: Birkhauser, 205-225.

***C6, 7, 8, 9, 10** MANDELBROT, B. B. 1985g. On the dynamics of iterated maps. Paper III: The individual molecules of the *M*-set: self-similarity properties, the N;**-2** rule, and the N;**-2** conjecture. Paper IV: The notion of "normalized radical" *R,* and the fractal dimension of the boundary of *R.* Paper V: Conjecture that the boundary of the *M*-set has a fractal dimension equal to 2. Paper VI: Conjecture that certain Julia sets include smooth components. Paper VII:

Domain-filling ("Peano") sequences of fractal Julia sets, and an intuitive rationale for the Siegel discs. *Chaos, Fractals and Dynamical Systems.* Eds. P. Fischer & W. Smith. New York: Marcel Dekker, 213-253.

H21* MANDELBROT, B. B. 1985l. Self-affine fractals and fractal dimension. *Physica Scripta:* **32, 257-260.

***C11** MANDELBROT, B. B., 1985n. Continuous interpolation of the complex discrete map $z \rightarrow \lambda(1-z)$, and related topics (On the dynamics of iterated maps, IX). Ed. N. R. Nilsson, *Physica Scripta:* **T 9**, 59-63.

MANDELBROT, B.B. 1986k. Multifractals and fractals (letter to the Editor). *Physics Today* (September issue): 11-12.

MANDELBROT, B.B. 1986p. Fractals and the rebirth of iteration theory. *The Beauty of Fractals,* by Heinz-Otto Peitgen & Peter H. Richter, New York: Springer, 151-160.

MANDELBROT, B. 1986r. Comment j'ai découvert les fractales. (Entretien avec Marc Lesort) *La Recherche* (Mars 1986): 420-424.

***H22,23,24** MANDELBROT, B. B. 1986t. Self-affine fractal sets, I: The basic fractal dimensions, II: Length and surface measurements, III: Hausdorff dimension anomalies and their implications. *Fractals in Physics* (Trieste, 1985). Eds. L. Pietronero & E. Tosatti. Amsterdam: North-Holland, 3-28.

MANDELBROT, B. 1987c. Propos à bâtons rompus. *Fractals: dimensions non entières et applications,* dirigé par G. Cherbit, Paris: Masson, 4-15.
• English translation: Sundry observations. *Fractals: Non Integer Dimensions and Applications.* Ed. G. Cherbit. Chichester & New York: John Wiley, 1991, 3-9.

MANDELBROT, B. B. 1987d. *Die fraktale Geometrie der Natur.* German translation of 1982F{*FGN*}, by R. & U. Zähle. Basel: Birkhauser.

MANDELBROT, B. B. 1987e. Fractals. *The Encyclopedia of Physical Science and Technology* : **5**, 579-593. San Diego, CA: Academic Press.

MANDELBROT, B. 1987i. *Gli oggetti frattali.* Traduction italienne de Mandelbrot 1975O, par R. Pignoni; préface par L. Peliti & A. Vulpiani. Torino: Giulio Einaudi.

MANDELBROT, B. 1987m, 1989m. *La geometria della natura.* Milano: Montedison. Roma: Edizioni Theoria.

MANDELBROT, B. B. 1987r. Towards a second stage of indeterminism in science (preceded by historical reflections), *Interdisciplinary Science Reviews*: **12**, 117-127.

MANDELBROT, B. 1987s. *Los objetos fractales.* Spanish translation of M 1975O by J. M. Llosa. Barcelona: Tusquets.

MANDELBROT, B. B. 1988c. An introduction to multifractal distribution functions. *Fluctuations and Pattern Formation* (Cargèse, 1988). Eds. H. E. Stanley & N. Ostrowsky. Dordrecht-Boston: Kluwer, 345- 360.

MANDELBROT, B. B. 1988f. Flare: A by-product of the study of a two-dimensional dynamical system, *IEEE Tr. on Circuits and Systems*: **36**, 1988, 768-769.

MANDELBROT, B. B. 1988m. Naturally creative. Interview by Mike Dibb. *Modern Painters* (London), Premier Issue (Spring), 52-53.

***H20** MANDELBROT, B. B. 1988p. Fractal landscapes without creases and with rivers. *The Science of Fractal Images.* Eds. H.-O. Peitgen & D. Saupe, New York: Springer, 1988, 243-260.

MANDELBROT, B. B. 1988s. People and events behind the science of fractal images. *The Science of Fractal Images.* Eds. H.-O. Peitgen & D. Saupe, New York: Springer, 1-19.

MANDELBROT, B. B. 1989a. The principles of multifractal measures. *The Fractal Approach to Heterogeneous Chemistry.* Ed. D. Avnir, New York: Wiley, 45-51.

MANDELBROT, B. B. 1989b. An overview of the language of fractals. *The Fractal Approach to Heterogeneous Chemistry.* Ed. D. Avnir. New York: Wiley, 3-9.

MANDELBROT, B. B. 1989e. A class of multifractal measures with negative (latent) values for the "dimension" $f(\alpha)$. *Fractals' Physical Origin and Properties* (Erice, 1988). Ed. L. Pietronero. New York: Plenum, 3-29.

MANDELBROT, B. B. 1989g. Multifractal measures, especially for the geophysicist. *Pure and Applied Geophysics*: **131**, 5-42 Also *Fractals in Geophysics.* Eds. C. H. Scholz & B. B. Mandelbrot. Boston: Birkhauser.

MANDELBROT, B. B. 1989h. Lewis Fry Richardson and prematurity in science. *The British Society for the History of Mathematics, Newsletter*: **12**, October, 2-4.

MANDELBROT, B. B. 1989l. Survol du langage fractal: *Les objets fractals* (3e édition). Paris: Flammarion, 185-240.

MANDELBROT, B. B. 1989m. Chaos, Bourbaki and Poincaré. *Mathematical Intelligencer:* **11** (Summer) 10-12.

MANDELBROT, B. B. 1989p. Temperature fluctuations: a well-defined and unavoidable notion. *Physics Today*. January : **42**, 71 & 73.

MANDELBROT, B. B. 1989r. Fractal geometry: What is it, and what does it do? *Proc. Royal Society* (London): **A423**, 3-16.

MANDELBROT, B. B. 1989s. Fractals: an art for the sake of Science. *Leonardo* : **22** (Special SIGGRAPH issue), 21-24.

MANDELBROT, B. B. 1989t. The fractal range of the distribution of galaxies: crossover to homogeneity and multifractals. *Large-scale structure and Motions in the Universe.* Eds. F. Mardirossian et al., Dordrecht-Boston: Kluwer, 259-279.

MANDELBROT, B. B. 1990d. New "anomalous" multiplicative multifractals: left-sided $f(\alpha)$ and the modeling of DLA. *Condensed Matter Physics, in Honor of Cyril Domb. Physica*: **A168**, 95-111.

MANDELBROT, B. B. 1990n. Fractals: a geometry of nature. *The New Scientist*: September 15, 1990, cover & pp. 38-43.

MANDELBROT, B. B. 1990r. Negative fractal dimensions and multifractals. *Statistical Physics 17, International IUPAP Conference.* Ed. C. Tsallis. *Physica:* **A 163**, 306-315.

MANDELBROT, B. B. 1990t. Limit lognormal multifractal measures. *Frontiers of Physics: Landau Memorial Conference* (Tel Aviv, 1988). Eds. E. A. Gotsman, Y. Ne'eman & A. Voronel New York: Pergamon, 309-340.

MANDELBROT, B. B. 1990w. Two meanings of multifractality, and the notion of negative fractal dimension. *Chaos/Xaoc: Soviet-American Perspectives on Nonlinear Science.* Ed. D. K. Campbell. New York: American Institute of Physics, 79-90.

MANDELBROT, B. B. 1991g. The gray and the green. *Fractal Forms.* Eds. E. Guyon & H. E. Stanley. Paris: Palais de la Découverte, 1991.

MANDELBROT, B. B. 1991k. Random multifractals: negative dimensions and the resulting limitations of the thermodynamic formalism. *Proc. of the Royal Society* (London): **A434**, 79-88. Also in *Turbulence and Stochastic Processes: Kolmogorov's ideas 50 years on.* Eds. J. C. R. Hunt, O. M. Phillips & D. Williams. London: The Royal Society, 1991.

MANDELBROT, B. B. 1991n. Fractal craft. (Letter to the Editor.) *The New Scientist*: September 14, 1991.

***C** MANDELBROT, B. B. 1991p. Fractals and the rebirth of experimental mathematics. *Fractals for the Classroom,* by H.-O. Peitgen, H. Jürgens, & D. Saupe, E. M. Matelski, T. Perciante, & L. E. Yunker. New York: Springer, 1991.

MANDELBROT, B. B. 1992g. Avant-propos: *Physique et structures fractales,* par J.-F. Gouyet. Paris: Masson, 1992.

MANDELBROT, B. B. 1992h. Plane DLA is not self-similar; is it a fractal that becomes increasingly compact as it grows? *Physica*: **A 191,** 95-107.

MANDELBROT, B. B. 1992s. Avant-propos: *Dieu joue-t-il aux dés?,* par I. Stewart. Paris: Flammarion, 1992, 7-13.

MANDELBROT, B. B. 1993f. Opinions. *Fractals:* **1**, 117-123. Reprint in a special issue on Symmetry: *Culture and Science.* **4**, 1993, 319-328.

MANDELBROT, B. B. 1993g. Fractals. *Chaos: The New Science.* (Gustavus Adolphus Nobel Conference XXVI.) Ed. J. M. Holte. Lanham MD: University Press of America, 1993, 1-27.

MANDELBROT, B. B. 1993n. A fractal's lacunarity, and how it can be tuned and measured. *Fractals in Biology and Medicine.* Eds. T. F. Nonnenmacher et al. Basel: Birkhauser, 8-21.

***C21** MANDELBROT, B. B. 1993s. The Minkowski measure and multifractal anomalies in invariant measures of parabolic dynamic systems. *Chaos in Australia* (Sydney, 1990). Eds. G. Brown & A. Opie. Singapore: World Publishing, 1993, 83-94. Slightly edited reprint: *Fractals* *and Disordered Systems*. Second edition. Eds. A. Bunde & S. Havlin. New York: Springer: 1995.

MANDELBROT, B. B. 1994d. Les fractales, l'art algorithmique et le test de Turing. *La science et la métamorphose des arts*, textes réunis par R. Daudel. Paris: Presses Universitaires de France, 39-52.

MANDELBROT, B. B. 1994h. Fractals as a morphology of the amorphous. Preface of *Fractal Landscapes from the Real World*, **by** William Hirst. Manchester, UK: Cornerhouse Publications, 1994.

MANDELBROT, B. B. 1994j. Comment on "Theoretical mathematics..." by A. Jaffe and F. Quinn, *Bulletin of the American Mathematical Society*: 1994, 193-196.

MANDELBROT, B. B. 1994q. Fractals, the computer, and mathematics education. *Proc. of the International Congress of Mathematics Education*, ICME-7 (Québec, 1992), Québec: Presses de l'Université Laval, 1994, 77-98.

MANDELBROT, B.B. 1995b. Statistics of natural resources and the law of Pareto. *Fractal Geometry and its Uses in the Geosciences and in Petroleum Geology*. Eds. C. C. Barton & P. La Pointe. New York: Plenum, 1-12. First appeared as Mandelbrot 1962n.

MANDELBROT, B.B. 1995k. Negative dimensions and Hölders, multifractals and their Hölder spectra, and the role of lateral preasymptotics in science. *J.P. Kahane meeting* (Paris, 1993). Eds. Aline Bonami & Jacques Peyrière. *J. of Fourier Analysis and its Applications 1995, special issue*: 409-432.

MANDELBROT, B. B. 1995l. The Paul Lévy I knew. *Lévy Flights and Related Phenomena in Physics* (Nice, 1994). Eds. M. F. Schlesinger, G. Zaslawsky, & U. Frisch (Lecture Notes in Physics). New York: Springer, 1995, ix-xii.

MANDELBROT, B. B. 1995n. Introduction to fractal sums of pulses. *Lévy Flights and Related Phenomena in Physics*. Eds. M. F. Shlesinger, G. Zaslawsky, & U. Frisch. New York: Springer, 110-123.

*C MANDELBROT, B. B. 1995s. The Minkowski measure and multifractal anomalies in invariant measures of parabolic dynamic systems. *Fractals and Disordered Systems*. Second edition. Eds. A. Bunde & S. Havlin. New York: Springer, 345-353.

MANDELBROT, B. B. 1995z. Measures of fractal lacunarity: Minkowski content and alternatives. *Fractal Geometry and Statistics*. Eds. C. Bandt, S. Graf & M. Zähle. Basel & Boston: Birkhauser, 12-38.

*E MANDELBROT, B. B. 1997E. *Fractals and Scaling in Finance: Discontinuity, Concentration, Risk*. New York: Springer-Verlag.

*FE MANDELBROT, B.B. 1997FE. *Fractales, hasard et finance*. Paris: Flammarion.

MANDELBROT, B.B. 1998e. Fractality, lacunarity and the near-isotopic distribution of galaxies. *Current Topics in Astrofundamental Physics*. Eds. N. Sanchez & A. Zichichi. Dordrecht: Kluwer, 585-603.

*N MANDELBROT, B.B. 1999N. *Multifractals & 1/f Noise: Wild Self-Affinity in Physics*. New York: Springer-Verlag.

MANDELBROT, B.B. 1999s. Multifractal walk down Wall Street. *Scientific American:* February issue, cover & 70-73.

MANDELBROT, B.B. 1999p. Renormalization and fixed points in finance, since 1962. *Statistical Physics 20, International IUPAP Conference* (Paris, 1998). Ed. D. Iagolnitzer. *Physica:* **A26,** 477-487.

MANDELBROT B. B. 2000p. Some mathematical questions arising in fractal geometry. *Development of Mathematics 2000*: Ed. J-P. Pier. Basel: Birkhäuser, 795-811.

MANDELBROT, B.B. 2001a. Scaling in financial prices: I. Tails and dependence. *Quantitative Finance:* **1,** 113-123.

MANDELBROT, B.B. 2001b. Scaling in financial prices: II. Multifractals and the star equation. *Quantitative Finance:* **1,** 124-130.

MANDELBROT, B.B. 2001c. Scaling in financial prices: III. Cartoon Brownian motions in multifractal time. *Quantitative Finance:* **1,** 427-440.

MANDELBROT, B.B. 2001d. Scaling in financial prices: IV. Multifractal concentration. *Quantitative Finance:* **1**, 641-649.

MANDELBROT, B. B. 2001e. Stochastic volatility, power-laws, and long memory. *Quantitative Finance:* **1**, 588-589.

MANDELBROT, B.B. 2001R. *Nel mondo dei frattali.* Roma: di Renzo.

***H** MANDELBROT, B.B. 2002H. *Gaussian Self-Affinity and Fractals: R/S, 1/f, Globality, Reliefs & Rivers,* New York: Springer-Verlag.

***C17** MANDELBROT, B.B. 2002w. Symmetry by dilation/reduction, fractals, and roughness, *Symmetry 2000.* Eds. I. Hargittai and T. Laurent, **I**, 133-141.

MANDELBROT, B.B. 2003b. Fractal sums of pulses and a practical challenge to the distinction between local and global dependence. *Long Range Dependent Stochastic Processes: Theory and Applications* (Bengalore, India, 2002). Eds. Govindan Rangarajan and Ming Ding. New York: Springer, 118-135.

MANDELBROT, B.B. 2003P. *Fractals and Multifractals in Finance.* Webbook.

MANDELBROT, B.B. 2003f. Multifractal power-law distributions, other "anomalies," and critical dimensions, explained by a simple example. *Journal of Statistical Physics:* **110**, 739-777.

MANDELBROT, B.B. 2003r. Heavy tails in finance for independent or multifractal price increments. *Handbook on Heavy Tailed Distributions in Finance.* Ed. Svetlozar T. Rachev (Volume 30 of *Handbooks in Finance, Senior Ed.*: William T. Ziemba): **1**, 1-34.

***C** MANDELBROT, B.B. 2004C (This book's self-reference). *Fractals and Chaos: the Mandelbrot Set and Beyond,* New York: Springer-Verlag.

MANDELBROT, B.B. 2004f (expected). Fractal sums of pulses: self-affine global dependence and lateral limit theorems.

MANDELBROT, B. B., CALVET, L & FISHER, A. 1997. *The Multifractal Model of Asset Returns; Large Deviations and the Distribution of Price Changes; The Multifractality of the Deutschmark/US Dollar Exchange Rate.* New Haven CT: Cowles Foundation Discussion Papers **1164, 1165** and **1166.** On Mandelbrot's Internet homepage.

***C19** MANDELBROT, B. B. & EVERTSZ, C. J. G. 1990. The potential distribution around growing fractal clusters. *Nature:* **378** (6296), front cover & pp. 143-145.

***C22** MANDELBROT, B. B. & EVERTSZ, C. J. G. 1991. Multifractality of harmonic measure on fractal aggregates, and extended self-similarity. *Physica*: **A177**, 386-393.

MANDELBROT, B. B. & EVERTSZ, C. J. G. 1991n. Exactly self-similar multifractals with left-sided $f(\alpha)$. *Fractals and Disordered Systems.* Eds. A. Bunde & S. Havlin. 323-346.

MANDELBROT, B. B., EVERTSZ, C. J. G. & HAYAKAWA, Y. 1990. Exactly self-similar "left-sided" multifractal measures. *Physical Review*: **A42**, 4528-4536.

MANDELBROT, B.B. & FRAME, M. 1999. The canopy and shortest path in a self-contacting tree. *The Mathematical Intelligencer:* **21(#2)**, 1999, 18-27.

MANDELBROT, B. B., GEFEN, Y., AHARONY, A., & PEYRIÈRE, J. 1985. Fractals, their transfer matrices and their eigen-dimensional sequences. *J. of Physics*: **A18**, 335-354.

MANDELBROT, B. B. & GIVEN, J. A. 1984. Physical properties of a new fractal model of percolation clusters. *Physical Review Letters*: **52**, 1853-1856.

MANDELBROT, B. B., KAUFMAN, H., VESPIGNANI, A., YEKUTIELI, I., & LAM, C.-H. 1995. Deviations from self-similarity in plane DLA and the infinite drift scenario. *Europhysics Letters*: **29**, 599-604.

MANDELBROT, B.B., KOL, B., & AHARONY, A., 2002. Angular gaps in radial diffusion-limited aggregation fractal dimensions and non-transient deviations from linear self-similarity. *Physical Review Letters*: **88**, 055501-1-4.

***H28** MANDELBROT, B. B. & MCCAMY, K. 1970. On the secular pole motion and the Chandler wobble. *Geophysical J.*: **21**, 217-232.

MANDELBROT, B.B. & NORTON, A. Fractal Surfaces Defined by Iteration of Rational Functions in the Quaternions, to appear.

MANDELBROT, B. B. & PASSOJA, D. E. Eds. 1984. *Fractal Aspects of Materials: Metal and Catalyst Surfaces, Powders and Aggregates.* Extended Abstracts of a MRS Symposium, Boston. Pittsburgh PA: Materials Research Society.

MANDELBROT, B. B., PASSOJA, D., & PAULLAY, A. 1984. The fractal character of fracture surfaces of metals. *Nature*: **308**, 721-722.

MANDELBROT, B.B. & RIEDI, R.H. 1995. Multifractal formalism for infinite multinomial measures. *Advances in Applied Mathematics:* **16**, 132-150.

MANDELBROT, B. B. & STAUFFER, D. 1994. Antipodal correlations and texture (fractal lacunarity) in critical percolation clusters. *J. of Physics*: **A27**, L237-L242.

MANDELBROT, B. B., STAUFFER, D. & AHARONY, A. 1993. Self-similarity of fractals: a random walk test. *Physica*: **A196**, 1-5.

***E21** MANDELBROT, B. B. & TAYLOR, H. M. 1967. On the distribution of stock price differences. *Operations Research*: **15**, 1057-1062.

***H11** MANDELBROT, B. B. & VAN NESS, J. W. 1968. Fractional Brownian motions, fractional noises and applications. *SIAM Review*: **10**, 422-437.

MANDELBROT, B. B., VESPIGNANI, A. & KAUFMAN, H., 1995a. Cross-cut analysis of large radial DLA: departures from self-similarity and lacunarity effects, *Europhysics Letters*: **32**, 199-204.

MANDELBROT, B. B., VESPIGNANI, A. & KAUFMAN, H., 1995b. The geometry of DLA: different aspects of the departure from self-similarity. *Fractal Aspects of Materials*. Eds. F. Family, P. Meakin, B. Sapoval & R. Wool. Pittsburgh, PA: Materials Research Society, 1995, 73-79.

MANDELBROT, B. B. & VICSEK, T. 1989. Directed recursive models for fractal growth. *J. Physics*: **A22**, L377-383

***H10** MANDELBROT, B. B. & WALLIS, J. R. 1968. Noah, Joseph and operational hydrology. *Water Resources Research*: **4**, 909-918.

***H12.13.14** MANDELBROT, B. B. & WALLIS, J. R. 1969a. Computer experiments with fractional Gaussian noises, Part 1, 2 & 3. *Water Resources Research*: **5**, 228-267.

***H27** MANDELBROT, B. B. & WALLIS, J. R. 1969b. Some long-run properties of geophysical records. *Water Resources Research*: **5**, 321-340. Reprinted in Barton & La Pointe, 1994, 41-64.

***H25** MANDELBROT, B. B. & WALLIS, J. R. 1969c. Robustness of the rescaled range *R/S* in the measurement of noncyclic long run statistical dependence. *Water Resources Research*: **5**, 967-988.

***H7** MANDELBROT, B. B. & WALLIS, J. R. 1969d. Operational hydrology using self similar processes. *Proc. of the Fifth International Conference on Operations Research*. Ed. John Lawrence. London: Tavistock.

MANDELBROT, B. B. & ZARNFALLER, F. 1959. *Five-place Tables of Certain Stable Distributions*. IBM Research Technical Report: RC-421.

• **NOTE**: In other items co-authored by Mandelbrot the first-listed author is one of the following: Apostel, Asikainen, Barral, Berger, Blumenfeld, Calvet, Cioczek-Georges, Coppens, Damerau, Evertsz, Fisher, Frame, Gefen, Gerstein, Given, Gutzwiller, Hovi, Hurd, Jaffard, Kahane, Kaufman, Laibowitz, Lovejoy, Musgrave, Riedi, Schaefer, Sholz, Shlesinger, Voldman, Wallis, Weitz, or Yekutieli.

MAY, R. 1976. Simple mathematical models with very complicated dynamics. *Nature*: **261**, 459-467.

McMULLEN, C.T. 1998. Hausdorff dimension and conformal dynamics. III: Computation of dimension. *American J. Mathematics*: **120**, 691-721.

MEAKIN, P. 1988. The growth of fractal aggregates and their fractal measures. in: *Phase Transitions and Critical Phenomena*, Eds. C. Domb and J. Lebowitz: **12**, 335-489.

METROPOLIS, N., STEIN, M. L. & STEIN, P. R. 1973. On finite limit sets for transformations on the unit interval. *J. of Combinatorial Theory*: **A15**, 25-44.

MINKOWSKI, H. 1911. *Gesammelte Abhandlungen*, Chelsea reprint.

MONTEL, P. 1927. *Leçons sur les familles normales de fonctions analytiques et leurs applications*, Paris: Gauthier-Villars.

MONTEL, P. 1957. *Leçons sur les récurrences et leurs applications*, Paris, Gauthier-Villars.

MUMFORD, D., SERIES, C., & WRIGHT, D. 2002. *Indra's Pearls; The Vision of Felix Klein*. Cambridge: the University Press.

MUSGRAVE, F. K. & MANDELBROT, B. B. 1989. Natura ex machina. *IEEE Computer Graphics and its Applications*: **9**, Jan. Cover & 4-7.

MUSGRAVE, F. K. & MANDELBROT, B. B. 1991. The art of fractal landscapes. *IBM J. of Research and Development*: **35** 1991, front and back covers & pp. 535-540.

MYRBERG, P. J. 1962. Sur l'itération des polynomes réels quadratiques. *J. des Mathématiques pures et appliquées*: **(9)41**, 339-351.

NEVALINNA, R. 1970. *Analytic Functions*, New York: Springer.

NIEMEYER, I., PIETRONERO, L., & WIESMANN, H.J. 1984. Fractal dimension of dielectric breakdown. *Physical Review Letters*: **52**, 1033-1036.

PAINLEVÉ, P. 1895. *Leçon d'ouverture faite en présence de Sa Majesté le Roi de Suède et de Norvège*, In Painlevé 1972, **1**, 200-204.

PAINLEVÉ, P. 1972-. *Oeuvres de Paul Painlevé*. Paris: Éditions du CNRS.

PEANO, G. 1980. Sur une coube, qui remplit une aire plane. *Mathematische Annalen*: **36**, 157-160. Translation in Peano 1973.

PEITGEN, H. O. & RICHTER, P. H., 1986. *The Beauty of Fractals*, New York: Springer.

PERES, Y. 1994. The self-affine carpets of McMullen and Bedford have infinite Hausdorff measure. *Mathematical Proc. of the Cambridge Philosophical Society*: **116**, 513-526.

PERES, Y. & KENYON R. 1996. Hausdorff dimensions of affine-invariant sets. *Israel J. Mathematics:* **94**, 157-178.

PIETRONERO, L. & TOSATTI, E. Eds. 1986. *Fractals in Physics*, Amsterdam: North-Holland.

POINCARÉ, H. 1916-. *Oeuvres de Henri Poincaré*, Paris: Gauthier Villars.

RIEDI, R. H. & MANDELBROT, B. B., 1997a. Inverse measures, the inversion formula, and discontinuous multifractals. *Advances in Applied Mathematics:* **18,** 50-58.

RIEDI, R.H. & MANDELBROT, B.B. 1997b. Inversion formula for continuous multifractals. *Advances in Applied Mathematics*: **19**, 332-354.

RIEDI, R. H. & MANDELBROT, B. B., 1998. Exceptions to the multifractal formalism for discontinuous measures. *Mathematical Proc. of the Cambridge Philosophical Society*: **123**, 133-157.

RUELLE, D. 1970. Repellers for real analytic maps. *Ergodic Theory and Dynamic Systems*: **2**, 99-107.

SALEM, R. 1943. Singular monotone functions. *Trans. American Mathematical Society*: **53**, 427-439.

SALEM, R., 1967. *Oeuvres Mathématiques*. Paris: Hermann.

SALEM, R. & ZYGMUND, A., 1945. Lacunary power series and Peano curves. *Duke Mathematical J.*: **12**, 569-578. Also in Salem 1967, pp. 336-345.

SCHAEFER, D. W., LAIBOWITZ, R. B., MANDELBROT, B. B. & LIU, S. H. Eds. 1986. *Fractal Aspects of Materials II*. Extended Abstracts of a Symposium, Boston. Pittsburgh PA: Materials Research Society.

SCHOLZ, C. H. & MANDELBROT, B. B. Eds. 1989. *Fractals in Geophysics*. Special issue of *Pure and Applied Geophysics*: **131 (1-2)**. Also issued as a book, Basel & Boston: Birkhauser.

SCHWARTZ, L. 1997. *Un mathématician aux prises avec le siècle*, Paris: Odile Jacob. Translation: *A Mathematician Grappling with his Century*. Basel & Boston: Birkhauser, 2001.

SCHWARZER, S., LEE, J., BUNDE, A., HAVLIN, S., ROMAN, H.E. & STANLEY, H.E., 1990. Minimum growth possibility of diffusion-limited aggregates. *Physical Review Letters*: **65**, 603-606.

SERIES, C. 1985. The modular surface and continued fractions. *J. London Mathematical Society*: (2) **31**, 69-80.

SERIES, C. 1985. The geometry of Markoff numbers. A geometric interpretation is found in *Mathematical Intelligencer*: **7**, 20-29.

SERIES, C. 1986. Geometrical Markov coding of geodesics on surfaces of constant negative curvature, *Ergodic Theory Dynamical Systems*: **6**, 601-625.

SHLESINGER, M. F., MANDELBROT, B. B. & RUBIN, R. J. Eds. 1984. *Fractals in the Physical Sciences* (Gaithersburg, 1983). Special issue of *J. of Statistical Physics*: **36**, 516.

SHISHIKURA, M. M. 1994. The boundary of the Mandelbrot set has Hausdorff dimension two. *Complex analytic methods in dynamical systems*: (Rio de Janeiro, 1992). *Astérisque*: **222**, 389-405.

SHISHIKURA, M. M. 1998. The Hausdorff dimension of the boundary of the Mandelbrot set and Julia sets. *Annals of Mathematics*: **147**, 225-267.

SHISHIKURA, M. M. 2000. Bifurcation of parabolic points, in Tan 2000.

SIEGEL, C. L., 1942. Iteration of analytic functions. *Annals of Mathematics*: **43**, 607-612.

SIERPIŃSKI, W. 1915. Sur une courbe dont tout point est un point de ramification. *Comptes rendus* (Paris): **160**, 302. More detail in Sierpińński 1974-, **II**, 99-106.

SIERPINSKI, W. 1974. *Oeuvres choisies.* Ed. S. Hartman et al. Warsaw: Éditions scientifiques.

STEIN, P. R. & ULAM, S. 1964. Non-linear transformation studies on electronic computers. *Rozprawy Matematyczne*: **39**, 1-66. Also in Ulam 1974, 401-484.

STONE, M. 1961. The revolution in mathematics. *American Mathematical Monthly*: **68**, 715-734.

SULLIVAN, D. 1979. The density at infinity of a discrete group of hyperbolic motions. *Institut des Hautes Etudes Scientifiques. Publications Mathematiques*: **50**.

TAN, L. 1984. In *Étude dynamique des polynomes complexes* Eds. Adrien Douady & John H. Hubbard, Vol. II, pp. 139-152, Publications Mathématique d'Orsay, Orsay, France.

TAN, L. Ed. 2000. *The Mandelbrot Set, Theme and Variations,* London Mathematical Society Lecture Notes: **274**. Cambridge University Press, Cambridge.

TANNERY, J. & MOLK, J. 1898. Élements de la théorie des fonctions elliptiques. Tome III, Paris: Gauthier-Villars.

THOMAS, P. B. & DHAR, D. 1994. The Hausdorf dimension of the Apollonian packing of circles. *J. of Physics*: **A27**, 2257-2268.

URBANSKI, M. 2003. Measures and dimensionalities in conformal dynamics. *Bull. American Mathematical Society*: **40**, 281-321.

VICSEK, T. 1989. *Fractal Growth Phenomena*. Singapore: World Scientific.

***H31** VOLDMAN, J. MANDELBROT, B.B., HOEVEL, L.W., KNIGHT, J. & ROSNFELD, P. 1983. Fractal nature of software-cache interaction. *IBM J. of Research and Development*: **27**, 164-170.

VON KARMAN, T. 1940. The engineer grapples with nonlinear problems. *Bull. of the American Mathematical Society*: **46**, 615-683.

WALLIS, J.R. & MANDELBROT, B.B. 1968. *2nd Symposium on the Use of Analog and Digital Computers in Hydrology,* Tucson, AZ, Publication 81 of the International Association of Scientific Hydrology: **2**, 738-755.

WALSH, J. L. 1956. *Interpolation and Approximation by Rational Functions in the Complex Domain,* Colloquium Publications: **20**. Providence, RI: American Mathematical Society.

WEGENER, A. 1915-1929-1966. *The Origins of Continents and Oceans.* The 1966 date is that of the Dover translation of the 1929 German edition.

WEIERSTRASS, K. 1872. Über continuirliche Functionen eines reellen Arguments, die für keinen Werth des letzteren einen bestimmten Differential-quotienten besitzen. Unpublished until Weierstrass 1895-, *Mathematische Werke*: **II**, 71-74.

WEIL, A. 1992. *A Mathematician's Apprenticeship.* Basel-Boston: Birkhaüser.

WEITZ, D. A., SANDER, L. M. & MANDELBROT, B. B. Eds. 1988. *Fractal Aspects of Materials: Disordered Systems.* Extended Abstracts of a MRS Symposium, Boston. Pittsburgh PA: Materials Research Society.

WHITTAKER, E.T. & WATSON, G.N. 1902-1927. *A Course of Modern Analysis* (4th edition). Cambridge: The University Press.

WITTEN, T.A. & SANDER, L.M. 1981. Diffusion-limited aggregation, a kinetic critical phenomenon, *Physical Review Letters*: **47**, 1400-1403.

Index